INTERNATIONAL HYDROLOGY SERIES

The **Intergovernmental Hydrological Programme** (IHP) was established by the United Nations Educational, Scientific and Cultural Organization (UNESCO) in 1975 as the successor to the International Hydrological Decade. The long-term goal of the IHP is to advance our understanding of processes occurring in the water cycle and to integrate this knowledge into water resources management. The IHP is the only UN science and educational programme in the field of water resources, and one of its outputs has been a steady stream of technical and information documents aimed at water specialists and decision-makers.

The **International Hydrology Series** has been developed by the IHP in collaboration with Cambridge University Press as a major collection of research monographs, synthesis volumes, and graduate texts on the subject of water. Authoritative and international in scope, the various books within the series all contribute to the aims of the IHP in improving scientific and technical knowledge of fresh-water processes, in providing research know-how and in stimulating the responsible management of water resources.

TITLES IN PRINT IN THIS SERIES

M. Bonell, M. M. Hufschmidt, and J. S. Gladwell *Hydrology and Water Management in the Humid Tropics: Hydrological Research Issues and Strategies for Water Management*

Z. W. Kundzewicz *New Uncertainty Concepts in Hydrology and Water Resources*

R. A. Feddes *Space and Time Scale Variability and Interdependencies in Hydrological Processes*

J. Gibert, J. Mathieu, and F. Fournier *Groundwater/Surface Water Ecotones: Biological and Hydrological Interactions and Management Options*

G. Dagan and S. Neuman *Subsurface Flow and Transport: A Stochastic Approach*

J. C. van Dam *Impacts of Climate Change and Climate Variability on Hydrological Regimes*

D. P. Loucks and J. S. Gladwell *Sustainability Criteria for Water Resource Systems*

J. J. Bogardi and Z. W. Kundzewicz Risk, *Reliability, Uncertainty, and Robustness of Water Resource Systems*

G. Kaser and H. Osmaston *Tropical Glaciers*

I. A. Shiklomanov and J. C. Rodda *World Water Resources at the Beginning of the Twenty-First Century*

A. S. Issar *Climate Changes during the Holocene and their Impact on Hydrological Systems*

M. Bonell and L. A. Bruijnzeel Forests *Water and People in the Humid Tropics: Past, Present and Future Hydrological Research for Integrated Land and Water Management*

F. Ghassemi and I. White Inter-Basin *Water Transfer: Case Studies from Australia, United States, Canada, China and India*

K. D. W. Nandalal and J. J. Bogardi *Dynamic Programming Based Operation of Reservoirs: Applicability and Limits*

H. S. Wheater, S. Sorooshian, and K. D. Sharma *Hydrological Modelling in Arid and Semi-Arid Areas*

J. Delli Priscoli and A. T. Wolf *Managing and Transforming Water Conflicts*

H. S. Wheater, S. A. Mathias, and X. Li Groundwater *Modelling in Arid and Semi-Arid Areas*

L. A. Bruijnzeel, F. N. Scatena, and L. S. Hamilton *Tropical Montane Cloud Forests*

S. Mithen and E. Black Water *Life and Civilization: Climate, Environment and Society in the Jordan Valley*

K. A. Daniell *Co-Engineering and Participatory Water Management*

R. Teegavarapu *Floods in a Changing Climate: Extreme Precipitation*

P. P. Mujumdar and D. Nagesh Kumar *Floods in a Changing Climate: Hydrologic Modeling*

G. Di Baldassarre *Floods in a Changing Climate: Inundation Modelling*

S. Simonović *Floods in a Changing Climate: Risk Management*

S. Hendry *Frameworks for Water Law Reform*

J. Martin-Ortega, R. C. Ferrier, I. J. Gordon, and S. Khan *Water Ecosystem Services: A Global Perspective*

Impact of Urban Floods on Water Quality

Edited by

Blanca Jiménez Cisneros

Division of Water Sciences, UNESCO

Alexandros K. Makarigakis

Natural Sciences Sector, UNESCO

International Hydrology Series

CAMBRIDGE
UNIVERSITY PRESS

CAMBRIDGE
UNIVERSITY PRESS

Shaftesbury Road, Cambridge CB2 8EA, United Kingdom

One Liberty Plaza, 20th Floor, New York, NY 10006, USA

477 Williamstown Road, Port Melbourne, VIC 3207, Australia

314–321, 3rd Floor, Plot 3, Splendor Forum, Jasola District Centre, New Delhi – 110025, India

103 Penang Road, #05–06/07, Visioncrest Commercial, Singapore 238467

Cambridge University Press is part of Cambridge University Press & Assessment, a department of the University of Cambridge.

We share the University's mission to contribute to society through the pursuit of education, learning and research at the highest international levels of excellence.

Published jointly by the United Nations Educational, Scientific and Cultural Organization (UNESCO), 7, Place de Fontenoy, 75007 Paris, France, and Cambridge University Press, University Printing House, Shaftesbury Road, Cambridge CB2 8BS, United Kingdom.

www.cambridge.org
Information on this title: www.cambridge.org/9781107118720

DOI: 10.1017/9781316340332

When citing this work, please include a reference to the DOI 10.1017/9781316340332

First published 2026

A catalogue record for this publication is available from the British Library

A Cataloging-in-Publication data record for this book is available from the Library of Congress

ISBN 978-1-107-11872-0 Hardback

Contents

Colour versions of the images can be found online at www.cambridge.org/jimenez-cisneros

Figures

Tables

Contributors

The contributors of the book are listed in alphabetical order.

Sheila Aboulouard
Deputy Director in charge of operations at the Direction for Environment and Sanitation Services of Val-de-Marne (Conseil départemental du Val-de-Marne)

Simon Berkhahn
Institute of Fluid Mechanics and Environmental Physics in Civil Engineering, Leibniz University Hannover, Germany

Béatrice Blanchet
Director at the Department of the Sanitation System and Networks, Syndicat Interdépartemental pour l'Assainissement de l'Agglomération Parisienne (SIAAP)

Daniel Castro-Fresno
University of Cantabria, GITECO Research Group, Spain

Susanne Charlesworth
Institute for Sustainability, Equity and Resilience, Centre for Agroecology, Water and Resilience, Coventry University, UK

Ana Deletic
Faculty of Engineering, Queensland University of Technology, Brisbane, Australia

Lothar Fuchs
Institute for Technical and Scientific Hydrology (itwh GmbH), Hannover, Germany

Patricia Gadd
Australian Nuclear Science and Technology Organisation, Sydney, Australia

Thomas Graf
Institute of Fluid Mechanics and Environmental Physics in Civil Engineering, Leibniz University Hannover, Germany

Kapil Gupta
Indian Institute of Technology Bombay, Mumbai, India

Roland I. Hall
Department of Biology, University of Waterloo, Canada

Hendrik Heijnis
School of Postgraduate Studies, Diponegoro University, Semarang, Indonesia

C. K. Jain
Centre for Science and Environmental Research, Roorkee, India

Haifeng Jia
School of Environment, Tsinghua University, Beijing, China.

Mingyi Jia
School of Energy and Environment, City University of Hong Kong

Blanca Jiménez Cisneros
National Water Commission, Embassy of Mexico in France, Engineering Institute of the National Autonomous University of Mexico, Mexico

Stefan Krämer
Institute for Technical and Scientific Hydrology (itwh GmbH), Hannover, Germany

Rakesh Kumar
National Institute of Hydrology, Roorkee, India

Paul Leahy
Environment Protection Authority Victoria, Melbourne, Australia
Royal Melbourne Institute of Technology, Melbourne, Australia

Anna Lintern
Department of Civil Engineering, Monash University, Melbourne, Australia

Alexandros K. Makarigakis
Division of Water Sciences, Natural Sciences Sector, United Nations Educational Scientific and Cultural Organization, France

David T. McCarthy
Faculty of Engineering, Queensland University of Technology, Brisbane, Australia

Lauren MacDonald
Department of Biology, University of Waterloo, Canada

Insa Neuweiler
Institute of Fluid Mechanics and Environmental Physics in Civil Engineering, Leibniz University Hannover, Germany

Aaron Peche
Institute of Fluid Mechanics and Environmental Physics in Civil Engineering, Leibniz University Hannover, Hannover, Germany

Vincent Rocher
Delegate Director for Innovation, Strategy and Environment, SIAAP

Jorge Rodriguez-Hernandez
University of Cantabria, GITECO Research Group, Spain

Antonio Lastra De La Rubia
Canal de Isabel II, Madrid, Spain

Robert Sämann
Institute of Fluid Mechanics and Environmental Physics in Civil Engineering, Leibniz University Hannover, Germany

Jean-Pierre Tabuchi
Retired and former Project Coordinator at SIAAP

Julian Wahl
Institute for Technical and Scientific Hydrology (itwh GmbH), Hannover, Germany

Zheng Wang
Center for Urban Runoff Control and Stream Restoration, School of Environment, Tsinghua University, Beijing, China

Brent B. Wolfe
Department of Geography and Environmental Studies, Wilfrid Laurier University, Waterloo, Canada

Changqing Xu
School of Environment, Tsinghua University, Beijing, China

Te Xu
Center for Urban Runoff Control and Stream Restoration, School of Environment, Tsinghua University, Beijing, China

Atun Zawadzki
Australian Nuclear Science and Technology Organisation, Sydney, Australia

Acknowledgements

The editors would like to thank all the authors for contributing to the improvement of the knowledge base for a topic that lacks research and should be further examined. They would further like to thank their teams and all the experts involved in the realisation of this book and, more specifically, Ms Kaixin Lee and Ms Langrun Song, whose unwavering dedication in coordinating all the authors was crucial in finalising the book.

Introduction

Blanca Jiménez Cisneros and Alexandros K. Makarigakis

Since the 1970s, 44% of all disaster events have been flood related (IPCC, 2022). Since antiquity, it is well known that floods can threaten lives, cause physical injuries, inundate properties and businesses, destroy belongings, prevent access to essential public services and damage agricultural fields, as well as public and private infrastructures. Between 2013 and 2022, an average of 5,034 deaths per year were attributed to floods (with data from *Statistica*, 2023). Although mortality has been reduced worldwide, figures remain at least one order of magnitude higher in low-income countries than in higher-income ones. Flooding is associated with increased water-borne diseases, it contributes to reduced food availability and increased food prices and it is amongst the most common climatic drivers for disasters, especially in poor parts of Asia, Africa and South and Central America (IPCC, 2022).

Unfortunately, this list above of negative impacts is not exhaustive: one missing impact is the effect of floods on water quality. This effect has been frequently overlooked in literature, most of all because the majority of negative effects have an immediate and significant impact on population and assets, but also because the effects on water quality are much more complex and less well understood. This book contributes to shed some light on the subject.

Floods are a natural phenomenon, and they become a disaster only when they occur in areas where population, infrastructure or ecosystems are exposed and damaged. It is to discuss if floods are a natural disaster or simply a disaster, as nowadays many experts concur that the term *natural disaster* should simply be replaced by the term *disaster*, since, for it to happen, a combination of a natural hazard or a human-made hazard impacting a vulnerable community, infrastructure or ecosystem must occur.

Floods usually start with an extreme precipitation, but this is not the only cause, nor is it the predetermining one, as the occurrence of a flood also requires a combination of factors related to hydrology, climate and human management of a basin (land use and land cover, urbanisation and economic development). There are many types of floods: pluvial floods, flash floods, river floods, groundwater floods, surge floods and coastal floods (Kovács, 2013; Ranasinghe et al., 2021; Weber, 2023). The IPCC (2022) opted to regroup them in three main categories: river, pluvial and coastal floods.

A fluvial or river flood occurs when the water level in a river, lake or stream rises and overflows onto the neighbouring land. It is caused by an excessive (amount and/or intensity) precipitation. In addition to precipitation, soil moisture, basin morphology, land-use and land-cover characteristics, hydraulic infrastructure and tidal levels (when available) also play a role. When the flood plain (the land adjacent to a river) is flooded, and houses and infrastructure were built on it, these assets and their inhabitants become vulnerable to floods.

Extreme rain events exceeding the capacity of natural and artificial drainage systems can lead to pluvial floods in cities, roadways, farmlands, industries, metro stations, subway tunnels and building basements. When pluvial floods occur in cities, they are called urban floods. Very intense rainfall, along with a high fraction of impervious surfaces, can result in flash floods in urban areas. Urban flooding can be linked to a major disaster, such as hurricanes, extreme rain events and intense hail precipitation. Frequent sewer blockages, sewer backups or sewer overflows are observed. Even small amounts of rain can overwhelm the urban drainage if it is deteriorated, poorly designed or built, or simply blocked with uncontrolled solid wastes (partly from: Ranasinghe et al., 2021; Seneviratnet et al., 2021).

Coastal floods occur in cities, infrastructure or land located along the shores. They can affect communities, farmlands, buildings, transportation routes, industries and, eventually, ecosystems. Coastal floods are caused by extreme total water levels which are the combination of the relative sea level, tides, storm surge and high wave setup at the shoreline. Coastal floods can affect port and harbour operations. Frequent inundation with saline water can also have significant impacts on water resources, crops, aquaculture and infrastructure (due to corrosion).

FLOODS AND CLIMATE CHANGE

According to the IPCC (2022) report, there is high confidence that climate change and projected socioeconomic development would increase exposure in inundation areas, resulting in a large increase in direct flood damages, and an estimated increase of

120% and 400% of population affected by river flooding at 2°C and 4°C warming, respectively. The highest numbers of people affected by river flooding are projected for countries in Southern, Eastern and Southeastern Asia. Under climate change scenarios, many regions are expected to experience more frequent and intense urban and flash floods, up to levels that can exceed current flood protection design standards. For coastal areas, the sea level rise will also bring, more frequent and intense floods. In Europe alone, coastal flood damage is projected to increase at least 10 folds by the end of the 21st century (IPCC, 2022).

FLOODS AND WATER QUALITY

Concerning water quality, heavy rain events mobilise chemical and biological pollutants from the paved ground and urban soil to water sources, endangering its quality. Floods intensify the mixing of floodwater with wastewater and the redistribution of pollutants. For coastal areas, the risks also include a higher salinisation of water sources (IPCC, 2022). Furthermore, floods damage infrastructure, including water and sanitation services, leaving population exposed to water quality problems. Unfortunately, little research on the impacts of floods on water quality has been conducted and, when available, it focuses almost exclusively on Europe, the United States and Canada.

ADAPTATION AND SOLUTIONS

This book analyses the state of the art on floods and their impact on water quality. It presents adaptation methods and solutions to the problems caused, varying from the control of floods to the control of its effects on water quality.

Chapter 1, "Urbanisation process and urban water challenges," describes urban water challenges, setting the scene for the rest of the book. Merging inputs from authors coming from three different regions (Asia, Latin America and Europe) give not only an interesting view on how cities impact water as a resource but also how water resources affect cities with a focus on floods and water quality. Following global economic and political decisions, the urbanisation process is dominating the way humans settle in, such a manner that there was a need to define the "urban water cycle" to better understand the phenomena and processes linking both cities and water. Consequently, the frequently overlooked link between urban floods and water quality is considered.

Chapter 2, "General Approaches on Urban Water challenges," describes current and future water challenges and presents general approaches on how to face the effects of urban flooding. It initially provides a brief description of the transition in the management of urban water challenges, from the initial sanitary engineering to the sustainable period. It also provides an evolution in terminology and types of urban systems and a presentation of strategies and examples of flood risk management in selected countries. Chapter 2 also refers to the different approaches used to conceptualise urban drainage, including low impact development (LID), sustainable urban design systems (SUDS), water-sensitive urban design (WSUD), ending with the Sponge Cities concept. These different practices have different priorities, application scopes, and management styles; decision-makers can choose amongst them according to the actual condition of their specific regions.

Chapter 3, "Integrating water quality in urban flood risk management: structural approaches," demonstrates how the use of structural measures, such as SuDS, provides multiple benefits and flexibility to reduce the effects of storm peaks while supporting cities' efforts in adapting to climate change effects. SuDS can become an integral part of WSUD and provide the opportunity for reducing water demand in cities where water is considered in planning and is part of the constructed environment.

Chapter 4, "Impacts of urban drainage management on water quality following extreme rainfall," examines, in particular, the main factors to be considered by a city to reduce the impact of a flooding event and presents the ways to manage the drainage system sustainably. Their presentation includes all management phases, from the project phase to the maintenance and cleaning activities and even the operation of the infrastructure during the flood.

"Online forecast model for urban flooding and induced water contamination" is the title of Chapter 5. The need of early warning systems to predict pluvial flood events in urban environments is discussed in this chapter. However, the nature of the threat during pluvial floods in urban areas is dual, as further to the large amounts of water the city has to receive and evacuate in a short period of time, it needs to deal with the release and transport of potentially hazardous substances. The chapter examines existing numerical models addressing the challenge and presents a fully coupled urban flow model along with a transport model to assess the migration of contaminants during and after a pluvial flood event.

In Chapter 6, the methods to identify the contaminant levels in river floods that occurred in the past are discussed. Riverine floods can result in the dispersal of contaminants throughout inundated areas, including urbanised landscapes and regions downstream of industrialisation. This chapter reflects on a missed opportunity that could support water quality managers and engineers in their efforts to build robust flood mitigation strategies. In particular, it presents how water quality and hydrologic data collected from sedimentary records by palaeoenvironmental scientists can be incorporated into models and into management frameworks.

Chapter 7 is very relevant to protect water supply users' health. It describes how to cope with water quality problems that occur on the supply of water during floods. This chapter provides insight to the basic principles to be followed during an emergency flood situation. Analysing the planning exercises to take place, the identification, development and maintenance of the water supply to be used and the quality control using available facilities it presents suggestions to manage the response and recovery part.

A case study on the impact of urban flooding on water quality in India, with special reference to Mumbai, is presented in Chapter 8. This situation described in here is common to many of the heavily populated cities of India. In many of these cities, the urban water table has dropped significantly, and raw sewage is discharged into the open stormwater drains. During the monsoon, raw sewage in stormwater drains overflows polluting water in the surroundings and resulting in outbreaks of the diseases such as gastroenteritis, malaria and dengue.

In June 2016, the Seine and the Marne, in the Paris region of France, had substantial flooding that was unusual at that time of the year as flooding usually occurs in the winter. Chapter 9 presents the impact of this flood on the operation of the sanitation system of the Paris urban area, including the subsequent impact on the quality of the receiving environment: the Marne and the Seine. One aspect that was retained from this experience is the relevance of having a coordinated and effective management between all sanitation operators together with a good communication system between them and all the stakeholders. With regard to the sanitation system, it was assessed that the treatment systems were able to process substantial volumes of the pollution received during floods limiting the further impact on the environment.

Chapter 10 discusses about "The Sponge City" initiative in China, where frequent urban flooding is a major concern. Pervious natural lands are gradually subsumed by the dramatically increasing impermeable concrete surface that prevents any water infiltration and results in urban floods. To deal with them, the concept of the Sponge City was developed. The Sponge City was initially defined as "natural storage, natural infiltration, and natural purification." Its purpose is to mitigate urban flooding and purify stormwater runoff. The Sponge City is also beneficial for ecological restoration and microclimate improvement. It is noted that the Sponge City concept is not purely "green" but an integration of green and grey infrastructures. This chapter describes in detail the concept and its implementation, presenting practical cases in which it has been applied.

The relevance and significance of this book has been evident during the 2024 Olympic Games in France, where the local government decided to use the Seine River, as a location to host some water related competitions. This visionary way to promote the city soon gave place to be an unexpected way to show to society the impact that extreme, or even intense, rain events can have on the quality of water bodies which are used for recreational, sportive and even drinking water purposes. In several occasions the competitions had to be rescheduled until the quality of water attaint an acceptable level for the organisers to allow its use. Similarly, in October 2024 a Depresión Aislada en Niveles Altos (DANA), or isolated depression at high levels, caused torrential rain and flooding in an Eastern and Southern Spain resulting to the loss of more than 200 lives and an unprecedented crisis in access to drinking water, adding stress to the already complicated situation the people found themselves in.

References

IPCC (2022). Climate Change 2022: Impacts, Adaptation and Vulnerability. In *Contribution of Working Group II to the Sixth Assessment Report of the Intergovernmental Panel on Climate Change* [H.-O. Pörtner, D. C. Roberts, M. Tignor, E. S. Poloczanska, K. Mintenbeck, A. Alegría, M. Craig, S. Langsdorf, S. Löschke, V. Möller, A. Okem, B. Rama (eds.)]. Cambridge University Press, Cambridge, UK and New York, NY, 3056 pp., doi:10.1017/9781009325844.

Kovács, J. (2013). "Flood Deposits." Encyclopedia of Natural Hazards. Encyclopedia of Earth Sciences Series: 325. doi:10.1007/978-1-4020-4399-4_137. ISBN 978-90-481-8699-0.

Ranasinghe, R., A. C. Ruane, R. Vautard, N. Arnell, E. Coppola, F. A. Cruz, S. Dessai, A. S. Islam, M. Rahimi, D. Ruiz Carrascal, J. Sillmann, M. B. Sylla, C. Tebaldi, W. Wang, and R. Zaaboul (2021). Climate Change Information for Regional Impact and for Risk Assessment. In *Climate Change 2021: The Physical Science Basis. Contribution of Working Group I to the Sixth Assessment Report of the Intergovernmental Panel on Climate Change* [V. Masson-Delmotte, P. Zhai, A. Pirani, S. L. Connors, C. Pean, S. Berger, N. Caud, Y. Chen, L. Goldfarb, M. I. Gomis, M. Huang, K. Leitzell, E. Lonnoy, J. B. R. Matthews, T. K. Maycock, T. Waterfield, O. Yelekci, R. Yu, B. Zhou (eds.)]. Cambridge University Press, Cambridge, UK.

Scottish sensory Center (SSC) (2013). www.ssc.education.ed.ac.uk/BSL/geography/floodplaind.html, consulted in September 2023.

Seneviratne, S. I., X. Zhang, M. Adnan, W. Badi, C. Dereczynski, A. Di Luca, S. Ghosh, I. Iskandar, J. Kossin, S. Lewis, F. Otto, I. Pinto, M. Satoh, S. M. Vicente-Serrano, M. Wehner, and B. Zhou (2021). Weather and Climate Extreme Events in a Changing Climate. In *Climate Change 2021: The Physical Science Basis. Contribution of Working Group I to the Sixth Assessment Report of the Intergovernmental Panel on Climate Change* [V. Masson-Delmotte, P. Zhai, A. Pirani, S. L. Connors, C. Pean, S. Berger, N. Caud, Y. Chen, L. Goldfarb, M. I. Gomis, M. Huang, K. Leitzell, E. Lonnoy, J. B. R. Matthews, T. K. Maycock, T. Waterfield, O. Yelekci, R. Yu, and B. Zhou (eds.)]. Cambridge University Press, Cambridge, UK and New York, NY, pp. 1513–1766, doi:10.1017/9781009157896.013.

Statista (2023). www.statista.com/statistics/1293207/global-number-of-deaths-due-to-flood/#:~:text=In%202022%2C%20floods%20caused%20the,the%20peak%20recorded%20in%201999, consulted in September 2023.

Weber, A. (2023). www.nrdc.org/bio/anna-weber/what-urban-flooding, consulted on September 2023, SSC 2023, Kovacz 2013.

World Bank (WB) (2023). www.worldbank.org/en/topic/urbandevelopment/overview#:~:text=Today%2C%20some%2056%25%20of%20the,people%20will%20live%20in%20cities. Last Updated: 03 April 2023, consulted in September 2023.

Zope, P. E., T. I. Eldho, and V. Jothiprakash (2016). Impacts of Land Use–Land Cover Change and Urbanization on Flooding: A Case Study of Oshiwara River Basin in Mumbai, India. *Catena* 145: 142–154.

1 Urbanisation Process and Urban Water Challenges

Changqing Xu, Mingyi Jia, Haifeng Jia, Alexandros K. Makarigakis, and Blanca Jiménez Cisneros

1.1 THE PROCESS OF URBANISATION IN THE WORLD

Urbanisation is the process of a population shift from rural areas to cities, often motivated by economic and social factors. In many countries, it is the national strategy for societal development and is a process of multi-layer structural change involving environment, economy, society, population, and change of system. Urbanisation is relevant to a range of disciplines, including engineering, geography, sociology, economics, urban planning, and public health. In many perceptions, the phenomenon has been closely linked to modernisation, industrialisation, and the sociological process of rationalisation. Urbanisation can be seen as a specific condition at a set time (e.g., the proportion of total population or area in cities or towns) or as an increase in that condition over time. So urbanisation can be quantified either in terms of, say, the level of urban development relative to the overall population or as the rate at which the urban proportion of the population is increasing. Urbanisation creates enormous social, economic, and environmental changes, which provide an opportunity for sustainability with the "potential to use resources more efficiently, to create more sustainable land use and to protect the biodiversity of natural ecosystems (Cohen, 2015)." It can also give rise to challenging pressures, within relatively small geographical areas, both by high demand for natural resources and by releasing back to the environment waste products that cannot be adequately treated (Jimenez, 2008, 2009b).

Urbanisation is not merely a modern phenomenon, but a rapid and historic transformation of human social roots on a global scale, whereby predominantly rural culture is being rapidly replaced by predominantly urban culture. The first major change in settlement patterns was the accumulation of hunter-gatherers into villages many thousand years ago. Village culture is characterised by common bloodlines, intimate relationships, and communal behaviour, whereas urban culture is characterised by distant bloodlines, unfamiliar relations, and competitive behaviour. This unprecedented movement of people is forecast to continue and intensify during the next few decades, mushrooming cities to sizes unthinkable only a century ago.

Urbanisation has manifested differently in various civilisations. For instance, in the 16th century while London and Paris had between 10,000 and 15,000 people, at the same time the Tenochtitlan, the capital of the Aztec Empire in Mexico, before it was conquered by the Spanish and "modernised," had 400,000 inhabitants. Paris and London were experiencing already pollution challenges and inequity in social development, while in Tenochtitlan pollution did not exist as everything was reclaimed, water was used wisely, and poverty and its associated problems were not existent (Jimenez, 2009a). This shows that cities in different models of social development can truly be different.

Indeed, in experiencing urbanisation, the developed and developing countries have different routes and modes of development. However, there is a tendency to follow in all of the economic factors defined by a limited number of social groups and countries. During the last century, global populations have urbanised rapidly: in 1900, 13% of people lived in urban environments; however, in 1950, 29% of people lived in urban environments (Dewan, 2013). Moreover, urbanisation is still increasing globally. The most urbanised region is Latin America with 81% of its population living in cities in 1987 (more than 260 million of people, Statistica, 2021). Asia is also a highly urbanised region. In China, for example, in the 21st century, urbanisation has entered a new developmental period, and the urbanisation rate (the percentage of urban population) increased from 12.5% in 1952 to 57.35% in 2016. Figure 1.1 represents the degree of urbanisation in China from 2005 to 2016 (National Bureau of Statistics, PRC, 2006–2017).

It is predicted that the urbanisation rate in China would continue to rise. The expectation is that, if unabated, urbanisation's negative impacts on the natural ecosystems and the environment would become widespread and much more severe (Jia et al., 2013; Dewan, 2013). Figure 1.2 (Department of Economic and Social Affairs of the United Nations Secretariat, 2014) and Figure 1.3 (Department of Economic and Social Affairs of the United Nations, 2018) show the percentage of world population from 1950 to 2050, by major cities, and the growth rates of urban agglomerations in 1990–2018, respectively.

If properly managed, urbanisation can lead to social and economic progress. First, it can increase food production due to high demand. When more people move from rural to urban areas in search of better job opportunities, food production significantly increases to feed the population. Besides, cities are tremendously efficient in certain ways. For instance, imagine 50 families living in 50 separate houses spread out evenly over a large piece of land. If these families were to stay in housing (apartments, flats, condos etc.) within the same building or in a complex of buildings, it would require less effort and resources to supply social

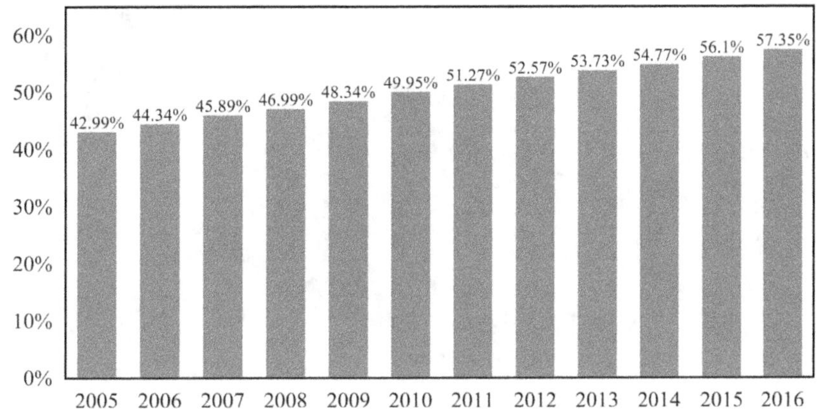

Figure 1.1 Degree of urbanisation in China from 2005 to 2016 (produced by authors with information from: National Bureau of Statistics of China, 2006–2017)

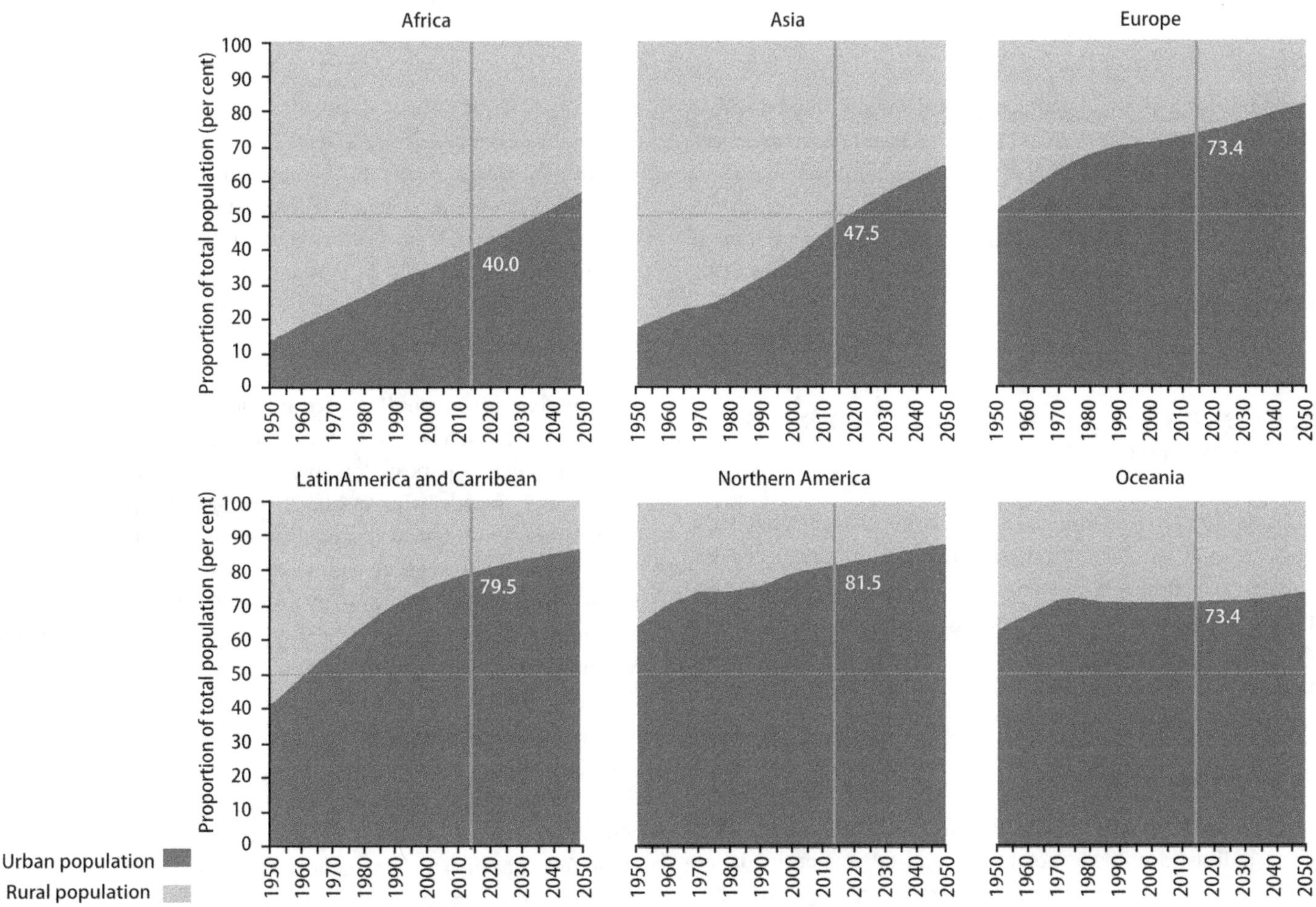

Figure 1.2 Urban and rural population graph as proportion of total population by major areas, 1950–2050 (source: Department of Economic and Social Affairs of the United Nations Secretariat, 2014)

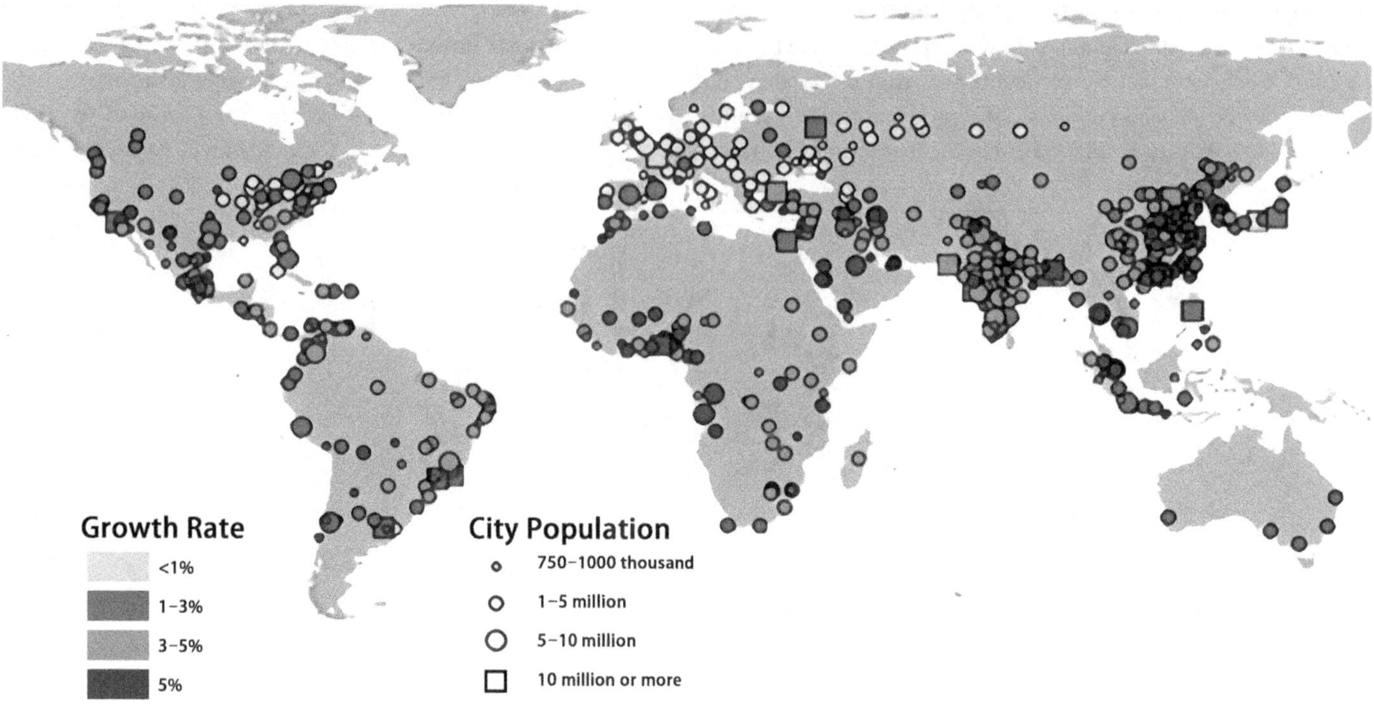

Figure 1.3 Growth rates of urban agglomerations, 1990–2018 (source: Department of Economic and Social Affairs of the United Nations Secretariat, 2018)

amenities such as water and energy. In addition, in urban areas, waste removal is efficient, and access to health, social services, cultural events, and education is easy. The same is true in the case of transportation. Cities tend to invest in mass transport making movement fluent and easy, which is not the case in rural areas, where transportation without private means is nearly impossible (Nowak, 1997).

1.2 URBAN WATER CHALLENGES WORLDWIDE

Urbanisation can lead to social and economic progress, but it also puts pressure on cities to provide infrastructure and social services. It thus brings negative impacts to natural ecosystems and the environment, especially urban hydrology, for which the negative effects include urban flooding, water shortage, water pollution, and other water-related ecological environment problems.

1.2.1 Urban Flooding

Over the past few years, many cities around the world have experienced disastrous floods, notably, at least five major floods ravaged Queensland and Victoria in Australia during 2010–2011 (DFO, 2012; Wilby and Keenan, 2012). The Indian commercial capital, Mumbai, experienced an unprecedented flood in 2005 that resulted in the estimated damages of around US$2 billion and 500 fatalities (Gupta, 2007; Ranger et al., 2011). Floods in New Orleans, the United States, in 2005 engulfed 80% of the city and resulted in enormous damage (Bizimana and Schilling, 2010). In 2012, Beijing, China, suffered a severe flood that resulted in a damage of about US$2 billion and 79 fatalities. Recently, in 2017, Texas-Houston experienced an unprecedented flood that caused severe waterlogging for the whole city and serious economic losses.

Excessive urban runoff caused by the dramatic increases of urbanisation, which is mainly caused by the increase of impervious areas (such as parking lots, buildings, and roads) (Marsalek et al., 2007; Kundzewicz et al., 2010), is causing flooding in urban areas and urban water quality deterioration in receiving water bodies and even economic losses (such as agricultural losses, urban and rural resident property losses, and hydraulic infrastructure losses). Urbanisation has called for a new hydrological concept called the urban water cycle (Figure 1.4). Globally, floods account for more than 55% of all fatalities with nearly 2.5 billion people affected (EM-DAT, 2012) and more than 30% of global economic losses from natural disasters. In China, according to a survey of 351 cities in China by the Ministry of Housing and Urban-Rural Development in 2010, about 62% of cities had encountered flooding events from 2008 to 2010. Among these cities, nearly 63% encountered more than three flooding disasters. During the periods of flooding, the standing waters were appeared in low-lying areas. The depth of

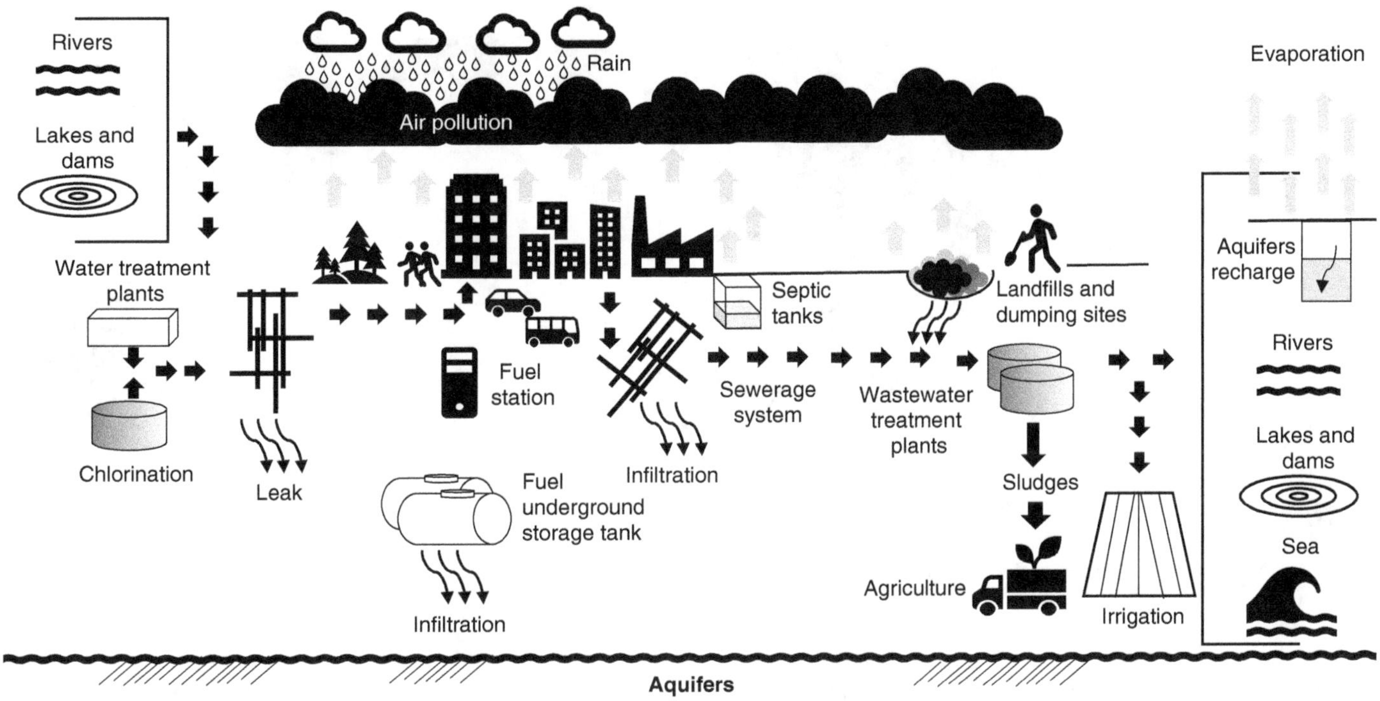

Figure 1.4 The urban water cycle, from Marsalek, 2007

standing water exceeded 50 cm in more than 74.6% of the cities, and the duration of standing water exceeded 0.5 hours in 78.9% of the cities (Jia et al., 2013).

Most of the available information on floods combines the effects of both urban and plain floods. Urban flooding has significant economic implications for affected neighbourhoods. In the United States, industry experts estimate that wet basements can lower property values by 10–25% and are cited among the top reasons for not purchasing a home (Center for Neighbourhood Technology, 2013). According to the US Federal Emergency Management Agency (FEMA), almost 40% of small businesses never reopen their doors following a flooding disaster.

A flood in Bangkok, Thailand, lasted for 175 days, took 815 lives, and caused about US$45.7 billion in damage (World Bank, 2011). Moreover, as reported, an average of 500,000 people per year are affected by floods in West Africa alone (Douglas et al., 2009). Meanwhile, the projected average annual population to be affected by river floods is around 21 million people worldwide and is likely to rise to 54 million by 2030. Because of the increasing damages and unacceptably high numbers of injuries, floods have become the public concern worldwide. Aggravated urban water challenges and vulnerabilities to flood risk pose great problems for policymakers worldwide. Figure 1.5 shows the number of occurrences of flood disasters by country from 1974 to 2003 (Guhasapir, 2004).

The vulnerability of megacities to floods, particularly in developing nations, is exacerbated due to rapid urban expansion, increasing concentration of population and property, rampant poverty, physical location, and poor-quality housing (Kubal et al., 2009; Suriya and Mudgal, 2012).

The degree of vulnerability to natural hazards is likely to increase exponentially (Sharma et al., 2011). A recent study indicated that, due to weaknesses in the adopted disaster preparedness system, communities' capacity to cope with extreme events is being reduced in many Asian megacities (Osti et al., 2011), and it is expected to worsen. Further, the dynamics of the socioeconomic system that results from global change, development practice, and political instability could lead to more frequent and disastrous floods if disasters risk management is not mainstreamed into the planning process of these megacities (Adger, 2006).

Many of the largest cities are exposed to more than one type of flood hazard. For example, the largest megacity in the world, Tokyo, is susceptible to cyclones and floods. Of the natural hazards analysed, estimates show that at least 233 cities across the globe are at a high risk of flooding, potentially affecting 663 million people. Some recent catastrophes include the inundations of Nîmes and Vaison-la-Romaine, the flooding of New Orleans, the flooding of Beijing, and the flooding in Rockhampton. Flood flows in urban environments have been studied relatively recently despite many centuries of experiencing flood events (Brown et al., 2011). Urban floods can exist even in elevated, mountainous areas, as was the case in Mexico City, with impacts not only on the infrastructure but also on the quality of the water that

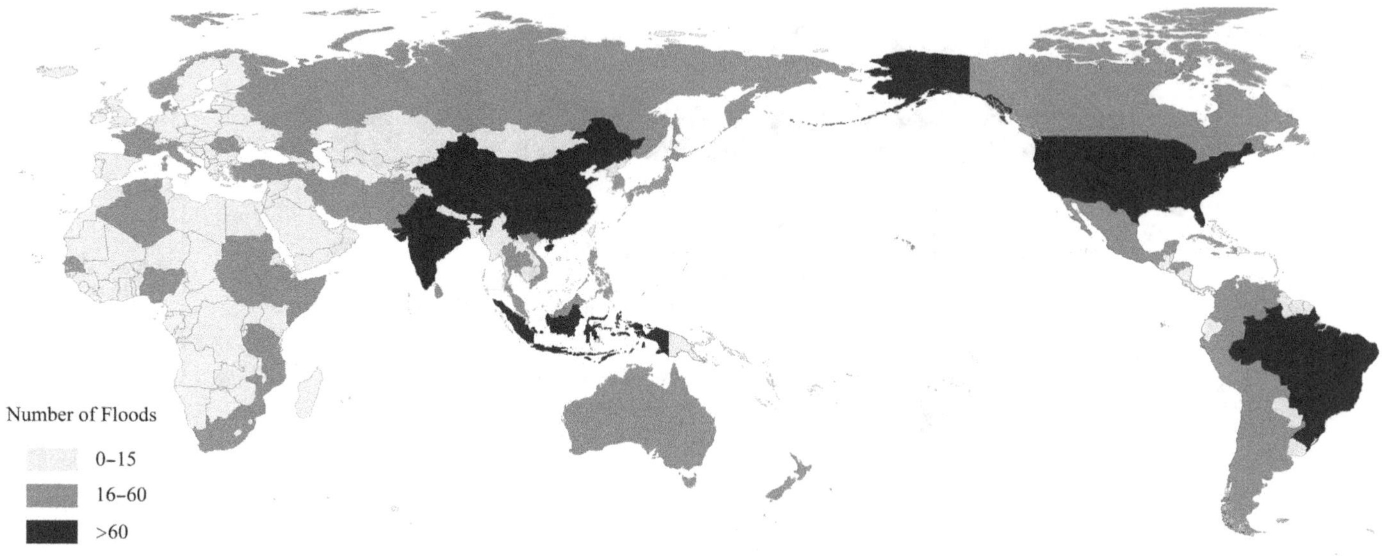

Figure 1.5 Number of occurrences of flood disasters by country: 1974–2003 (Guhasapir, 2004)

is used as supply (Jiménez Cisneros et al., 2017). Some recent research has even considered criteria for safe evacuation of individuals in flooded areas.

Floods are the most costly and debilitating natural hazards in the world. To save lives and property, flood management has been heavily dominated by engineering solutions. Although a degree of success has been reported, flood still remains an inordinate threat to people, and it is blamed for environmental degradation across the country (DFO, 2012).

1.2.2 Water Shortage

As previously mentioned, with the rapid development of urbanisation, the number of people living in the urban areas increased significantly and the water consumption rose sharply. Therefore, the groundwater has been over-mined, and the groundwater funnels increased. At the same time, the increased impervious area has cut off the water to replenish the groundwater and so on, causing the urban water shortage. For some countries, the percentage of people lacking adequate water supplies exceeds well over half of their total populations. As a result, many, especially the very young, die (Loucks and Jia, 2012). In addition, the supply shortage has resulted in conflicts between countries, created prolonged drought, closing of industrial units, shifting of local inhabitants and so on. In the past decade, progress has been made in providing more people with access to clean drinking water and basic sanitation (UNESCO, 2009). But a major effort is still required to extend these essential conditions to those still without them; the vast majority of them are poor and cannot pay the costs of these basic services.

How can one allocate water supplies optimally among all demands that impact the quality of, or even the existence of, life – both human and ecosystem life – in times of critical resource scarcity (UNDESA, 2014)? A general precise answer that fits all circumstances is never clear. By 2025, an estimated 3.4 billion people will be living in countries defined as water scarce. Many in those countries seem to be able to survive on as little as 3 L per day per person although WHO recommends a minimum of 100 L/person per day to have enough water to drink and provide the supply for other basic water needs. However, it takes about 3,000 L of water to produce a daily ration of food (Loucks and Jia, 2012), about 1,000 times what it is minimally need for dietary purposes. A substantial portion of our food comes from irrigated lands. On average over 70% of total freshwater use in the world is devoted to irrigation. Over the next 30 years, about 70% of grains used in cereal production are expected to come from irrigated land (UNESCO, 2009).

The physical evidence of increasing periods of water scarcity can be found almost everywhere in the world. Water scarcity (Figure 1.6) affects rich and poor countries alike; however, the lower income class, in both cases, is the one who suffers the most. Nearly three billion people live in water scarce conditions (over 40% of the world's population), and this situation could worsen if current population growth trends continue, and if the melting of some of the major sources of water – the glaciers – continues too (Loucks and Jia, 2012). The manifestations of pervasive water poverty include millions of deaths every year due to malnourishment and water-related diseases, political conflict over scarce water resources, extinction of freshwater species, and degradation of aquatic ecosystems. Roughly half of all the world's wetlands have already been lost and dams have seriously altered the flow of roughly 60% of the world's major river basins (Davis, 2007). The situation only worsens with time and urbanisation. In China, over 20% of the world's population is found but there

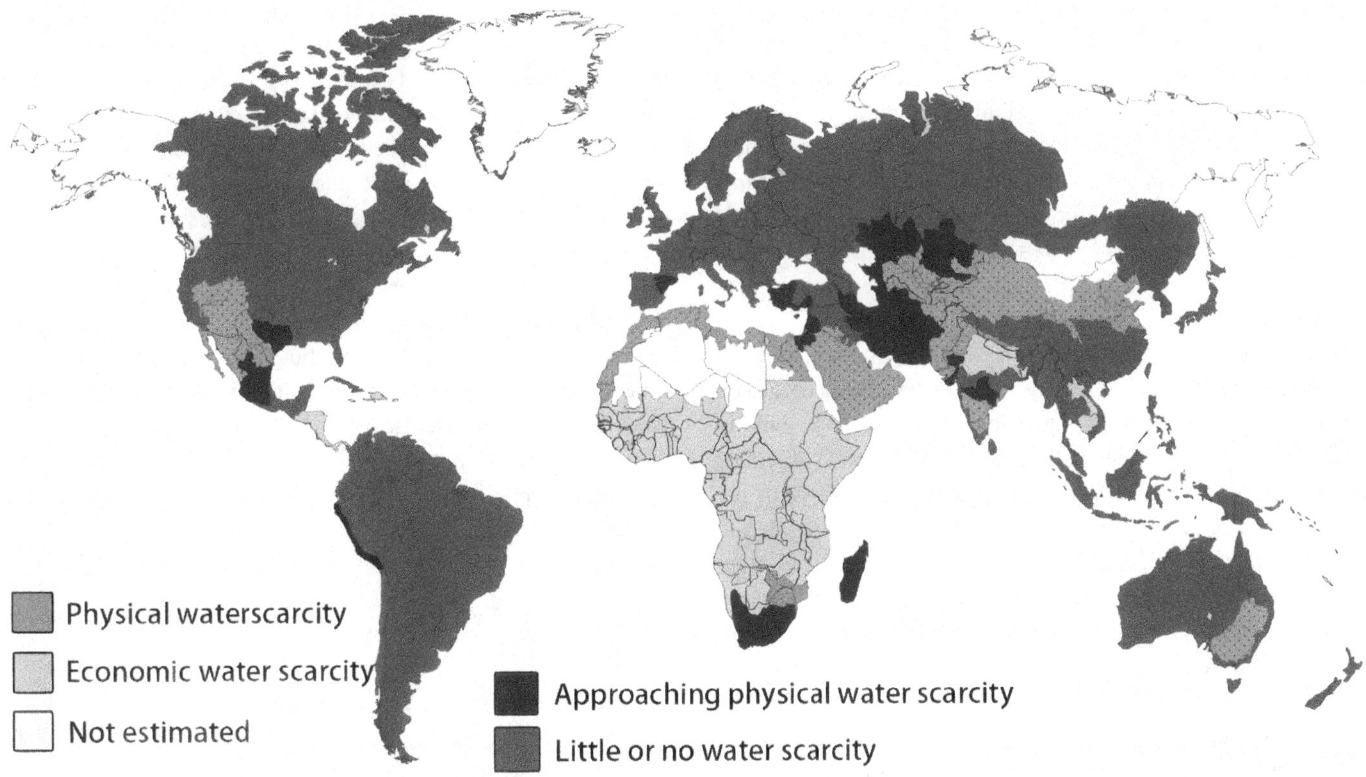

Figure 1.6 Global physical and economic water scarcity (Loucks and Jia, 2012)

is only about 7% of the world's freshwater runoff and approximately 400 of 600 cities are facing a water shortage challenge. China had a per capita water of 2079 m^3, while the world average was 6225 m^3. In addition, the economic losses caused by urban water shortage was 200 billion yuan/year (University of Seoul & Seoul Waterworks Authority, 2017), which is about 27 billion US$/year. Water pollution, over-exploitation of underground water, and low efficiency of water usage along with water shortages have continuously deepened the imbalance between water supply and demand in China (Loucks and Jia, 2012).

Water shortage has brought about many negative impacts on the environment, including lakes, rivers, wetlands, and other freshwater resources. The resulting water overuse that is related to water shortage, often located in the areas of irrigation agriculture, harms the environment in several ways including increased salinity, nutrient pollution, and the loss of floodplains and wetlands (United Nations Development Programme, 2006). Furthermore, water shortage makes flow management in the rehabilitation of urban streams problematic (Lawrence et al., 2014). Through the last hundred years, more than half of the Earth's wetlands have been destroyed and have disappeared (WWF, 2013). These wetlands are important not only because they are the habitats of numerous inhabitants such as mammals, birds, fish, amphibians, and invertebrates, but they support the growing of rice and other food crops as well as provide water filtration and protection from storms and flooding. Freshwater lakes such as the Aral Sea in central Asia have also suffered. Once the fourth largest freshwater lake, it has lost more than 58,000 km^2 of area and vastly increased in salt concentration over the span of three decades (WWF, 2013). Subsidence, or the gradual sinking of landforms, is another result of water shortage. The US Geological Survey estimates that subsidence has affected more than 44,029 km^2 in 45 US states, 80% of it due to groundwater usage. In some areas east of Houston, Texas, the land has dropped by more than 9 feet (2.74 metres) due to subsidence. Brownwood, a subdivision near Baytown, Texas, was abandoned due to frequent flooding caused by subsidence and has since become part of the Baytown Nature Center.

1.2.3 Water Pollution

In the urbanisation process, many types of open land (such as forest and agricultural land) as well as water bodies are transformed into impervious surfaces. This process alters the water balance by promoting polluted runoff at the expense of infiltration (Jia et al., 2013). Usually, urban development negatively impacts water quality in three ways. First, many of the new materials and components used in land development contribute to higher pollutant loads during the episodes of rainfall and the subsequent runoff. Second, the natural filtering action of wild vegetation (trees, bushes, tall grasses, etc.) is replaced by concrete, asphalt, and rooftops, which offer little means for water quality improvement (Jia et al., 2013). Larger impervious areas lead to higher levels of more diverse pollutants, diminished pollutant removal during overland flow, reduced infiltration, and greater runoff peak flows, which, in turn, can increase stream erosion. Most traditional pollutants found in urban runoff include suspended solids, nutrients, and heavy metals. Table 1.1 shows the typical pollutant concentrations in urban runoff for different land uses (Jia et al., 2013).

In urban areas, greater overland flows from impervious areas increase the erosion rates of nearby soils and the amount of dust on impervious surfaces. Loss of vegetation eliminates the possibility of attenuation of suspended solids and results in total suspended solids levels that range from tens to hundreds of milligrams per litre in urban runoff (Jia et al., 2013). Fertilisers are applied to managed grass and other landscaped areas. If not managed properly, nutrients readily wash from these lawns. Nitrate deposition from automobile NOx emissions also becomes part of the runoff flows. In high-density residential areas, wastes from pets (if not cleaned up) and other urban animals are washed away by rainfall and add significant organic and pathogen loadings. Moreover, many toxic compounds also exist in urban runoff. Heavy metals, such as lead and zinc, are released from automobiles through brake and tire wear, as are corroded and weathered paints, coatings, decorative flashing, galvanised materials, and structural components (Councell et al., 2004; Li et al., 2012). Motor oil, fuel, gear fluids, and coolants from vehicles are readily washed from impervious surfaces. Numerous toxic hydrocarbons and their oxidised products, which may persist in the environment, are also present in these fluids. Polycyclic aromatic hydrocarbons (PAHs) may be found in leaking fluids, incompletely combusted fuel in atmospheric deposits, and asphaltic products used in roofing and pavements. Various herbicides and insecticides are frequently used to control weeds and damage from insects in green urban areas. Again, without careful, controlled application, excess pesticides will leach from the application areas. Some of these pollutant sources are unavoidable with current practices, but the development focussed on the water runoff issues can reduce the impacts on the environment (Jia et al., 2013). The impacts of cities and floods on water quality have been described by Jimenez and Rose (2009), just as, under the framework of climate change, the effects of extreme rain events and floods on water quality have been presented by Jiménez Cisneros et al. (2014).

The third way cities impact water quality is through the leakages of their many underground networks. These networks of course convey "potable" water, but there are other networks transporting sewage, and, in some cities, even oil and gasoline. Underground tanks used to store different kind of fluids are as well a source of pollution to aquifers that are used to fed cities. Pollution seepage to aquifers is increased during urban floods (Jimenez and Rose, 2009).

Table 1.1 *Typical pollutant concentration in urban runoff for different land uses*

	Median event mean concentration for land use		
Pollutant	Residential	Mixed	Commercial
Chemical oxygen demand ($mg{\cdot}L^{-1}$)	73	65	57
Total suspended solids ($mg{\cdot}L^{-1}$)	101	67	69
Total Kjeldahl nitrogen ($mg{\cdot}L^{-1}$)	1.90	1.29	1.18
Nitrate and nitrite ($mg{\cdot}L^{-1}$)	0.74	0.56	0.57
Total phosphorus ($mg{\cdot}L^{-1}$)	0.38	0.26	0.20
Total lead ($mg{\cdot}L^{-1}$)	0.14	0.11	0.10
Total zinc ($mg{\cdot}L^{-1}$)	0.13	0.15	0.23

1.2.4 Other Water-Related Ecological Environment Problems

1.2.4.1 ECOLOGICAL DEGRADATION

Urban development results in multiple stressors that can degrade aquatic ecosystems by altering the hydrology, habitat, and water quality of urban rivers (Figure 1.7) (Cappiella et al., 2012). Results of the previous investigation found that no single environmental factor was universally important in explaining why the health of rivers declines as levels of urban development increases. Changes in the condition of the aquatic biologic communities from urban development were related in varying degrees to alterations in hydrology, habitat, and water quality.

For the change in hydrology caused by urban development, it is obvious that urban development could typically increase the amount of water entering a stream after a storm and decrease

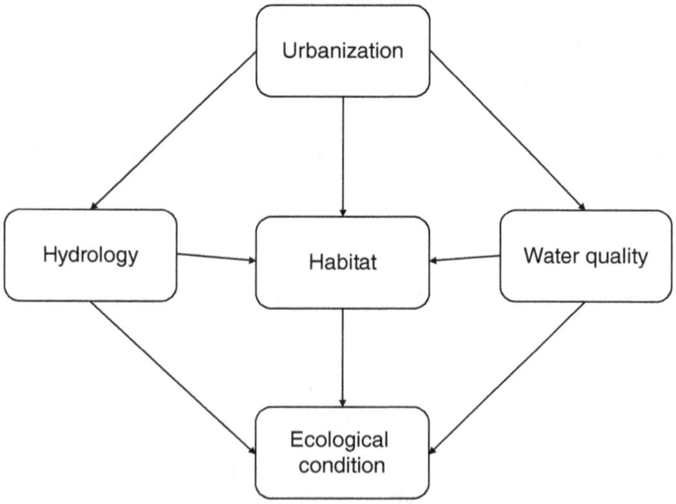

Figure 1.7 Stressors of aquatic ecosystem degradation by urbanisation (Jia et al., 2013)

the time that it takes for the water to travel over altered land surfaces before entering the stream. These factors can lead to stream flashiness and altered stream channels. Additionally, rapid runoff reduces the amount of water available to infiltrate the soil and recharge the aquifers, which often results in lower sustained stream flows (Jia et al., 2013). Habitats can be severely degraded when urban development occurs along riverbanks, such as where a stream has been straightened by channelisation or where manmade structures have replaced natural riparian vegetation. Furthermore, urban development that occurs throughout a watershed can result in a degraded habitat within a stream channel through flow alteration and sediment erosion. Water quality deterioration is another important factor that contributes to ecological degradation. As described earlier, the concentration of contaminants, including nitrogen, chloride, insecticides, and PAHs, increased with urban development. The changed chemistry brings negative impacts to the aquatic biota, as most contaminants have toxic effects on biologic communities.

Under the impact of changes in hydrology, habitat, and water quality caused by urban development, ecological degradation is observed. Generally, it may result in a shift in the species composition of the algal, invertebrate, and fish communities. The most consistent change in any of the biological communities was the loss of sensitive invertebrate species and a shift to a community with a higher percentage of species more tolerant to physical and chemical stressors (Jia et al., 2013).

A loss in the numbers of aquatic insect species that occurred in the orders Ephemeroptera (mayflies), Plecoptera (stoneflies), and Trichoptera (caddisflies) (EPT) was a common response in all study areas where urban development occurred in forested watersheds (Jia et al., 2013). The number of EPT species is a biological-condition metric that is used in many biomonitoring programmes across the country because it is sensitive to stressors from environmental degradation (Wang and Kanehl, 2003); reduction of more than 50% of EPT species was observed in some study areas as the percentage of urban development increased in the watersheds from low to high levels (Cappiella et al., 2012).

Studies have indicated that modifications in stream ecology occur when the proportion of impervious area is only a few percentage. Once the impervious portion reaches 10–30%, researchers see major declines in habitat and water-quality indicators (Wang et al., 2001). In addition to aquatic ecosystem degradation, terrestrial ecosystems are also affected by urbanisation. Converting "natural" land, such as forest, to residential or commercial use drastically changes its ecological characteristics. Vegetation is uprooted and replaced by impervious surfaces – rooftops, driveways, roadways, parking lots, and sidewalks. Even the remaining vegetated areas in residential developments tend to be smoothed, and small water storage depressions are eliminated. Diverse plant species are replaced with monoculture grasses that are regularly cut short and may be inundated with fertilisers and pesticides. In this way, the ecosystem service values decline. Ecological degradation in urban areas also aggravates the heat island effect of cities (Jia et al., 2013).

1.2.4.2 URBAN HEAT ISLANDS

Urban heat island (UHIs) is another ecological environment problem caused by urbanisation. Urban heat island is an urban area or metropolitan area that is significantly warmer than its surrounding rural areas due to human activities. The temperature difference usually is larger at night than during the day, and is most apparent when winds are weak. Urban heat island is most noticeable during the summer and winter. This effect was first described for Mexico City (Jauregui, 1997). The main cause of the UHI effect is from the modification of land surfaces (Solecki et al., 2005). The UHIs are formed when industrial and urban areas replace and reduce the amount of land covered by vegetation or open soil (Huang and Lu, 2015). In rural areas, the ground helps regulate temperatures by using a large part of the incoming solar energy to evaporate water in vegetation and soil (Joshi et al., 2015). This evaporation, in turn, has a cooling effect. However, in cities, where less vegetation and exposed soil exists, the majority of the Sun's energy is absorbed by urban structures and asphalt (Ren et al., 2015). During the day, cities experience higher surface temperatures because urban surfaces produce less evaporative cooling. Additional city heat is given off by vehicles and factories, as well as industrial and domestic heating and cooling units. Together, these effects can raise city temperatures by or 1°C up to 10°C (Figure 1.8) (Jimenez, 2009b; Kershaw et al., 2010). There are concerns raised about possible contribution from UHIs to global warming. Research in China (Huang and Lu, 2015; Ren et al., 2015) and India (Joshi et al., 2015) indicates that UHI effect contributes to climate warming by about 30%.

Aside from the effect on temperature, UHIs can produce secondary effects on local meteorology, including the altering of local wind patterns, the development of clouds and fog, the humidity, and the rates of precipitation. The extra heat provided

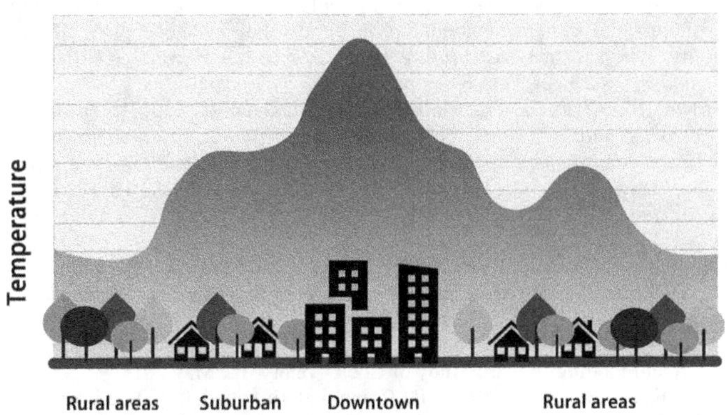

Figure 1.8 The temperatures of rural, commercial, urban residential, and suburban residential areas

by the UHI leads to greater upward motion, which can induce additional shower and thunderstorm activity. In addition, the UHI creates during the day a local low-pressure area where relatively moist air from its rural surroundings converges, possibly leading to more favourable conditions for cloud formation (Chiel et al., 2008). Rainfall rates downwind of cities are increased between 48% and 116%. Partly, as a result of this warming, monthly rainfall is about 28% greater between 20 miles (32 km) to 40 miles (64 km) downwind of cities, compared with upwind. Some cities show a total precipitation increase of 51% (Goddard Space Flight Center, 2008). One of the main effects of the heat island phenomenon in cities is to form well-localised heavy rains in cities producing an amount of water that cannot be handled by urban sewers (Jiménez Cisneros et al., 2014). Research has been done in a few areas suggesting that metropolitan areas are less susceptible to weak tornadoes due to the turbulent mixing caused by the warmth of the UHI.

UHIs also impair water quality. Hot pavement and rooftop surfaces transfer their excess heat to stormwater, which then drains into storm sewers and raises water temperatures as it is released into streams, rivers, ponds, and lakes. Additionally, increased urban water body temperatures lead to a decrease in diversity in the water (Chen et al., 2011). In August 2001, rains over Cedar Rapids, Iowa led to a 10.5°C (18.9°F) rise in the nearby stream within one hour, which led to a fish kill. Since the temperature of the rain was comparatively cool, it could be attributed to the hot pavement of the city. Similar events have been documented across the American Midwest, as well as Oregon and California (Paul and Gene, 2011). Rapid temperature changes can be stressful to aquatic ecosystems. Permeable pavements may mitigate these effects by percolating water through the pavement into subsurface storage areas where it can be dissipated through absorption and evaporation (USEPA, 2013).

References

Adger, W. (2006). Vulnerability. *Global Environmental Change*. 16(3): 268–281.

Bizimana, J. P., Schilling, M. (2010). Geo-information technology for infrastructural flood risk analysis in unplanned settlement: A case study of informal settlement flood risk in the Nyabugogo flood plain, Kigali city, Rwanda. In *Geospatial Techniques in Urban Hazard and Disaster Analysis*. Springer: Dordrecht, pp. 99–124.

Brown, R., Chanson, H., McIntosh, D., Madhani, J. (2011). *Turbulent Velocity and Suspended Sediment Concentration Measurements in an Urban Environment of the Brisbane River Flood Plain at Gardens Point on 12–13 January 2011*. Hydraulic Model Report No. CH83/11. Brisbane, Australia: The University of Queensland, School of Civil Engineering (CH83/11), p. 120.

Cappiella, K., Stack, W. P., Fraley-McNeal, L., Lane, C., McMahon, G. (2012). *Strategies for Managing the Effects of Urban Development on Streams*. Circular 1378. Reston: US Geological Survey.

Center for Neighborhood Technology (2013). The Prevalence and Cost of Urban Flooding, A Case Study of Cook County, IL, May 2013.

Chen, F., Kusaka, H., Bornstein, R., Ching, J., Grimmond, C. S. B., Grossman-Clarke, S., Loridan, T., Manning, K. W., Martilli, A., Miao, S., Sailor, D., Salamanca, F. P., Taha, H., Tewari, M., Wang, X., Wyszogrodzki, A., Zhang, C. (2011). The integrated WRF/urban modelling system: Development, evaluation, and applications to urban environmental problems. *International Journal of Climatology*. 31(2): 273–288.

Chiel, C., Heerwaarden, V., Vilà-Guerau de Arellano, J. (2008). Relative humidity as an indicator for cloud formation over heterogeneous land surfaces. *Journal of the Atmospheric Sciences*. 65(10): 3263–3277.

Cohen, B. (2015). Urbanization, city growth, and the new united nations development agenda. *Cornerstone, The Official Journal of the World Coal Industry*. 3(2): 4–7.

Councell, T. B., Duckenfield, K. U., Landa, E. R., Callender, E. (2004). Tire-wear particles as a source of zinc to the environment. *Environmental Science & Technology*. 38(15): 4206–4214.

Dartmouth Flood Observatory (DFO) (2012). Global register of major flood events. www.dartmouth.edu/~floods/Currentt.htm. Accessed 10 February 2012.

Davis, C. (2007). *The Multiple Dimensions of Water Scarcity*. World Resources Institute, http://earthtrends.wri.org/updates/node/264

Department of Economic and Social Affairs of the United Nations (2018). World Urbanization Prospects: The 2018 Revision. Available at: https://population.un.org/wup/Maps/. Accessed 7 September 2023.

Department of Economic and Social Affairs of the United Nations Secretariat (2014). World urbanization prospect-2014 revision. ISBN 978-92-1-123195-3.

Dewan, A. M. (2013). *Floods in a Megacity: Geospatial Techniques in Assessing Hazardous, Risk and Vulnerability*. ISBN 978-94-007-5875-9 (eBook), DOI 10.1007/978-94-007-5875-9, New York, London: Springer Dordrecht Heidelberg. Library of Congress Control Number: 2013930017.

Douglas, I. (2009). Climate change, flooding and food security in South Asia. *Food Security*. 1(2): 127–136.

EM-DAT (2012). Global flood records, 1950–2011. Centre for Research on the Epidemiology of Disasters (CRED), Université Catholique de Louvain, Belgium. www.emdat.be.

Goddard Space Flight Center (2008). NASA Satellite Confirms Urban Heat Islands Increase Rainfall Around Cities. National Aeronautics and Space Administration. Archived from the original on 12 June 2008. Retrieved 2009-07-17.

Guhasapir, D., Hargitt, D., Hoyois, P. (2004). Thirty years of natural disasters, 1970–2003: The numbers. Centre for Research on the Epidemiology of Disasters. ISBN: 2-930344-71-7.

Gupta, K. (2007). Urban flood resilience planning and management and lessons for the future: A case study of Mumbai, India. *Urban Water Journal*. 4(3): 183–194.

Huang, Q., Lu, Y. (2015). The effect of urban heat island on climate warming in the yangtze river delta urban agglomeration in China. *International Journal of Environmental Research and Public Health*. 12(8): 8773–8789.

Jauregui, E. (1997). Heat island development in Mexico City. *Atmospheric Environment*. 31(22): 3821–3831.

Jia, H. F., Yao, H. R., Yu, S. L. (2013). Advances in LID BMPs research and practices for urban runoff control in China. *Frontiers of Environmental Science and Engineering*. 7(5): 709–720.

Jimenez, B. (2008). Wastewater risks in the urban water cycle. In *Urban Water Security: Managing Risks*. Jimenez, B. and Rose, J. Ed., pp. 151–200. Taylor and Francis Group. ISBN: 978-0-415-48567-8.

Jimenez, B. (2009a). Coming to terms with nature: Water reuse new paradigm towards integrated water resources management. Chap. 2.20 in *Water and Health Vol II theme, Encyclopedia of Life Support*. Grabow, W. Ed., pp. 398–428. System EOLSS Encyclopedia of Life Support Systems, Publishers Co Ltd UNESCO. ISBN: UNESCO 93-3-103999-7.

Jimenez, B. (2009b). Risks and integrated management of the urban water cycle in megacities of the developing world: Mexico city. In *Chapter in Water and Urban Development Paradigms*. Feyen, S. J. and Neville, K. M. Eds., pp. 387–397. CRC Press Taylor and Francis Group UNESCO, London, UK. ISBN: 978-0-415-483346.

Jimenez, B., Rose, J. (Ed.) (2009). Urban Water Security: Managing Risks. Urban Water Series, Vol 5, CRC Press Taylor and Francis Group and UNESCO, 308 pp. The Netherlands. ISBN: 978-0-415-48567-8. DOI 13:978-0-203-88162-0 (In English), China Water & Power Press and UNESCO, 364 pp. China. ISBN: 978-7-5170-2658-7. [In Chinese].

Jiménez Cisneros, B., Navarro González, I., Montes Rojas, R. (2017). Water quality risk associated with climate change: three Mexico City case studies, II Adaptation and Integral Management Chapter 3 in Climate Change-Sensitive Cities: Building Capacities for Urban Resilience, Sustainability, and Equity. Delgado Ramos G. C. (Ed.) pp. 81–132 PNUD. E- E-ISBN: 978-607-02-9973-5 [In Spanish].

Jiménez Cisneros, B. E., Oki, T., Arnell, N. W., Benito, G., Cogley, J. G., Döll, P., Jiang, T., Mwakalila, S. S. (2014). Freshwater resources. In *Climate Change (2014) Impacts, Adaptation, and Vulnerability. Part A: Global and Sectoral Aspects*. Contribution of Working Group II to the Fifth Assessment Report of the Intergovernmental Panel on Climate Change [Field, C. B., V. R. Barros, D. J. Dokken, K. J. Mach, M. D. Mastrandrea, T. E. Bilir, M. Chatterjee, K. L. Ebi, Y. O. Estrada, R. C. Genova, B. Girma, E. S. Kissel, A. N. Levy, S. MacCracken, P. R. Mastrandrea, and L. L. White Eds.]. Cambridge University Press, Cambridge, UK and New York, NY, pp. 229–269. http://ipcc-wg2.gov/AR5/images/uploads/WGIIAR5-Chap3_FGDall.pdf.

Joshi, R., Raval, H., Pathak, M., Prajapati, S., Patel, A., Singh, V., Kalubarme, M. H. (2015). Urban heat island characterization and isotherm mapping using geo-informatics technology in Ahmedabad City, Gujarat State, Indi. *International Journal of Geosciences*. 6(3): 274.

Kershaw, T. J., Sanderson, M., Coley, D., Eames, M. (2010). Estimation of the urban heat island for UK climate change projections. *Building Services Engineering Research and Technology*. 31(3): 251–263.

Kubal, C., Haase, D., Meyer, V., Scheuer, S. (2009). Integrated urban flood risk assessment – adapting a multi-criteria approach to a city. *Natural Hazards and Earth System Sciences*. 9(6): 1881–1895.

Kundzewicz, Z. W., Hirabayashi, Y., Kanae, S. (2010). River floods in the changing climate-observations and projections. *Water Resources Management*. 24(11): 2633–2646.

Lawrence, J. E., Pavia, C. P. W., Kaing, S., Bischel, H. N., Luthy, R. G., Resh, V. H. (2014). Recycled water for augmenting urban streams in mediterranean-climate regions: A potential approach for riparian ecosystem enhancement. *Hydrological Sciences Journal*. 59: 488–501.

Li, W., Shen, Z. Y., Tian, T., Liu, R. M., Qiu, J. L. (2012). Temporal variations of heavy metal pollution in urban stormwater runoff. *Frontiers of Environmental Science and Engineering*. 6(5): 692–700.

Loucks, D. P., Jia, H. F. (2012). Managing water for life. *Frontiers of Environmental Science and Engineering*. 6(2): 255–264.

Marsalek, J., Jimenez, B., Malmquist, P., Karamouz, M., Goldenfum, J., Chocat, B. (2007). Urban Water Cycle Processes and Interactions, Urban Water Series, Taylor & Francis Group, 152 pp. The Netherlands. [152 pp in English ISBN: 978-92-3-104010-3 y 160 pp. and in Chinese ISBN: 978-7-5170-2583-2].

National Bureau of Statistics, People's Republic of China (2006–2017). National Report on Economic and Social Development in 2016 (in Chinese). Available from; www.gov.cn/xinwen/2017-02/28/content_5171643.htm.

Nowak, J. (1997). Neighborhood initiative and the regional economy. *Economic Development Quarterly*. 11: 3–10.

Osti, R., Hishinuma, S., Miyake, K., Inomata, H. (2011). Lessons learned from statistical comparison of flood impact factors among southern and eastern Asian countries. *Journal of Flood Risk Manage*. 4(3): 203–215.

Paul, A. T., Gene, M. (2011). *Physics for Scientists and Engineers*. Macmillan. 686. ISBN 978-1-4292-0124-7.

Ranger, N, Hallegatte, S., Bhattacharya, S., Bachu, M., Priya, S., Dhore, K., Ra fique, F., Mathur, P., Naville, N., Henriet, F., Herweijer, C., Pohit, S., Corfee-Morlot, J. (2011). An assessment of the potential impact of climate change on flood risk in Mumbai. *Climate Change*. 104: 139–167.

Ren, Y., Parker, D., Ren, G., Dunn, R. (2015). Tempo-spatial characteristics of sub-daily temperature trends in mainland China. *Climate Dynamics*. 46: 2737–2748 (2016). https://doi.org/10.1007/s00382-015-2726-7

Sharma, A., Surjan, A., Shaw, R. (2011). Overview of Urban Development and Associated Risks. In *Community, Environment and Disaster Risk Management*. Sharma, A., Surjan, A. and Shaw R Eds. Bingley: Emerald Group Publishing, pp. 1–16.

Solecki, W. D., Rosenzweig, C., Parshall, L., Pope, G., Clark, M., Cox, J., Wiencke, M. (2005). Mitigation of the heat island effect in urban New Jersey. *Global Environmental Change Part B: Environmental Hazards*. 6(1): 39–49.

Statistica (2021). www.statista.com/statistics/699089/urbanization-in-latin-america-and-caribbean/, consulted on 3 May 2021.

Suriya, S., Mudgal, B. V. (2012). Impact of urbanization on flooding: The Thirusoolam subwatershed – a case study. *Journal of Hydrology*. 412–413: 210–219.

United Nations Development Programme (2006). *Human Development Report 2006: Beyond Scarcity–Power, Poverty and the Global Water Crisis*. Basingstoke, United Kingdom: Palgrave Macmillan.

United Nations Department of Economic and Social Affairs (UNDESA) (2014). Water Scarcity. International Decade for Action 'Water for Life' 2005–2015. Last update: 24th, November 2014. Available from: www.un.org/waterforlifedecade/scarcity.shtml.

United States Environmental Protection Agency (USEPA) (2013). Cool Pavement Report (PDF). Available from: www.epa.gov/heat-islands.

University of Seoul & Seoul Waterworks Authority, Seoul Metropolitan Government, 16–17, August (2017). The 13th International Conference on Sustainable Water Environment (ICSWE 2017).

UNESCO (2009). *Water in a Changing World, the 3rd United Nations World Water Development Report: (WWDR-3)*. Paris: UNESCO Publishing.

Wang, L. Z., Kanehl, P. (2003). Influences of watershed urbanization and instream habitat on macroinvertebrate in cold water streams. *Journal of the American Water Resources Association*. 39(5): 1181–1196.

Wang, L. Z., Lyons, J., Kanehl, P., Bannerman, R. (2001). Impacts of urbanization on stream habitat and fish across multiple spatial scales. *Environmental Management*. 28(2): 255–266.

Wilby, R. L., Keenan R. (2012). Adapting to flood risk under climate change. *Progress in Physical Geography*. 36(3): 348–378.

World Bank (2011). *The World Bank Supports Thailand's Post-Floods Recovery Effort*. The World Bank. http://go.worldbank.org/TCFEHXJML0. Accessed 20 February 2012.

World Wildlife Fund (WWF) (2013). "Water Scarcity. Threats." Retrieved 20 October 2013. Available from: www.worldwildlife.org/threats/water-scarcity.

2 General Approaches on Urban Water Challenges

Mingyi Jia, Changqing Xu, and Haifeng Jia

2.1 TRANSITION OF URBAN WATER CHALLENGES MANAGEMENT

The development process of urban stormwater management in both the developed and the developing countries can be divided into four major periods: municipal sanitary engineering period, stormwater quantity regulation period, stormwater quality regulation period, and sustainable development period (Thomas and Andrew, 2003). The emergence, development, and succession of different stages are mainly driven by the contradictions faced by the historical stage of urban development, the needs of urban residents, as well as the progress of science and technology.

2.1.1 Municipal Sanitary Engineering Period

Urban drainage is a very old field, dating back to at least 3000 BC, with a primary focus on conveyance of water away from urban areas (Burian and Edwards, 2002). As the essential infrastructure of cities, urban drainage systems were built during the urbanisation process. The traditional philosophy for urban water management was to make stormwater or sewage go as far away and as fast as possible. In the beginning of the establishment, all cities need to consider and resolve the city's water supply and drainage problems. In the early stages, cities use the same type of the drainage as countryside, which is excavating ditches to collect and discharge urban sewage and stormwater. Although excessive sewage or rainfall would cause overflow and affect the city's road traffic and urban health, to a certain extent, ditches can meet the basic needs of urban drainage.

By the early nineteenth century, some developed countries (e.g., the United States of America) began to enter the stage of industrialisation, resulting in the migration of a large percentage of the rural population into the city areas, which accelerated urbanisation. However, the fast-growing cities did not have proper sewers and relied on contaminated wells within the city confines for drinking water supply (Thomas and Andrew, 2003). Large and medium cities are facing a series of urban problems, such as serious urban pollution, deteriorating environmental quality, infectious diseases, and plague, including cholera in 1832, 1849, and 1866 and typhoid in 1848 (Thomas and Andrew, 2003). The considerations of urban sanitation and the development of engineering technology, urban water supply, and drainage have gradually become a public utility and start developing rapidly. The city's ditch gradually replaced by the municipal drainage network, the city's domestic sewage, rain began to flow through the pipeline into the adjacent water body.

By the 1950s, most cities had established a relatively complete urban drainage system. Also, during this period, the source of drinking water in cities also shifted from well to cleaner river water. To ensure the safety of drinking water and prevent pollution, governors of cities began focussing on treating wastewater. However, treating domestic sewage and clean stormwater runoff together in a combined sewer system is not economical. Thus, to solve this problem, the modern urban drainage infrastructure was introduced, consisting of an efficient drainage system with catch basins and pipes leading to the nearest stream. The rapid economic growth and urban expansion after the wars contributed to the wide application of the separate sewer system, which also resolves the urban sanitation problems (Thomas and Andrew, 2003).

2.1.2 Stormwater Quantity Regulation Period

With the development and expansion of the city, the urban impervious surfaces are increasing, which changed the flow of rainfall in urban areas. This change exacerbated sewer line clogging, downstream floods, channel erosion, and other problems. The main objective of this period is to alleviate or solve the above-mentioned series of urban rain and flood problems caused by city expansions. In the late 1950s, the US National Meteorological Administration (NMA) released a series of basic data on rainfall maps, rainfall intensity-time-frequency curves across the country, and by the 1960s the data were already widely used and played a huge role (Thomas and Andrew, 2003).

Due to the deepening of cognition and the limitation of objective conditions, the mode and specific measures of water regulation are also in constant development and improvement. However, the period of water quantity regulation can be generally divided into stormwater detention basin and stormwater master plan.

A stormwater detention basin is an underground structure designed to manage excess stormwater runoff on a developed site, often in an urban setting. This type of best management practice (BMP) may be selected when there is insufficient space on the site to infiltrate the runoff or build a surface facility such as a detention basin or retention basin (Malcom, 1980). Detention vaults manage stormwater quantity flowing to nearby surface waters. They help prevent flooding and can reduce erosion in rivers and streams. They do not provide treatment to improve water quality, though some are attached to a media filter bank to remove pollutants (Malcom, 1980).

In the past, the core issue of stormwater master planning was the water balance analysis and hydrological processes simulation. During 1980–1990s, the improvement of hydrology and hydraulics model as well as the development of computer technology together, resolved these issues. In the 1970s, hydrological and hydraulics models were mainly used on large computers but became commonly available for PC use in the mid-1980s (Thomas and Andrew, 2003). By running the hydrological models at the basin scale, people can find solutions to the regulation of different stages of floods. The progress of these technical means makes the planning and overall coordination of the basin scale possible.

2.1.3 Stormwater Quality Regulation Period

The water quantity control phase of urban stormwater management came from the needs of urban life as well as the wishes to avoid urban flooding, whereas the emergence and development of stormwater quality management period are mainly driven by strict norms, standards, and the gradually improved legal and regulatory systems (Shaver et al., 1988). Therefore, before discussing the development of the water quality management period, it is necessary to first introduce the history of water pollution control. For example, in the United States, the Federal Water Control Act, enacted by the US Congress in 1948, is the nation's first law on water pollution control. It is also the predecessor of the Clean Water Act. By the late 1980s, the environmental protection agencies in cities began to focus on non-point source pollution, and some studies have suggested that the city's rainwater runoff is the main source of non-point source pollution. These have led to the change of stormwater management's focus from quantity control to quality control.

The introduction and implementation of the single largest daily maximum load capacity (TMDL) for a river and lake water bodies has also promoted the development of water quality management at watershed scale. The problem with this model is that when the flood management is extended to a basin scale, the processing capacity of information increases significantly, as it includes several components such as the development of the model itself, land use mode, traffic mode. and the coordination among different interest groups (such as developers, landowners, and other departments). So, although TMDL solves the provides a solution to a certain extent, it could prove a difficult and time consuming process. This led to experts thinking of other more economical and efficient alternative measures to solve water quality management problems (Shaver et al., 1988).

2.1.4 Sustainable Development Period

Urbanisation creates an environment that is ultimately not good for any living beings on the Earth. After realising that the original methods are not environmentally friendly, people start a multifaceted green revolution in stormwater management. Through a combination of structural and institutional practices, functional, environmentally friendly, sustainable, and beautiful living environments can be created. There have been many popularly named approaches that address some or all of this including low impact development (LID), green infrastructure, better site design, smart growth, conservation development, and sustainable development. They are innovative and integrate set of methodologies and engineering systems for urban runoff control (Davis, 2005; Kirk et al., 2006; Dietz, 2007; Jia et al., 2014). It basically consists of a network of interconnected linear and patch areas that seeks to sustain life and enrich the quality of life by mimicking "acceptable" hydrology; enhancing natural diversity and beauty; balancing ecological preservation/conservation with economic growth and development; building systems that are sustainable and maintainable; and working at a small, integrated scale with accumulated results.

The introduction of the concept of sustainable development shifts the urban stormwater management from the original passive responses to urban rain and flood problems to more active responses, which is identifying the root causes of the urban rain and flood problems before happens.

2.2 APPROACHES ON URBAN WATER CHALLENGES

2.2.1 The Evolution of Terminology Surrounding Urban Drainage

The management of urban stormwater has become increasingly complex over recent decades. Consequently, terminology describing the principles and practices of urban drainage has become increasingly diverse. In recent decades, the urban drainage and related literature has seen the development and adoption of a range of "new" terms that attempt to describe the evolution towards a more holistic approach. There has been rapid growth in the use of terms such as LID (Department of Environmental Resources, 1999), sustainable urban drainage systems (SUDS) (CIRIA, 2000), water-sensitive urban design (WSUD) (Whelans et al., 1994; Wong, 2007), BMPs (Schueler, 1987), and sponge city (MHURD, 2014). All of them involve the scientific management of the urban rainwater system to protect and utilise natural resources. But different practices have different priorities, application scopes, and management styles; decision-makers should choose different practices according to the actual condition of specific regions. This study makes a brief introduction of the abovementioned terminologies.

The original intent of LID was to achieve a "natural" hydrology by use of site layout and integrated control measures. Natural hydrology referred to a site's balance of predevelopment runoff, infiltration, and evapotranspiration volumes, achieved through a "functionally equivalent hydrologic landscape" (USEPA, 2000). Low impact development discouraged the then common practice of large end-of-catchment solutions, because of their inability to meet this catchment-wide hydrologic restoration. The term LID has been most commonly used in North America and New Zealand. Perhaps the most influential early use of the term was in Prince George's County Department of Environmental Resources, Maryland, USA in the early 1990s). The term LID was used to distinguish the site-design and catchment-wide approach from the common stormwater management approach at that time, which typically involved conveyance to large end-of-pipe detention systems. In contrast, LID was characterised by smaller scale stormwater treatment devices such as bioretention systems, green roofs, and swales, located at or near the source of runoff. Similar approaches were applied elsewhere, notably in New Zealand, guided by the publication of the Low Impact Design Manual (Shaver, 2000) by the then Auckland Regional Council. The basic principle of LID is to simulate the natural hydrological process of rainwater. Its core concept is to minimise the impact of the original natural hydrological process during the process of land development and utilisation. This approach, through a series of small, micro site design, can modify the hydrological processes associated to rainwater, such as permeability, filtration, evaporation, and interception and thus greatly reducing the burden of urban drainage pipe network (MHURD, 2014). Low impact development is suitable for use in residential areas, parking lots, commercial areas, squares, and waterfront areas. It is also suitable for the construction of urban new areas or urban renewal projects. In recent years, LID strategies gain much attention than other solutions.

The concept of the sustainable drainage triangle (quantity, quality, and habitat/amenity) was initially set out by D'Arcy (1998), and it is believed to be Jim Conlin of Scottish Water who first (October 1997) coined the term sustainable urban drainage systems to describe stormwater technology. Around this time, the principles of sustainable urban drainage more generally were also being fleshed out (Butler and Parkinson, 1997). A major set of guidance documents was published in 2000, with similar but separate design manuals for Scotland and Northern Ireland, and England and Wales (CIRIA, 2000), and it was there that the term sustainable urban drainage systems was formalised. The most authoritative guide to SUDS is currently the SUDS Manual (CIRIA, 2007) which aims to provide "comprehensive advice on the implementation of SUDS in the UK." Some parties omit "urban" from the term, referring to sustainable drainage systems (or SuDS) instead, although the meaning is essentially the same and this is now the terminology used in legislation effectively specifying the use of SuDS in England (Flood and Water Management Act, 2010). In UK practice, SUDS consist of a range of technologies and techniques used to drain stormwater/surface water in a manner that is (arguably) more sustainable than conventional solutions. They are based on the philosophy of replicating as closely as possible the natural, predevelopment drainage from a site, consistent with the previously described principles behind LID. Typically, SuDS are configured as a sequence of stormwater practices and technologies that work together to form a management train. In Scotland, SuDS have been mandatory in most new developments since 2003 (WEWS, 2003). Although they were introduced to improve water quality in receiving waters, these systems are also required to safely convey design storms for a range of recurrence intervals up to 200 years. In England and Wales, SuDS are aimed more at water quantity than quality control (Defra, 2011), although they are deemed to comply with water quality standards provided they comprise appropriate components of a defined treatment train.

The WSUD began to be used in the 1990s in Australia, with the first known reference to it being by Mouritz (1992) and then shortly after in a report prepared for the Western Australian Government by Whelans et al. (1994). The objectives of WSUD are to manage the water balance (considering groundwater and stream flows, along with flood damage and waterway erosion), to maintain and wherever possible to enhance water quality (including sediment, protection of riparian vegetation, and minimising the export of pollutants to surface and groundwater), to encourage water conservation (minimising the import of potable water supply, through the harvesting of stormwater and the recycling of wastewater, and reductions in irrigation requirements), and to maintain water-related environmental and recreational opportunities (Tim et al., 2014).

The United States first proposed the BMPs in response to the city's rain-flood problems and then it is used in the United States and Canada to describe a type of water pollution control (Ice, 2004). Historically the term has referred to auxiliary pollution controls in the fields of industrial wastewater control and municipal sewage control, while in stormwater management (both urban and rural) and wetland management, BMPs may refer to a principal control or treatment technique as well and can be classified into different kinds according to different classification criteria. The first classification is based on the control approach of BMPs, and it can be divided into three categories: source control, process control, and end control. The second classification is based on whether the practice has visible structure or not, and it can be divided into engineering practices and non-engineering practices (National Research Council, 2008).

The idea of "sponge city" emerged in the West and is gradually gaining popularity in China. Following the philosophy of "solving a city's problems with its own resources," it aims at storing rainwater, especially during heavy rainfall, to prevent floods and release it for the benefit of residents during dry times (MHURD, 2014). This will also ease a city's water shortage. Sponge cities save water and, therefore, are more environmentally friendly. Some cities in foreign countries have not only adopted the idea of sponge gardens but also built sponge squares. Unlike metaled squares that prevent water from replenishing the underground water table, these sponge squares have plants and well-designed pipelines that help store large volumes of water under the Earth's surface. Chinese cities can benefit by changing its metaled or tile-bedded squares to sponge squares. In more ways than one, a sponge city will be a smart city. Urban management officials can install devices across a city to collect real data on rainfall, ground and underground flows of water, and amounts of pollutants, which they can use to deal with emergencies. A sponge city follows the philosophy of innovation: a city can solve water problems instead of creating them. In the long run, sponge cities will reduce carbon emission in cities and help fight climate change. The idea of green cities will rule in the future and sponge cities will be part of it (MHURD, 2014).

2.2.2 Types of Urban Drainage Systems

In urban areas, the types of drainage systems include combined drainage systems and separate drainage systems. There is much disagreement as to the advantages and disadvantages of combined drainage systems and separate drainage systems. For example, the Code for the Design of Outdoor Wastewater Engineering (GB50014-2006) states that the type of drainage system in cities could be selected according to local conditions; however, separate drainage systems were recommended for new development areas (Ministry of Construction of the PRC, 2006). So, in China, combined drainage systems mainly exist in older towns, while new towns and some new residential areas employ separate systems. For urban runoff control, it is related to the stormwater drainage system of separate systems and the combined system. Conventional management of urban stormwater aims to move runoff away from a developed area as quickly as possible – from impervious surfaces to storm drains and finally to stream discharge. According to previous research on the impacts of urbanisation on peak flow of runoff, it was concluded that the peak flow would increase 1.5 to 6 times when a natural watershed with an area of 1 km^2 becomes urbanised (Hao et al., 2005).

To avoid loss of life and damage to property, higher drainage standards and larger scale engineered drainage systems (e.g., culverts, stormwater pumping stations, and channelised streams and rivers) need to be constructed and maintained to quickly convey increased stormwater runoff from urban areas to downstream receiving rivers. However, on one hand, such construction and maintenance bring huge financial burdens to society. On the other hand, it is difficult to determine proper drainage standards and then plan, design, and construct safe urban drainage systems under uncertain conditions of climate change and urbanisation patterns (Jia et al., 2013).

2.2.3 Typical LID Facilities for Urban Runoff Source Control

Facing the urban water challenges, the role of urban runoff source control by sponge facilities become more and more important. The functions of sponge facilities include infiltration, storage, regulation, transmission, decontamination, and purification. The combined application of various technologies can realise the target of total runoff control, peak runoff control, runoff pollution control, and rainwater resource utilisation. In practice, according to the local natural and social conditions, the suitable sponge facilities should be chosen to meet different requirements. Various types of "sponge" technologies include a number of different forms of LID facilities. Through their function, these facilities include bioretention facilities, green roof, permeable pavement, grass swales, and other small-type auxiliary facilities. This study will introduce some typical LID facilities.

2.2.3.1 BIORETENTION FACILITIES

Bioretention is the process in which contaminants and sedimentation are removed from stormwater runoff. A typical structure profile of bioretention facility is shown in Figure 2.1. Stormwater is collected into the treatment area which usually consists of a grass buffer strip, sand bed, ponding area, organic layer or mulch layer, planting soil, and plants (Brown and Clarke, 2007). First, the runoff passes over or through a sand bed, which slows the runoff's velocity, and then it distributes evenly along the length of the ponding area, which consists of a surface organic layer and/or groundcover and the underlying planting soil. The ponding area is graded, its centre depressed. Water is ponded to a depth of 15 cm (5.9 in.) and gradually infiltrates the bioretention area or is evapotranspired. The bioretention area is graded to divert excess runoff away from itself. Stored water in the bioretention area planting soil exfiltrates over a period of days into the underlying soils.

2.2.3.2 GREEN ROOF

A green roof or living roof is a roof of a building that is partially or completely covered with vegetation and a growing medium, planted over a waterproofing membrane (Getter and Rowe, 2006; Vandermeulen et al., 2011). A typical structure of green roof is shown in Figure 2.2. It may also include additional

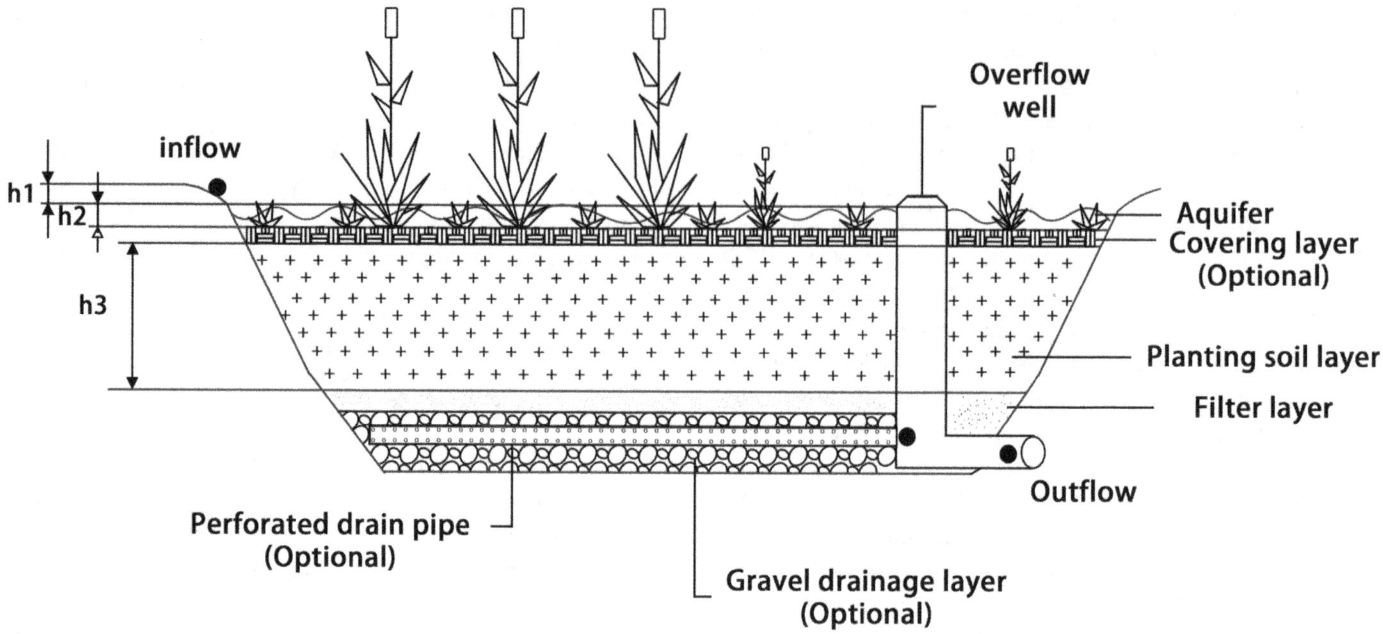

Figure 2.1 Structure of bioretention facilities

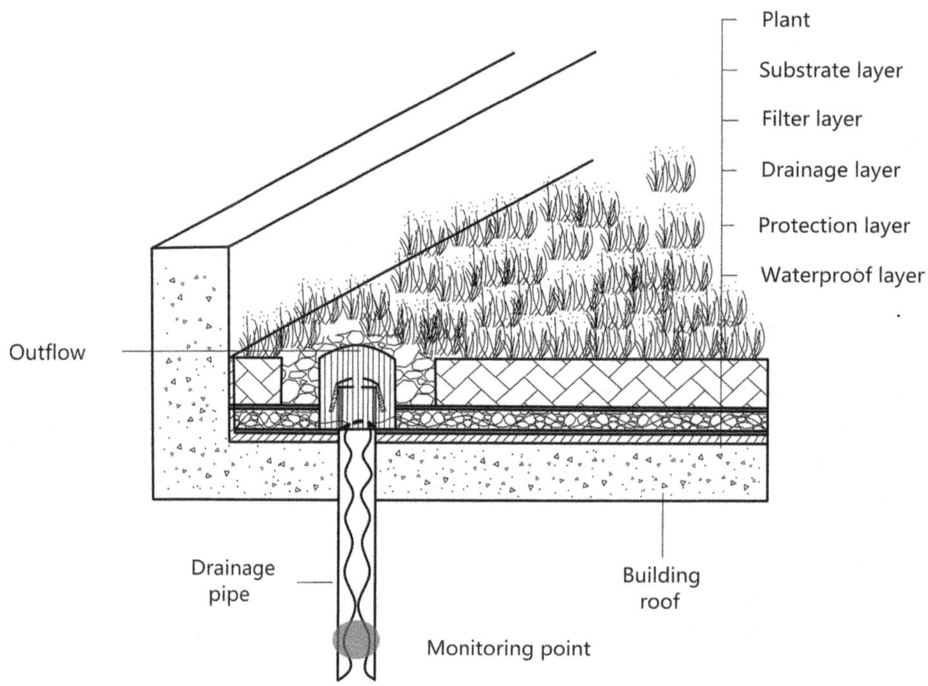

Figure 2.2 Structure of a green roof

layers such as a root barrier and drainage and irrigation systems. Container gardens on roofs, where plants are maintained in pots, are not generally considered to be true green roofs, although this is debated. Rooftop ponds are another form of green roofs which are used to treat grey water (Bastien et al., 2010; Woods Ballard et al., 2015). Green roof is one of the main measures of the LID management strategy. Green roofs serve several purposes for a building, such as absorbing rainwater, providing insulation, creating a habitat for wildlife, increasing benevolence (Kosareo and Rise, 2007) and decreasing stress of the people around the roof by providing a more aesthetically pleasing landscape, and helping to lower urban air temperatures and mitigate the heat island effect (Vandermeulen et al., 2011). They effectively utilise the natural functions of plants to filter water and treat air in urban and suburban landscapes (Volder and Dvorak, 2014). There are two types of green roof: intensive roofs, which are thicker, with a minimum depth of 12.8 cm (5.0 in), and can support a wider variety of plants but are heavier and require more maintenance, and extensive roofs, which are shallow, ranging in depth from 2 cm (0.79 in) to 12.7 cm (5.0 in), lighter than intensive green roofs, and require minimal maintenance (Volder and Dvorak, 2014).

With green roofs, water is stored by the substrate and then taken up by the plants from where it is returned to the atmosphere through transpiration and evaporation. Green roofs not only retain rainwater but also moderate the temperature of the water and act as natural filters for any of the water that happens to run off. Many green roofs are installed to comply with local regulations and government fees, often regarding stormwater runoff management (Getter et al., 2009). In areas with combined sewer-stormwater systems, heavy storms can overload the wastewater system and cause it to flood, dumping raw sewage into the local waterways. Green roofs decrease the total amount of runoff and slow the rate of runoff from the roof. It has been found that in certain condition, they can retain up to 70–90% of rainwater in summer and ~25–40% in winter, gradually releasing it back into the atmosphere via condensation and transpiration, while retaining pollutants in their soil (Carter and Keeler, 2008). Often, phosphorus and nitrogen are in this category of environmentally harmful substances even though they are stimulating the growth of plants and increase the agricultural productivity. When these substances are added to a system, it can create mass biological activity since they are considered limiting factors of plant growth and by adding more of them to a system, it allows for more plant growth (Saadation et al., 2013).

2.2.3.3 PERMEABLE/POROUS PAVEMENT

Permeable paving (also known as pervious or porous pavement) are surfaces that allow water to pass through voids in the paving material or between pavers while providing a stable, load-bearing surface (Booth and Leavitt, 1999). A typical structure profile of a permeable pavement is shown in Figure 2.3. This allows stormwater to filter through the soil below the paved surface, reducing the numerous environmental issues associated with water runoff. Just as drinking water can be filtered to remove impurities, the soil particles filter rainwater percolating through soil on its way to surface waters and to groundwater aquifers. This important step in the natural process of water purification is bypassed when rainwater falls on impermeable pavement surfaces or roofs and is carried directly through storm drainage systems into waterways (Pratt et al., 1989). Pollution carried in rainwater runoff is another concern, especially in urban areas. Stormwater flowing across streets, parking lots, and sidewalks picks up contaminants associated with air pollution particles, spilled oil, detergents, solvents, dead leaves, pesticides, fertiliser, and bacteria from pet waste. Natural filtration of water through soil is the simplest way

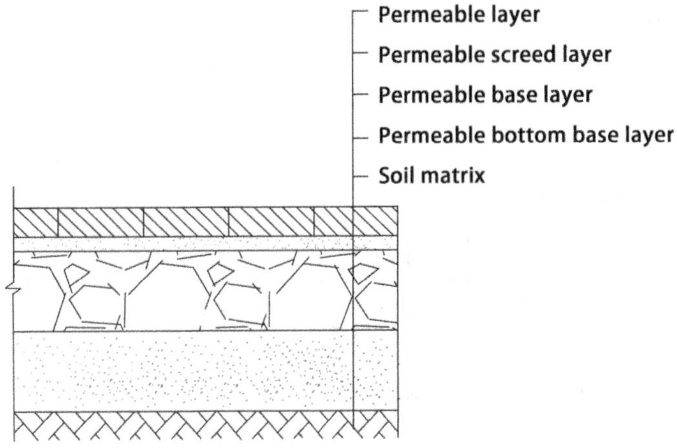

Figure 2.3 Structure of a permeable pavement

to control these pollutants, and is a direct advantage of permeable pavement (Morgenroth and Visser, 2011).

Permeable paving surfaces have been demonstrated as effective in managing runoff from paved surfaces (Brattebo and Booth, 2003). Large volumes of urban runoff causes serious erosion and siltation in surface water bodies. Permeable pavers provide a solid ground surface, strong enough to take heavy loads, like large vehicles, while at the same time they allow water to filter through the surface and reach the underlying soils, mimicking natural ground absorption. They can reduce downstream flooding and stream bank erosion and maintain base flows in rivers to keep ecosystems self-sustaining. Permeable pavers also combat erosion that occurs when grass is dry or dead, by replacing grassed areas in suburban and residential environments (Volder and Watson, 2009).

2.2.3.4 GRASS SWALES

Grass swales are vegetated open channels that are designed to manage the runoff by reducing the depth of flow and velocity through the channel (Zhou et al., 2010). It can be used to partially manage the first one inch of rainfall on-site using a grass swale designed to the required geometry and slope to maintain the design volume flow depth and velocity. Besides, it can reduce pollutant loads to meet water quality targets (total maximum daily loads or TMDLs. Also, grass swales can be blended into the landscape and drainage infrastructure design for many sites (Jia et al., 2015). The structure of two typical grass swales are presented in Figure 2.4.

Grass swales are primarily applicable for land uses such as roads, highways, and residential development. Key constraints with grass swales are including available space, site topography, water table, contributing drainage area, hotspot land uses, floodplains, proximity to utilities, community factors, baseflow, and non-stormwater discharges (Zuo et al., 2017). Unlike other stormwater practices, grass swales are designed based on a peak rate of flow. Designers must demonstrate both channel conveyance and treatment capacity.

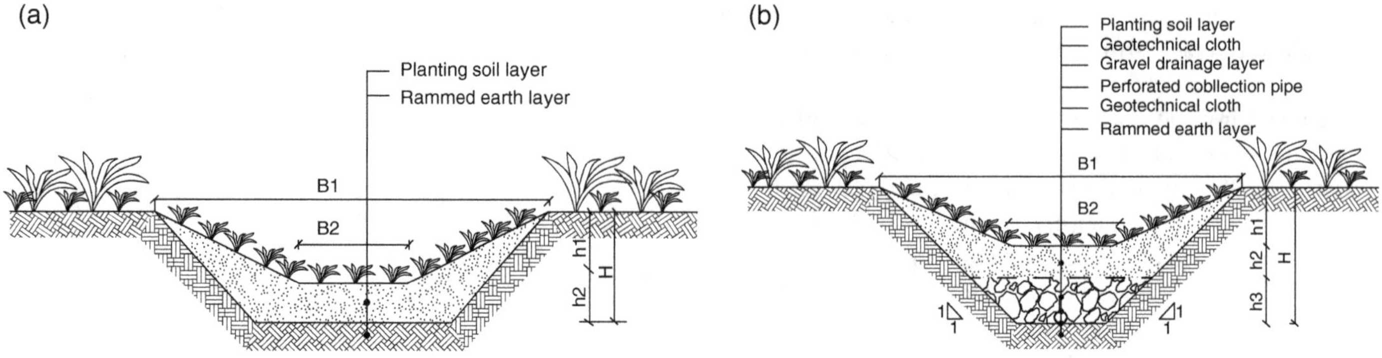

Figure 2.4 Structure of typical grassed swales

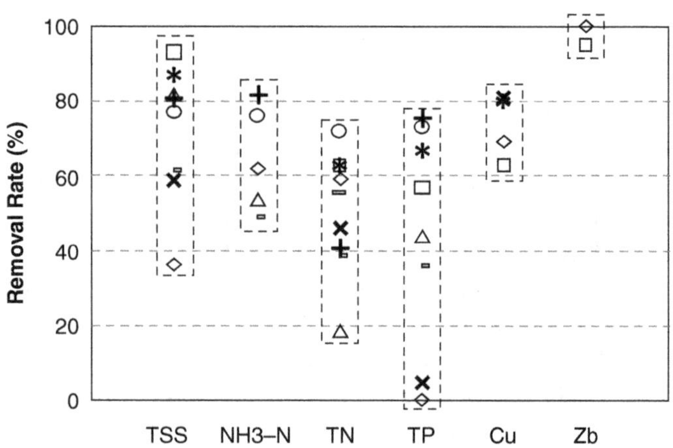

Figure 2.5 Pollutant removal efficiency of bioretention cell worldwide

2.2.3.5 OTHER FACILITIES

Besides, LID includes some mini-type treatment facilities, such as dry wells, tree boxes, filter strip, constructed wetland, vegetated buffers, lever spreaders, rain barrels, cisterns, and infiltration trenches, according to different situations. All of them interact with each other by reducing impervious area, increasing rainwater infiltration, and utilising rainwater sources to realise the sustainable rainwater management.

As mentioned earlier, LID and BMP practices perform well in flood control, rainwater harvesting, and water quality purification (Jia et al., 2013). Figure 2.5 presented the pollutant removal efficiency of bioretention cell worldwide (Xu et al., 2017). The removal efficiency capacities of TSS, NH_3-N, TN, TP, Cu, and Zn from those reported studies are 45% to 93%, 49.31% to 82.21%, 19% to 63%, 5% to 75.33%, 63% to 81%, and 95%, respectively (Center for Watershed Protection, 2007; Brown et al., 2011; Trowsdal and Simcock, 2011; Chen et al., 2013; Mangangka et al., 2015). In addition to the enhancement of runoff quality control efficiency, the LID and BMP practices can achieve good runoff quantity control efficiencies. As reported by Jia et al. (2015), the peak flow and stormwater runoff volume of the LID and BMP practices were reduced by an average of 62.2% and 66.0%, respectively. In particular, the peak flow rates of bioretention cell and grassed swale were approximately 50–84% and 17–79%, respectively and the stormwater runoff volume and the reduction rate were approximately 47–80% and 9–74%, respectively (Jia et al., 2015).

2.3 MANAGEMENT OF FLOOD RISK

2.3.1 The Strategies of Flood Risk Management

The flooding system includes the physical process of flooding, the inhabitants of floodplains, their infrastructures and ecosystems, and the people and organisations in the public and private sectors that influence or are subject to flooding and its impacts (Hall et al., 2003). Flooding systems are continuously changing because of natural and socioeconomic processes, as well as activities that are intended to reduce flood risk. Flood risk management is the process of data and information gathering, risk assessment, appraisal of options, and making, implementing, and reviewing decisions to reduce, control, accept, or redistribute risks of flooding. Integrated flood risk management explicitly recognises the interrelationships between all risk management measures, and their analysis, costs, and effectiveness, within changing social, economic, and environmental contexts. It can be helped by improving access to data and models, sharing and communicating risk analyses, involving a wider range of stakeholders, and coordinating risk management actions (Samules et al., 2006).

Flood risk management is aimed at managing whole flooding systems, be they catchments or coastlines, in an integrated way that accounts for all of the potential interventions that may alter flood risk. In support of this aim, the science and technology of flood management has made tremendous progress in the last half century. Process-based, parametric, and statistical models describing key elements of the flooding system (i.e., loads, defence response, inundation, and impacts) are now available and are continuing to be developed. The potential exists for an integrated description of the whole flooding system from causes to consequences, including risk management measures. Without appropriate decision support tools, however, "flood risk managers" understandably struggle to handle the complexities inherent in integrated management of the flooding system and the hazard it represents (Samules et al., 2006).

Rather few of the changes that impact on the flooding system will be in the control of the institutions or individuals with responsibility for flood risk management. Table 2.1 lists the changes that flood risk managers could realistically hope to have significant influence over, which will be referred to as potential "flood risk management actions." Table 2.1 also lists the current roles with the most capacity to implement change. Note that these roles are not all concentrated within one governmental department or agency, or indeed within the public sector (Hall et al., 2003).

It is useful to think of change on a continuous range – from those (relatively few) changes that flood risk managers have more or less absolute power over and all of the resources they need to implement, to changes over which flood risk managers have no power. One objective of integrated flood risk management is to extend the range of changes over which flood risk managers and other stakeholders have some control. In free societies with limited economic resources, there is a limit to the extent of statutory and publicly funded powers. Rather, the scope of intentional change may, for example, be extended through publicity campaigns or economic instruments (often operated through the private sector) such as insurance pricing, aimed at changing public behaviour and hence flood risk (Kunreuther 2000; Ryland 2000). In summary, integrated flood risk management is characterised by:

1) A broad definition to the flooding system and scope of flooding impacts.
   Arbitrary subdivision of the flooding system, for example, due to geographical boundaries or administrative divisions, is avoided. Temporal and spatial interactions in system performance are accounted for.
2) Continuous management of flood system performance.
   Consideration of one or a few design events is replaced by the consideration of a whole range of system behaviours and the development of appropriate management responses. There is a commitment to ongoing monitoring and control of the system at time intervals appropriate to the system dynamics.
3) Tiered analysis and iterative decision-making.
   Flood risk management cascades from high-level policy decisions, based on outline analysis, to detailed designs

Table 2.1 *Flood risk management actions and roles with control over those actions*

Risk management action	Effect of action	Role
Development control in floodplains	Limit to construction of buildings and infrastructure in the floodplain, hence controlled increase in vulnerability	Planning authorities
Improving flood resistance of buildings	Reduced flood damage	Property developers and building owners
Increasing public awareness of temporary measures to reduce flood impact on building contents	More effective public action to reduce flood damage to building contents	Building occupants
Flood insurance	Redistribution of the cost of damage across the population and through time	Insurance companies
Increasing storage in catchments and reducing the rate of runoff (source control)	Reduced flood severity	Property and infrastructure developers, planning authorities, farmers
Urban drainage	Reduced probability of flooding	Water authorities, local authorities, highway authorities
Flood defence planning, design, construction, operation, and maintenance	Reduced probability of flooding	Environment agency, local authorities
Real-time flood forecasting and warning	Reduced flood impact (if followed by appropriate action by the public)	Environment Agency, Metrological Office
Emergency repair of flood defences	Reduced probability of flooding	Environment Agency, emergency services
Evacuation of people in flood events	Reduced public safety and health impacts of flooding	Emergency services
Post-flood recovery and reconstruction	Reduced social, health, and economic impacts of flooding	Local authorities, insurers

and projects, which require more detailed analysis. High-level policy and plans provide the framework and common understanding within which more detailed actions are implemented.

4) Consideration of the widest possible set of management actions that may have some impact on flood risk.
This includes measures to reduce the probability of flooding and measures to reduce flood impact (vulnerability).
5) Development of integrated strategies that combine a range of flood risk management actions and implement them in a programmed way.
Management strategies are developed following consideration of both effectiveness, in terms of risk reduction, and cost. They will involve coordinating the activities of more than one organisation and multiple stakeholders.
6) Evolving within current policy framework.
Integrated flood risk management will remain an abstract concept unless it is placed within the current policy and administrative context. This involves making the best use of existing policy instruments and actively identifying opportunities to influence policy change. It may involve reacting opportunistically to policy, administrative, or regulatory reviews and changes that are initiated for non-flood-related reasons (Samules et al., 2006).

Although simply stated, integrated flood risk management, like sustainable development, is a complex endeavour. Indeed, the fragmented approaches that prevailed in the past can be seen as an attempt to reduce the complexity of the problem by subdividing it into manageable roles and allocating them to different individuals or organisations. Integrated flood risk management in practice requires decision-making ability that is beyond the information-processing capacity of any individual. It therefore requires a framework within which diverse activities can be enacted, which will help to ensure that they are complementary and based on a common understanding of key principles.

Water management faces major challenges over the coming decades, with existing social, ecological, and technical water subsystems subject to emerging global threats such as climate change, urbanisation, and depletion of resources (Butler et al., 2016). Managing flood risk in urbanised areas is likely to need a range of solutions, both structural and non-structural including improved spatial planning, coordination between different stakeholders, and greater community awareness of flooding hazards. To ensure the ongoing reliability, resilience, and sustainability of service provision, implementing strategies needs a long-term approach and countries need to build individually tailored solutions to suit their distinct water management challenges and reduce the risk of major impacts of flooding in their urban populations (Norwegian Ministry of Foreign Affairs, 2013).

2.3.2 Flood Risk Management in Selected Countries

Flood risk management strategies have developed in the past under the influence of societal debate and the power of coalitions. Hence, they are a result of a cultural process. In different countries, these cultural processes differ, as histories are very different indeed. Hereafter, a brief analysis is examined on the way river flood risk management is evolving in different parts of the world, in different socioeconomic and cultural historical contexts. This study will introduce some typical countries which have good experience in flood risk management.

2.3.2.1 ENGLAND AND WALES

In the UK over the twentieth century, three stages can be discerned during which certain coalitions were strong, influencing the approach to water management. From the 1930s to about 1970, there was a strong drainage coalition for agriculture, benefiting from the post-war need to secure food production, with a policy for flood prevention. From 1970 to the early 1990s, economic reasons predominated, leading to a flood defence coalition to protect people and property. From the mid 1990s onwards, there has been a gradual shift towards flood risk management, under the influence of an environmental movement. This also includes enhanced attention to social responsibility, community involvement, and public awareness. Further important background to inform a revision of flood risk management policy came from the Foresight "Future Flooding" project run from the Office of Science and Technology over an 18-month period in 2002–04. This project investigated drivers, responses, and scenarios for flood risk over a timescale of about 100 years. The flood risks were analysed at a scale of a 10-km grid for four socioeconomic scenarios, which were linked to standard IPCC global emissions scenarios and simulations of future climate by the UK Met Office Hadley Centre. The scenarios represent different policy frameworks for the country and the project considered flooding from all causes: urban storms, river, estuarine, and coastal flooding. Drivers of flood risks were identified and ranked under these scenarios and the potential flood damage estimated for the 2080s.

2.3.2.2 GERMANY

In Germany, a paradigm change can be observed since the floods of the Oder in 1997 and the Elbe in 2003. This paradigm change is based on the understanding that absolute protection against floods is unachievable. Therefore, the approach to managing flood risks is shifting away from only protection to holistic risk management by including spatial planning. This change of paradigm in practice was already introduced in theory during the Decade of Natural Hazards. In practice, this changing view is reflected by the wealth of plans which are being drafted or already implemented in Germany. Because of the federal structure of the country, plans are being made at various levels. There is a Flood Protection Act at the federal level; there are Water Management Plans in most lander states (e.g., the Saxony Water Management Plan) which address the issue of flood risk, and most large communities have an important role in decisions on land use. The communities should enforce land use zoning according to flood hazard zones at local level. This means that on paper flood risk management is well organised. In practice, however, it is difficult to change things, as there is no single managing entity responsible for the whole. Moreover, things may be arranged differently in different lander, which is the logical consequence of the federal structure. For any change to work practice, cooperation is a prerequisite, both vertical and horizontal. This means that, in the special context of the federal republic of Germany, informal cooperative processes are much more important than the formal ones.

2.3.2.3 THE NETHERLANDS

The Netherlands has a long tradition of flood risk management, which gradually developed towards a flood defence approach in the 1950s. The 1953 storm surge, which caused the death of some 1,850 people, was an important policy catalyst for better protection. It soon led to the installation of the so-called Delta Committee (DC), which performed many studies and came up with (independent) advice to the government. The advice focussed on the coast, but in its wake, the polders on the former floodplain (flood-prone) areas along the large rivers (e.g., Rhine and Meuse) were dealt with similarly. The DC advised basing the safety standard on flood risk, indeed as a combination of flood probability and flood consequence, in balance with the construction costs for realising a certain flood probability.

The DC advised basing the safety standard on flood risk, indeed as a combination of flood probability and flood consequence, in balance with the construction costs for realising a certain flood probability. Because of insufficient knowledge and to keep things simple, in practice at that time an approach was adopted based on safety standards for many flood-prone areas. Later, after a revision of criteria and a structural institutional reform of the number (and responsibilities) of water boards (back to a number of only 53 so-called dike rings), four major safety levels were adopted by Parliament, for example, a safety level of 1: 1,250; 1: 2,000; 1: 4,000; or 1: 10,000 (dependent on location of, and population and investments inside, those dike rings, and a potential flooding of flood-prone areas by salt or fresh floodwater). These safety levels relate to probabilities of exceeding design conditions. Both the safety levels (with a differentiation into four levels only) and the location of the dike rings are now specified in the Law on Flood Defence (1997). The state of the defences is monitored constantly and evaluated in relation to the latest measurements and scientific insights on river discharge every 5 years. Whether the standards for probability of exceeding of design conditions are still up to date and in tune with economic development is not yet a point of consideration.

2.3.2.4 USA

The USA has national flood standards and a floodplain management programme. Floods have been characterised as a menace to national welfare since the 1930s, after massive damage from the 1927 Mississippi floods. This led to the passing of the Flood Control Act in 1936, making flood risk management a national concern. On the basis of this Act, many engineered structures have been built, for example, embankments, floodways. With these, it is estimated that some $18 billion of damage have been prevented, in particular during the exceptional floods in the Mississippi basin in 1993. Since the 1960s, non-structural measures have been applied additionally, especially in the form of regulatory measures and insurance. For the implementation of such measures, the designation of 1% flood probability zones provided the basis. This has led to a 1% flood probability having become the accepted standard all over the USA. Although the practice of flood risk management is growing in the United States, the term is still not widely used, recognised, or understood.

Today, in the US, the three central approaches guiding floodplain management are: a risk-based approach used by the US Army Corps of Engineers on new studies; a simple standard approach for the National Flood Insurance Program (NFIP) and related land use; and a social goal-based approach to avoid disasters. Following the great Mississippi River flood of 1993, the United States has closely examined its floodplain management policies and procedures, conducting several regional and national assessments and has reacted to those assessments. Based on the post-flood assessments, there is growing consensus that the national goals should be to reduce flood damage, while protecting and enhancing the natural environment and continuing economic growth. However, despite the good progress in flood risk management, flood risk often increases as there are still large developments in floodplains; they remain attractive areas for industrial and housing development and floods do not seem to be a big concern for local decision-makers. Consequently, there are still counteracting policies on development in floodplains. It may thus be concluded that the implementation of a sound coherent flood risk management for the USA is hampered by the lack of one single Flood or Water Act, which integrates these three approaches or resolves the contradictions in their underlying philosophies (Loucks, 2003)

2.3.2.5 BANGLADESH

Some 50 years ago, the people of Bangladesh simply used to build above the expected floodwater level. This suited a population of only 40 million people, which also greatly depended on floods. But since then, population has grown to in excess of 100 million, requiring a substantial increase in agricultural production. From the time of independence, controlling floods has been a national goal. However, it is important to understand the definition of a flood in the social context of the country. If an area goes under and remains under water for some time, this is considered as inundation. When this causes damage to property and crops, as well as disrupting communication and bringing harmful effects to human beings and to flora and fauna, it is flood. It is essential to understand that inundation supports fishing, navigation, soil fertility, ecosystems, groundwater recharge, and so on. Recently, the aim of controlling floods has been broadened to a more integrated flood risk management. This policy shift was catalysed by the 1987–88 floods, which was followed by the Flood Action Plan as coordinated by the World Bank. It culminated in an integrated Water Management Plan in 2004, specifying 11 guiding principles and applying a portfolio of measures and instruments, partly related to flood risk management. In summary, Bangladesh's approach to flood management today is to safeguard life and livelihoods; minimise potential flood damage; improve agroecological conditions for enhanced crop production; meet the needs of fisheries, navigation, communications, and public health; promote commerce and industry; and create flood-free land for a better living environment.

When looking back, a major shift in strategy is discerned on how to deal with floods: from adaptation, through flood control, to flood management. In Bangladesh, floods are so severe that full flood control is impossible. Thus flood protection measures focus on the urban areas only where the population is most vulnerable compared to the rural areas where inundation is either a benefit or only a nuisance. The relative importance of flood prediction and warning in the case of dangerous floods arises from the impracticability of full control of flood. In practice, Bangladesh has a good flood prediction, although the subsequent warning requires further improvement, as it is hampered by the low level of development. However, evacuation and shelter management for coastal floods is a success story which has saved many lives.

References

Booth, D. B., Leavitt, J., 1999. Field evaluation of permeable pavement systems for improved stormwater management. *Journal of the American Planning Association*. 65(3), 314–325.

Brattebo, B. O., Booth, D. B., 2003. Long-term stormwater quantity and quality performance of permeable pavement systems. *Water Research*. 37, 4369–4376.

Brown, R., Chanson, H., McIntosh, D., Madhani, J., 2011. *Turbulent Velocity and Suspended Sediment Concentration Measurements in an Urban Environment of the Brisbane River Flood Plain at Gardens Point on 12–13 January 2011*. Hydraulic Model Report No. CH83/11. Brisbane, Australia: The University of Queensland, School of Civil Engineering (CH83/11): 120.

Brown, R., Clarke, J., 2007. *Transition to Water Sensitive Urban Design; The Story of Melbourne, Australia*. Melbourne: 538 14 T.D. Fletcher et al. Facility for Advancing Water Biofiltration, and National Urban Water Governance Program.

Burian, S., Edwards, F., 2002. Historical perspectives of urban drainage. In: *Proceedings of 9th International Conference on Urban Drainage*. Portland, Oregon: American Society of Civil Engineers.

Butler, D., Parkinson, J., 1997. Towards sustainable urban drainage. *Water Science and Technology*. 35(9), 53–63.

Butler, D., Sarah, W., Chris, S., Maryam, A., Kegong, D., Raziyeh, F., Fu, G., 2016. Reliable, resilient and sustainable water management: The Safe & SuRe approach. *Global Challenges*. Wiley on Line Library. https://doi.org/10.1002/gch2.1010

Carter, T., Keeler, A., 2008. Life-cycle cost–benefit analysis of extensive vegetated roof systems. *Journal of Environmental Management*. 87(3), 350–363.

Center for Watershed Protection. 2007. National pollutant removal performance database. Vol. 3. Ellicott City, Maryland.

Chen, X. L., Peltier, E., Sturm, B. S. M., Young, C. B., 2013. Nitrogen removal and nitrifying and denitrifying bacteria quantification in a stormwater bioretention system. *Water Research*. 47, 1691–1700.

CIRIA, 2000. *Sustainable Urban Drainage Systems – Design Manual for Scotland and Northern Ireland*. Dundee, Scotland: CIRIA Report No. C521.

CIRIA, 2007. *The SUDS Manual*. Dundee, Scotland: CIRIA Report No. C697.

Davis, A. P., McCuen, R. H., 2005. *Stormwater Management for Smart Growth*. New York: Spring Verlag. ISBN: 9780387260488.

Defra, 2011. National standards for sustainable drainage systems – designing, constructing, operating and maintaining drainage for surface runoff. Retrieved from www.defra.gov.uk/consult/files/suds-consult-annexa-nationalstandards-111221.pdf.

Demuzere, M., Orru, K., Heidrich, O., Olazabal, E., Geneletti, D., Orru, H., Bhave, A. G., Mittal, N., Feliu, E., Faehnle, M., 2014. Mitigating and adapting to climate change: Multi-functional and multi-scale assessment of green urban infrastructure. *Journal of Environmental Management*. 146, 107–115.

Department of Environmental Resources, 1999. *Low-Impact Development: An Integrated Design Approach*. Maryland: Department of Environmental Resources, Prince George's County.

Dietz, M. E., 2007. Low impact development practices; a review of current research and recommendations for future directions. *Water, Air, and Soil Pollution*. 186, 351–363.

Fletcher, T. D., Shuster, W., Hunt, W. F., Ashley, R., Butler, D., Arthur, S., … Viklander, M., 2014. SUDS, LID, BMPs, WSUD and more – The evolution and application of terminology surrounding urban drainage. *Urban Water Journal*. 12(7), 525–542. https://doi.org/10.1080/1573062X.2014.916314

Flood and Water Management Act, 2010. Retrieved from www.legislation.gov.uk/ukpga/2010/29/contents

Getter, K. L., Rowe, D. B., 2006. The role of extensive green roofs in sustainable development. *HortScience*. 41(5), 1276–1285.

Getter, K. L., Rowe, D. B., Robertson, G., Cregg, B. M., Andresen, J. A., 2009. Carbon sequestration potential of extensive green roofs. *Environmental Science & Technology*. 43(19), 7564–7570.

Hall, J. W., Meadowcroft, I. C., Sayers, P. B., Bramley, M. E., 2003. Integrated flood risk management in England and Wales. *Natural Hazards Review*. 4(3), 126–135.

Hao, T. W., 2005. Effect on natural water system of urban construction and sustainable utilization of wastewater discharge system. *Water & Wastewater Engineering*. 31(11), 39–42 (in Chinese).

Ice, G., 2004. History of innovative best management practice development and its role in addressing water quality limited waterbodies. *Journal of Environmental Engineering*. 130(6), 684–689.

Jia, H. F., Ma, H. T., Sun, Z. X., Yu, S. Ding, Y. W., Liang, Y., 2014. A closed urban scenic river system using stormwater treated with LID-BMP technology in a revitalized historical district in China. *Ecological Engineering*. 71, 448–457.

Jia, H. F., Wang, X. W., Ti, C. P., Zhai, Y. Y., Field, R., Tafuri, A. N., Cai, H. H., Yu, S. L., 2015. Field monitoring of a LID-BMP treatment train system in China. *Environmental Monitoring and Assessment*. 187, 373.

Jia, H. F., Yao, H. R., Yu, S. L., 2013. Advances in LID BMPs research and practices for urban runoff control in China. *Frontiers of Environmental Science and Engineering*. 7(5), 709–720.

Kirk, B., Roseen, R., Etnier, C., 2006. The big picture – evaluating stormwater BMPs through the life cycle lens. In: *Proceedings of the 5th Annual Storm Conference*, Denver, CO.

Kosareo, L., Ries, R., 2007. Comparative environmental life cycle assessment of green roofs. *Building and Environment*. 42, 2606–2613.

Kunreuther, H., 2000. Insurance as cornerstone for public-private sector partnerships. *Natural Hazards Review*. 1(2), 126–136.

Li, H., Davis, A., 2008. Urban particle capture in bioretention media I: Laboratory and field studies. *Journal of Environmental Engineering*. 136(6), 409–418.

Loucks, D. P., 2003. Managing America's rivers: Who's doing it? *International Journal of River Basin Management*. 1(1): 21–31.

Malcom, H. R., 1980. A Study of Detention in Urban Stormwater Management, Water Resources Research Institute of the University of North Carolina, Report No. 156, July 1980.

Mangangka, I. R., Liu, A., Egodawatta, P., Goonetilleke, A., 2015. Performance characterisation of a stormwater treatment bioretention basin. *Journal of Environmental Management*. 150, 173–178.

Ministry of Construction of the PRC, 2006. *General Administration of Quality Supervision*, Inspection and Quarantine of the PRC, Code for Design of Outdoor Wastewater Engineering (GB50014–2006). Beijing: China Planning Press (in Chinese).

Ministry of Housing and Urban-Rural Development (MHURD), 2014. Technical guidelines on Sponge City construction – Low impact development stormwater management system (in Chinese).

Morgenroth, J., Visser, R., 2011. Aboveground growth response of Platanus orientalis to porous pavements. *Arboriculture & Urban Forestry*. 37(1), 1–5.

Mouritz, M., 1992. *Sustainable Urban Water Systems; Policy & Professional Praxis*. Perth, Australia: Murdoch University.

National Research Council, 2008. *Urban Stormwater in the United States*. Washington, DC: National Academies Press (in Chinese).

Norwegian Ministry of Foreign Affairs, 2013. Challenges in flood risk management in urban areas of river deltas in South and South East Asia. Wednesday 6 – Thursday 7 November. Conference report, Bangkok.

Pratt, C. J., Mantle, J. D. G., Schofield, P. A., 1989. Urban stormwater reduction and quality improvement through use of permeable pavements. *Water Science and Technology*. 21(8/9), 769–778.

Ryland, H. G., 2000. A piece of the puzzle: Insurance industry perspective on mitigation. *Natural Hazards Review*. 1(1), 43–49.

Saadatian, O., Sopian, K., Salleh, E., Lim, C. H., Riffat, S., Saadatian, E., Toudeshki, A., Sulaiman, M. Y., 2013. A review of energy aspects of green roofs. *Renewable and Sustainable Energy Reviews*. 23, 155–168.

Samules, P., Klijn, F., Dijkman, A. J., 2006. An analysis of the current practice of policies on river flood risk management in different countries. *Irrigation and Drainage*. 55, 141–150.

Schueler, T. R., 1987. *Controlling Urban Runoff: A Practical Manual for planning and designing urban BMPs*. Washington, DC: Washington Metropolitan Water Resources Planning Board.

Shaver, H. E., 2000. Auckland Regional Council Technical Publication No. 124, April 2000. ISSN 1175 205 XP.

Shaver, H. E., Roesner, L. A., Urbonas, B., Sonnen, M. B., 1988. Stormwater management issues, in *Design of Urban Runoff Quality Controls*, American Society of Civil Engineers (ASCE).

Tim, D., William, S., William, F. H., Richard, A., David, B., Scott, A., Trowsdale, S. A. Sylvie, B., Annette, S. D., Jean-Luc, B. K., Peter, S. M., Gilles, R., Mathias, U., Danielle, D., Maria, V., 2015. SUDS, LID, BMPs, WSUD and more – The evolution and application of terminology surrounding urban drainage. *Urban Water Journal*. 12(7), 525–542.

Thomas, N. D., Andrew, J. R., 2003. Municipal stormwater management-2nd edition. ISBN 1-56670-584-3, CRC Press LLC, Lewis Publishers is an imprint of CRC Press LLC.

Trowsdale, S. A., Simcock, R., 2011. Urban stormwater treatment using bioretention. *Journal of Hydrology*. 397, 167–174.

United States Environmental Protection Agency (USEPA), 2000. Washington, D.C. Field Evaluation of Permeable Pavements for Stormwater Management, Olympia, Washington. Fact Sheet. Document No. EPA-841-B-00-005B.

Vandermeulen, V., Verspecht, A., Vermeire, B., Van Huylenbroeck, G., Gellynck, X., 2011. The use of economic valuation to create public support for green infrastructure investments in urban areas. *Landscape and Urban Planning*. 103(2), 198–206.

Volder, A., Dvorak, B., 2014. Event size, substrate water content and vegetation affect storm water retention efficiency of an un-irrigated extensive green roof system in Central Texas. *Sustainable Cities and Society*. 10, 59–64.

Volder, A., Watson, W. T., Viswanathan, B., 2009. Potential use of pervious concrete for maintaining existing mature trees during and after urban development. *Urban Forestry & Urban Greening*. 8(4), 249–256.

WEWS, 2003. *Water Environment and Water Services Act*. Edinburgh, Scotland: Scottish Government.

Whelans, C., Maunsell, H. G., Thompson, P., 1994. *Planning and Management Guidelines for Water Sensitive Urban (Residential) Design*. Perth, Western Australia: Department of Planning and Urban Development of Western Australia.

Wong, T. H. F., 2007. Water sensitive urban design; the journey thus far. *Australian Journal of Water Resources*. 110(3), 213–222.

Xu, C., Hong, J., Jia, H., Liang, S., Xu, T., 2017. Life cycle environmental and economic assessment of a LID-BMP treatment train system: A case study in China. *Journal of Cleaner Production*. 149, 227–237.

Zhou, J., Erdal, Z. K., McCreanor, P. T., Montalto, F., 2010. Sustainability. *Water Environment Research*. 82, 1376–1395.

Zuo, J., Pullen, S., Rameezdeen, R., Bennetts, H., Wang, Y., Mao, G., Zhou, Z., Du, H., Duan, H., 2017. Green building evaluation from a life-cycle perspective in Australia: A critical review. *Renewable and Sustainable Energy Reviews*. 70, 358–368.

3 Integrating Water Quality in Urban Flood Risk Management

Structural Approaches

Susanne Charlesworth, Daniel Castro-Fresno, and Jorge Rodriguez-Hernandez

3.1 INTRODUCTION

In previous chapter, it has been established that the removal of vegetation and its replacement with hard infrastructure such as impermeable surfaces (e.g., roads, pavements and roofs) as well as the constraining of water falling and flowing in and on urban areas to pipes, gutters and concrete-lined channels results in flooding and pollution events (Charlesworth et al., 2003; Shuster et al., 2005). Flooding is reflected in the urban storm hydrograph, with its flashy peak, reduced time to peak, increased volumes and discharge (Hung et al., 2018; see Figure 3.1).

The impacts described earlier also arguably result in problems with first flush whereby pollutants adhering to hard surfaces are removed in the first 10% of a storm, and contaminate the receiving water body (Lee and Bang, 2000). According to Al Mamun et al. (2020), there is some uncertainty in the occurrence and intensity of the effect since it may be site and storm specific associated with catchment topography, the structure of the drainage system, type of climate and the kinds of contaminants. Kim et al. (2003) illustrated the first flush phenomenon using a series of pollutographs (the concentration of the pollutant with time) which showed reducing concentrations of dissolved and particulate associated metals with time as the storm passed. Figure 3.2 illustrates this with an example from Bertrand-Krajewski et al. (1998), showing the rise in concentration of total suspended solids (TSS) before the storm peak as pollutants are removed from paved surfaces during the first 10 minutes of the storm.

However, the presence of this phenomenon and its impact are both catchment and climate dependent, as well as there are different behaviours for different pollutants (Deletic, 1998). It has been found that the majority of these pollutants are associated with industry, air pollution and traffic, with the emphasis in recent years changing with the increased importance of emerging pollutants (EPs) or contaminants of emerging concern such as personal care products (PCPs), medicines and pesticides (Jimenez, 2008; Ellis et al., 2013; Krishnakumar et al., 2022; Kumar et al., 2023), and the decreased importance of Pb with the introduction of unleaded petrol. For example, Charlesworth et al. (2003) found that the concentration of Pb in Birmingham, UK, street dusts over nearly 30 years had reduced almost 26-fold, leading some countries to drop Pb from their atmospheric monitoring programmes, and instead concentrate on Zn and Cu. Figure 3.3 illustrates the complex pathways of a variety of pollutants, their sources and deposits in the urban area. Thus, the integration of flood risk management in the urban environment with consideration of water quality improvement is of great importance since the two are inextricably linked.

3.2 CONTEXT OF USING STRUCTURAL SOLUTIONS TO ACHIEVE SUSTAINABLE WATER QUALITY IMPROVEMENTS IN URBAN STORMWATER

This chapter focusses on *structural* means of managing urban flooding-associated pollution, mainly particulate matter. The source and the mode of transport of such contaminants, and their combination, will define the pollutant load and hence procedures for its management. Further, the particle size of polluted solid material transporting potentially toxic elements is also important, since the majority of the load is carried by particles <63 μm, particulate associated pollutants (PAPs). Studies have found that whilst the <63 μm fraction may make up just 6% of the sediment load, they carry up to 50% of the total transported pollutant load (Loganathan et al., 2013). Knowing the partition of contaminants between PAPs and dissolved elements results in different processes for their management. For example, polluted atmospheric particulates will be washed out during heavy rainfall to join dissolved or particulate-associated contaminants in the drainage system, as well as microbiota, salts, nutrients, oil and grease.

Urban flooding will therefore contain many pollutants that are being transported around the environment in various pathways, differing in chemical species and forms. There have been

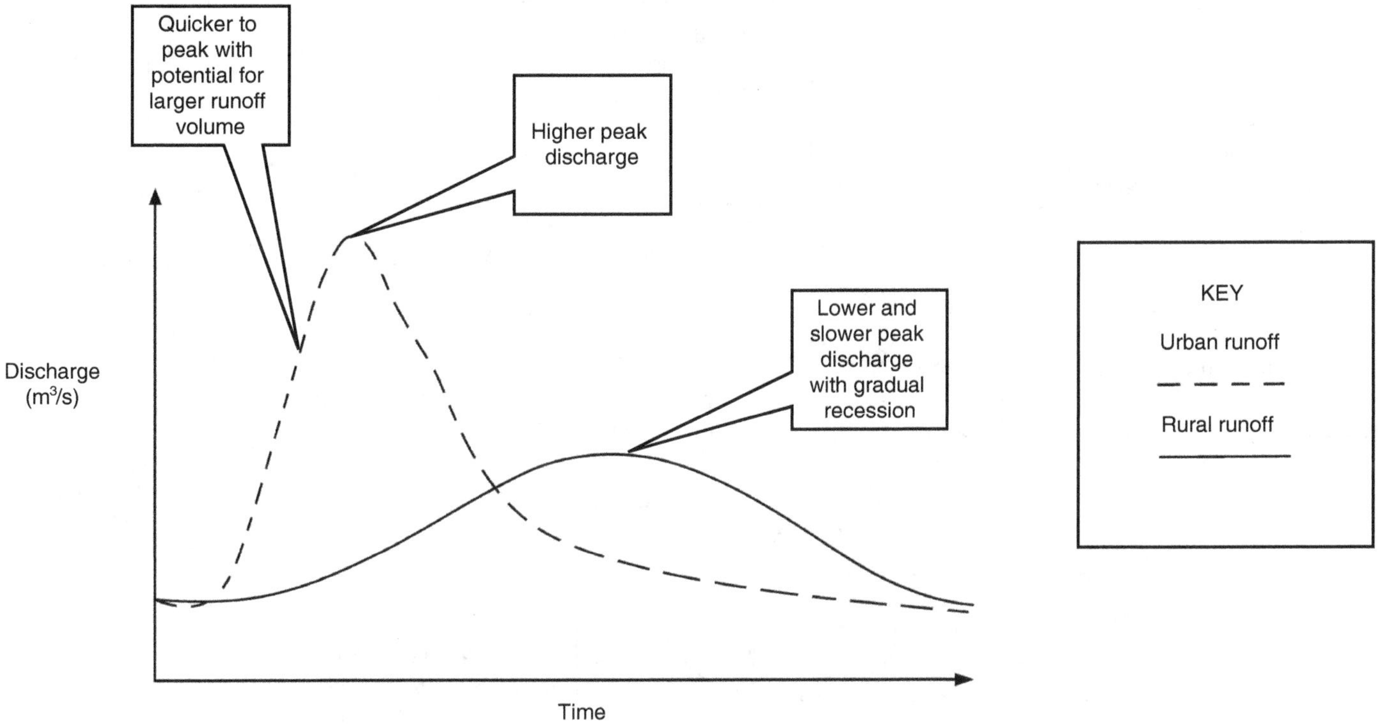

Figure 3.1 Comparison of urban and rural storm hydrographs (after Woods Ballard et al., 2007)

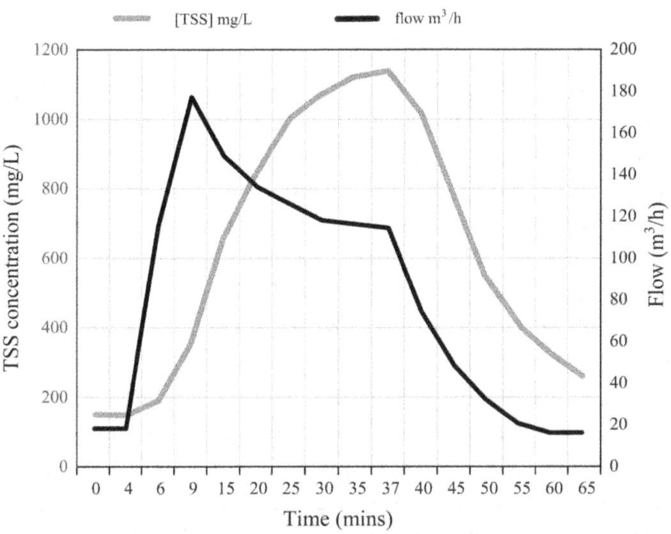

Figure 3.2 Example of a hydrograph with pollutograph illustrating the first flush of TSS during the first 10 minutes of the storm (taken from: Bertrand-Krajewski et al., 1998)

representations of these pathways (see Figure 3.3) including the "cascade" proposed by Charlesworth and Lees (1999), whereby PAPs are moved from their source to a site of deposition, which may be temporary, for example, in a gully pot or catch pit by the road gutter, or more permanent, for example, accumulating as lake or pond sediment. Charlesworth et al. (2011) proposed an urban geochemical cycle combining the cascade, the addition of natural materials and the production of urban sediments including street dusts and soil.

The EU's Water Framework Directive (2000/60/EC) in combination with the EU's Directive (2008/105/EC), amended by Directive 2013/39/EC, identify between them 45 priority substances (PS), of which 20 individual elements or groups of chemical species are considered to be "priority hazardous substances (PHS)." The latter category includes Cd, tributyltin, Hg and their compounds, but not lead or nickel, which are PSs. Also included under PHS are the PAHs benzo(a)pyrene; benzo(b)fluoranthene; benzo(g,h,i)perylene; benzo(k)fluoranthene and indeno(1,2,3-cd) pyrene. This combination of substances in the PS and PHS lists illustrate the potential chemical "cocktail" which can be carried by urban stormwater and which any management strategy must seek to address (Warwick, 2017; Lapointe et al., 2022).

Conventional drainage, comprising pipes, gutters and roadside gully pots, does not function efficiently when dealing with heavy storms and associated pollution since it has not been designed with these scenarios in mind. Bolognesi et al. (2008) found that the efficiency of gully pots was inversely proportional to the rate of water inflow and directly proportional to the size and specific gravity of sediment particles. As a result, during heavy rain, any polluted sediment trapped in gully pots can overflow onto the road including the overlying liquor which contains dissolved pollutants (Charlesworth and Lees, 2001).

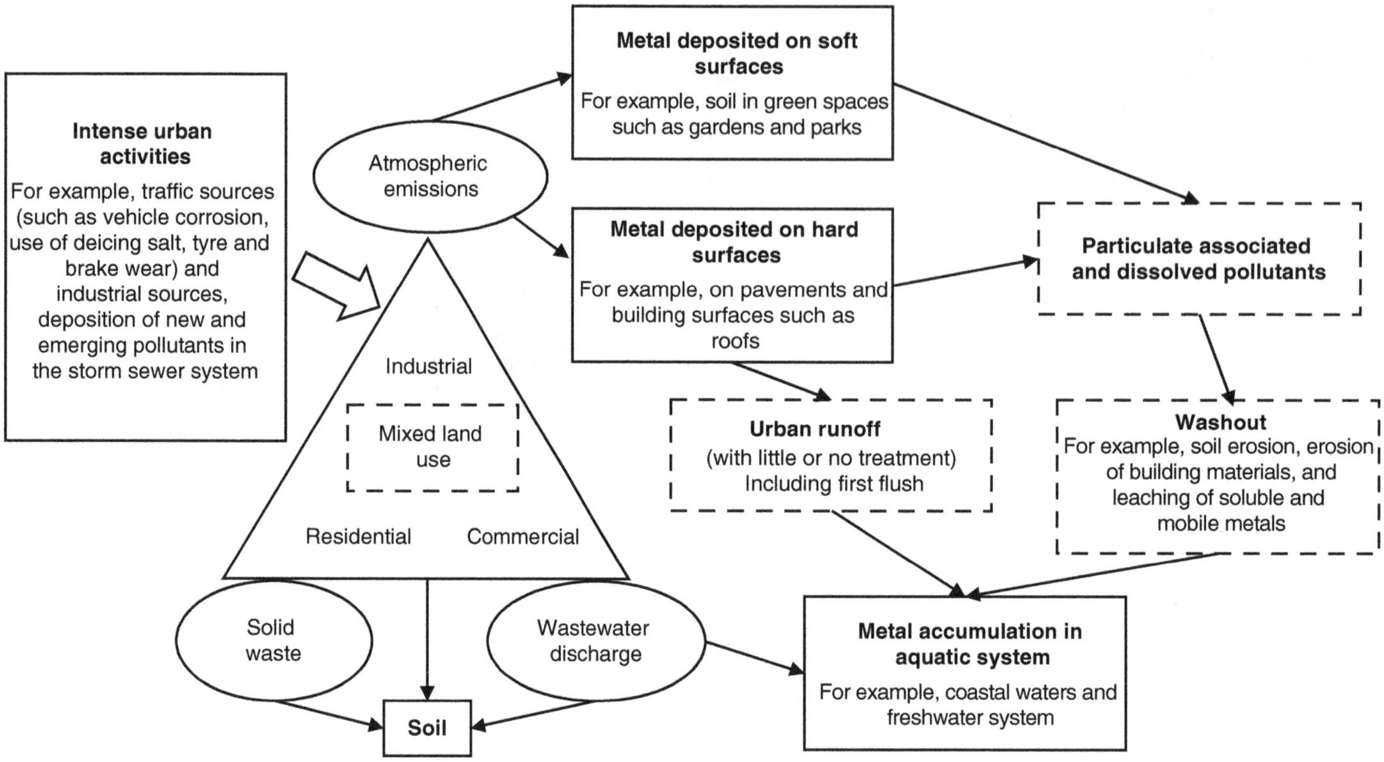

Figure 3.3 Pathways of pollutants in urban environments (after Wong et al., 2006; Escarameia et al., 2006)

Many cities worldwide have a legacy of combined sewer systems (CSS), in which excess surface water and sewage flow together in the same pipework. During storm events, large volumes of water pass through the wastewater treatment works unnecessarily, and when the system cannot cope, a toxic mixture of human waste and polluted stormwater floods homes, industrial premises and infrastructure. Such mixtures are called combined sewer overflows (CSOs); very often this kind of flooding facilitates the polluted water reaching the receiving watercourse as well, which causes diffusion or non-point pollution incidents impacting negatively on the environment beyond the city, particularly considering the predicted impacts of climate change (Botturi et al., 2021) which will exacerbate the incidences of CSOs due to more intense rainfall.

After the 2007 floods in the UK, Water UK (2008) stated that "bigger pipes are not the solution to bigger storms" that "sewers and drains are not flood defences" and that the way forward was through "sustainable drainage" and what it called "sacrificial areas." Sustainable drainage systems (SuDS) mimic natural drainage whereby water infiltrates into the ground, possibly held in temporary storage and can then be conveyed slowly to the receiving water body, be that a stream or river, pond or lake, or groundwater or the sea. Sacrificial areas include the floodplain on which much construction has already taken place, but rivers and their floodplains can be returned to their natural state as far as possible in the river restoration approach (RESTORE, 2013) by reinstating meanders, constructing backwaters and re-vegetating banks and riverbeds. Restoring or rehabilitating rivers in this way reduces flow velocity, encouraging PAPs to settle out and giving aquatic and riparian vegetation the time to trap and/or biodegrade any contaminants. Nature-based solutions (NbS) utilise vegetation to reduce water volume and bioremediate contaminants by taking up and metabolising nutrients and micronutrients. They offer similar co- and multiple benefits as SuDS, which includes NbS as well as hard infrastructure (e.g., raingardens). According to the 2018 World Water Development Report (WWAP/UN-Water 2018), NbS have the potential to address current and future water resources management issues, contributing to the achievement of the 2030 Agenda for Sustainable Development, particularly the water related Sustainable Development Goal, SDG6. However, the UN 2023 report on progress admitted that the ability to provide sustainable management of water and sanitation for all by 2030 was "alarmingly off-track." This lack of progress may be due to the many examples where NbS do not appear in policies, financing or management interventions, possibly owing to the absence of sufficient evidence of their efficacy by improved assessment, monitoring and evaluation. However, all of these strategies can be part of the much wider, holistic approach of water-sensitive urban design (WSUD), or low impact development (LID) (Lashford et al., 2022) and are also incorporated in the "Sponge City" concept (Lashford et al., 2019; Song, 2022),

Table 3.1 *SuDS devices and their water quality and quantity benefits (adapted from Dickie et al., 2010; Woods Ballard et al., 2007, 2015; Charlesworth et al., 2013)*

	Devices	Description	Process	Role	Water quantity benefits	Water quality benefits
Green roofs		Partial or complete coverage of a roof with vegetation such as *Sedum* sp., if extensive, or can be planted with shrubs and trees, if intensive.	Control of runoff close to source, some storage of water and filtering out of pollutants. Can provide other benefits such as reduction of the urban heat island, carbon sequestration.	Source control	✓✓	✓✓
Water Butts		Small-scale (compared to rainwater harvesting), off-line storage devices designed to capture and store runoff for use in gardens or other domestic uses.	Potable water demand reduction. Some water storage can reduce the storm peak.	Source control	✓✓✓	✓✓
Rainwater Harvesting		Large-scale water collection system from impermeable surfaces for non-potable water use such as toilet flushing.	Potable water demand reduction. Some water storage can reduce the storm peak.	Source control	✓✓✓	✓✓
Pervious Pavements		Pervious surfaces allowing water to infiltrate, replacing traditionally impermeable pavements.	Water is stored and released gradually during which water quality is also improved. Can be used in permeable and impermeable ground by incorporating outflow and overflow components.	Source control, in car parks and used for some roads.	✓✓✓	✓✓✓
Bioretention		Depressions filled with a sand/ soil mixture and planted with vegetation. Water enters, trickling via a filter layer to a perforated pipe at the base.	Stores water, releasing it gradually. The vegetation provides improvements in water quality, as can the filter layer to a certain extent.	Source control, in open spaces, next to roads and car parking.	✓✓✓	✓✓✓

Table 3.1 *(cont.)*

	Devices	Description	Process	Role	Water quantity benefits	Water quality benefits
Rain Garden		Vegetated area into which runoff infiltrates, is attenuated and can be stored or taken up by plants.	Stores runoff, filters out pollutants and recharges groundwater.	Near roads, in residential developments, throughout urban areas.	✓✓✓	✓✓✓
Swales		These are shallow vegetated ditches which can run parallel to hard surfacing. They allow runoff to trickle down the sides and into the base. Water is conveyed in a controlled manner to another SuDS device or the receiving watercourse.	Can both treat and attenuate runoff. Can be used in both permeable and impermeable grounds as long as the latter is underdrained properly.	In open spaces, next to roads and car parks.	✓✓✓	✓✓✓

Key: ✓ Low impact
✓✓ Medium impact
✓✓✓ High impact

allowing advantage to be taken of water as a "liquid asset" (Semadini-Davies et al., 2008) rather than a "wasted waste" (Charlesworth, 2010) and viewing it as "opportunity water" (Chocat et al., 2007) instead. Water-sensitive urban design is further discussed in Section 3.6.

The SuDS approach is therefore a *structural solution* and includes individual devices with permeable surfaces such as green roofs and pervious pavements, infiltration trenches and basins, swales, filter drains and strips, detention basins, retention ponds and wetlands (see Table 3.1 for details of individual devices or interventions; Charlesworth et al., 2013). These devices can be implemented alone, that is, in isolation, to address a specific environmental problem or set of problems, and/or end of pipe solutions whereby a single device is installed at the end of a conventionally drained catchment to address any final issues around flooding or pollution. Designed in combination, they provide a treatment, or management train, which offers a far better performance in terms of degradation of pollutants, attenuation of the storm peak, but also benefits in terms of amenity and biodiversity (Bastien et al., 2010; Woods Ballard et al., 2015). If designed properly (Watkins and Charlesworth, 2014) they can go some way to remove stormwater from conventional sewer systems, thus reducing the amount of treatment necessary, and also reduce floodwaters emanating from water treatment works.

This chapter therefore considers the ways in which individual SuDS devices, management trains and the holistic WSUD approach can address suspended solids and biodegradable pollution problems mainly associated with urban flooding and its impacts on water quality in cities worldwide.

3.3 FLOOD RESILIENCE AND WATER QUALITY BENEFITS OF SUDS

The role of SuDS is to deal with excess stormwater as near to the source as possible. In an ideal world, no rainwater would leave the building or its surroundings. This approach is exemplified by the Management Train (see Figure 3.4), whereby various SUDs devices are designed together to reduce flooding, improve water quality and provide amenity and biodiversity (Charlesworth, 2010). The focus should therefore be on source control, with whatever surface water is left progressing to site and regional controls. Within the management train, water infiltrates, is evaporated or retained, with various processes occurring during its passage to the receiving water body which improves water quality and reduces water quantity.

Table 3.2 shows the processes occurring in different types of SuDS structures used to address pollution problems using filtration and adsorption. Filtration occurs in almost all such systems with the exception of ponds. Adsorption removes, among other substances, heavy metals, which are pollutants transported in association with particulates and that are not present in all suspended sediment (SS). Scholes et al. (2008) and Lundy (2017) classified the individual removal processes detailed in Table 3.2 as physical, biological and physico-chemical as presented in Table 3.3.

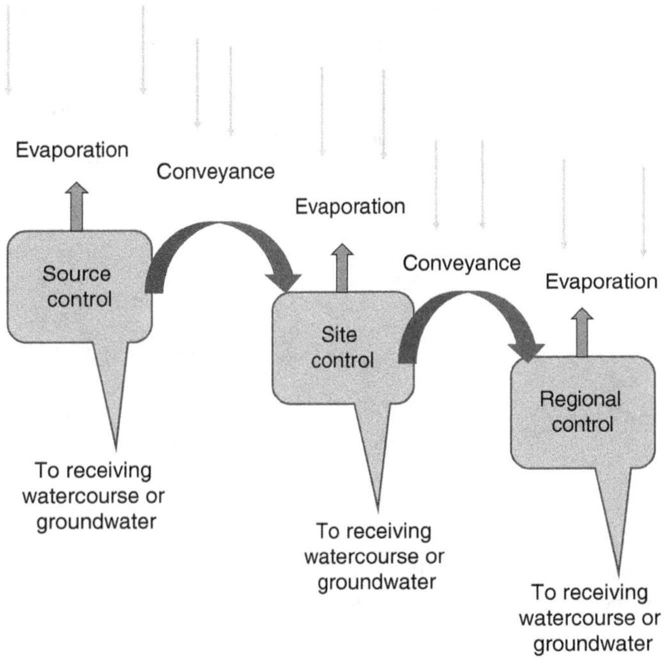

Figure 3.4 SuDS Management Train (after Charlesworth, 2010)

For the SuDS devices, similar to NbS, depending on biological processes occurring in vegetation, seasonality greatly affects their ability to improve water quality. Plant die back during autumn, and winter can potentially release contaminants into the system, for example, plants in wetlands (see Weis and Weis, 2004) or leaf drop of street trees. The United States Environmental Protection Agency (USEPA) (2009) recommends harvesting vegetation before the winter from stormwater ponds in order to remove it before die-off, preventing excess nutrients, bacteria and oxygen-consuming materials being released downstream. However, they do not provide recommendations for the disposal of this harvested vegetation should it prove to be contaminated. Maintenance may also be problematic, for example, grass cutting can have the potential to generate contaminated mowings which then have to be disposed of (Wilson et al., 2004; Berwick, 2017). A laboratory-based study by Charlesworth et al. (2016) on the uptake of contaminants in the leaves and roots of grass types recommended for use in swales found that *Agrostis canina* (velvet bent), *Poa pratensis* (smooth-stalked meadow grass) and *Agrostis capillaris* syn. *tenuis* (browntop bent) accumulated heavy metals in their shoots; therefore, maintenance practices such as mowing had the potential to remove the parts containing contaminants. In general, it is recommended that the grass cuttings are removed

Table 3.2 *Processes occurring in SuDS device to reduce potential pollutants. The most important processes have been* italicised.

Runoff pollutant	Swales	Infiltration basin	Surface flow (SF) constructed wetland	Sub-surface flow (SSF) constructed wetland	Balancing pond	Sedimentation pond
Suspended solids (SS)	*Filtration*	*Filtration* Settling	*Filtration* Settling	*Filtration*	*Settling*	*Settling*
Total heavy metals	*Filtration* Adsorption	*Filtration* Adsorption Settling Plant uptake Precipitation	*Filtration* Adsorption *Settling* Plant uptake	*Filtration* *Adsorption* Plant uptake	Adsorption *Settling*	Adsorption *Settling* Plant uptake Precipitation
Total organics	*Filtration* Adsorption	*Filtration* Adsorption Settling Biodegradation Volatilisation	*Filtration* Adsorption Settling Biodegradation Volatilisation	*Filtration* *Adsorption* *Biodegradation*	*Adsorption* Settling Biodegradation Volatilisation	*Adsorption* *Settling* Biodegradation Volatilisation
Nutrients	Plant uptake	Plant uptake	*Plant uptake*	*Plant uptake*	Plant uptake	Plant uptake
Oil and grease	*Filtration* Adsorption	*Filtration* Adsorption Settling Biodegradation	*Filtration* Adsorption Settling Biodegradation	*Filtration* *Adsorption* Biodegradation	Adsorption *Settling* Biodegradation	Adsorption Settling Biodegradation

Table 3.3 *Processes occurring in SuDS devices (adapted from Scholes et al., 2008; Wood-Ballard et al., 2007, 2015; and Humphrey et al., 2023).*

	Physical	Physicochemical	Biological
Process	*Settling of particulates*	*Adsorption to substrate*	*Uptake by plants and algae*
SuDS device	Infiltration basins Retention ponds Extended detention basin Sedimentation tank Detention basin Lagoons	Pervious paving Soakaways Infiltration trench Infiltration basin SSF constructed wetlands	SSF constructed wetlands
Pollutant Removed	Nutrients Total suspended sediments Metals Organic matter Biological oxygen demand	Hydrocarbons Metals Pesticides	Metals Nutrients
Process	*Filtration through the substrate*	*Flocculation*	*Degradation by microbiota*
SuDS device	Porous asphalt Pervious paving Soakaways Infiltration trench Infiltration basin SSF constructed wetlands	Water harvesting tanks	Infiltration basin SSF constructed wetlands Geotextile if included in permeable pavement systems (PPS)
Pollutant Removed	Total suspended sediments Hydrocarbons Organic matter Biological oxygen demand	None	Nutrients Hydrocarbons Pesticides Organic matter Biological oxygen demand

Table 3.3 *(cont.)*

	Physical	Physicochemical	Biological
Process	*Volatilisation*	*Precipitation*	*Denitrification*
SuDS device	SF wetland Infiltration basin	Sedimentation pond Infiltration basin	
Pollutant Removed	Pesticides	Nutrients Metals	Nutrients
Process	*Photolysis*	*Avoidance*	*Trapping*
SuDS device	Any device which allows UV light to penetrate, for example, surface of ponds	None	SF constructed wetlands Swales
Pollutant Removed	Hydrocarbons	Chloride	Litter Sediment

Table 3.4 *Classification of some SuDS devices according to their performance in reducing pollution.*

Runoff constituent	Swales	Infiltration basins	SF constructed wetlands	SSF constructed wetlands	Balancing ponds	Sedimentation ponds
Suspended solids and associated heavy metals	Good	Good	Good	Good	Moderate	Good
Dissolved heavy metals	Moderate–Good	Moderate–Good	Good	Good	Poor	Poor–Moderate
Oil and grease	Good	Moderate–Good	Good	Good	Moderate	Moderate
Nutrients	Poor	Poor	Moderate–Good	Good	Poor	Poor–Moderate

to prevent clogging of outlets and clumping on the bottom of the device (Colwell, 2000; Berwick, 2017). Maintenance is discussed further in Section 3.5.

Table 3.3 details the specific processes occurring in individual SuDS devices according to these three categories. Some of these processes occur together, depending on the device used. For instance, the removal of pollutants in vegetated devices such as green roofs and walls or filter strips and swales may include aspects of all three categories given in Table 3.3, mainly biological, in that they may utilise contaminants such as nitrate and phosphate as nutrients and also heavy metals such as Zn and Cu as essential to their metabolic function, vegetation also encourages the physical settling of particulates due to the reduction in flow through it, giving time for biodegradation and also the trapping of PAPs in their stems and roots. In the case of swales, Allen et al. (2015) modified a trapping efficiency equation for sediment transport over a grassed surface which had been developed by Deletic (2005) and concluded that bed deposition and trapping efficiency were found to decrease progressively over multiple runoff events. Extrapolating from field results, a tentative estimation was proposed that detention efficiency in a swale after service of 25 years was 3% of the initial inflow.

Table 3.4 classifies individual devices according to their performance in addressing various pollutants carried in urban stormwater, and Table 3.5 lists potential removal efficiencies. All of the devices listed remove PAPs and SS well; however, dissolved heavy metals and nutrients were not addressed as well apart from in the wetlands. Oil and grease were dealt with reasonably well in all of the devices listed.

Geotextiles are generally recommended for inclusion in PPS initially to prevent blocking of the sub-base (generally a 5–40-mm aggregate) by the bedding layer of particles with smaller size (3–10-mm gravel). However, it was found by Coupe et al. (2003) that the geotextile retained and treated pollutants by filtering out the PAPs, as well as dissolved contaminants, microorganisms and hydrocarbons (Charlesworth et al., 2014). Permeable pavement systems depend on infiltration and hence filtering of the inflowing stormwater. The unique role of the geotextile is a fundamental support in this regard in that a hydraulic head is developed above the geotextile, slowing the water down and thus allowing particulates to settle out, but also the geotextile fibres provide the medium on which a biofilm can develop (Bayon et al., 2015). This biofilm is responsible for biodegrading pollutants such as hydrocarbons and nutrients, but it is also responsible for the physical entrapment of PAPs such as metals. Biofilms are a microecosystem comprising bacteria, fungi and protozoa held together by a slimy extra polymeric substance, with the first photomicrographs of them taken by Coupe et al. (2003); further details can also be found in Charlesworth et al. (2014). These microbes are similar to those found in soils, and thus are naturally occurring in the environment. There were some studies in which suitable microorganisms were inoculated into PPS in an attempt to improve

Table 3.5 *SuDS devices and their potential pollutant removal efficiencies (after Wilson et al., 2004)*

	Percentage removal of urban stormwater pollutant					
SuDS device	TSS	TP	TN	NOx	Metals	Hydrocarbons
Dry detention pond	61	19	31	9	26–54	–
Wet retention pond	68±10	55±7	32±11	34±21	26–65	–
Detention basins	65–90	20–50	20–30	–	40–90	30–60
Treatment ponds	75–90	30–50	30–50	–	50–80	30–60
Infiltration trench	70–80	60–80	25–60	–	60–90	–
Pervious pavement	60–95	50–80	65–85	–	60–99	70–90
Bioretention	80	65–87	49	15–16	43–97	–
SF constructed wetland	71±35	56±35	19±29	40±68	0–57	50–80
Swale	60–90	25–80	30–90	13±31	40–90	40–90
Filter strip	54–84	−25–20	−27–20	−16	25–40	70–90
Infiltration basins	45–75	60–70	55–60	–	85–90	–
Filter drains	50–85	–	–	–	50–80	30–70
SuDS in general	*45–95*	*20–80*	*10–90*	–	*25–90*	*30–90*

Key: TSS = total suspended solids; TP = total phosphorus; TN = total nitrogen; NO_x = nitrogen oxides.

their development, but it was found that it was better to allow them to develop naturally (Bond and Pratt, 1998; Pratt et al., 1999; Newman et al., 2002). It was found that biodegradation of the retained hydrocarbons achieved by these microorganisms was more efficient with the addition of fertilisers. Spicer et al. (2006) produced a self-fertilising geotextile system which, whilst it performed extremely effectively as microbial activity increased during the experiments, was nonetheless expensive to produce commercially, and has therefore not been brought to market. Whilst the inclusion of a single geotextile is generally recommended, Paul and Tota-Maharaj (2015) investigated the utility of having two geotextile layers in a set of test rig PPS experiments. They found that water quality wise, there were distinct advantages, in having two layers of geotextile, as they enhanced particulate removal and biochemical treatment performance. Sañudo-Fontaneda et al. (2018) undertook laboratory-based experiments to assess the utility of position and number of geotextiles in highway filter drain (HFD) test rigs, finding that where the geotextile was located had a significant effect on hydraulic performance. The presence of a geotextile tended to delay time to peak; with a geotextile higher in the test rig (at 50 mm from the surface) peak flow was longer than the rig with a lower geotextile (at 150 mm from the surface). Test rigs without a geotextile at all had the shortest time to peak.

Geotextile thickness is an important factor in supporting the generation of biomass, since they provide larger volumes of fabric for microorganisms to grow on, although the manufacturing of the fabric influences this ability as well. For example, needle-punched fabrics can be thicker than the ones produced by heat-sealing, which makes the former more suited to support microbiological growth (Bayon et al., 2015). However, until experiments carried out by Bayon et al. (2015), it was not known what the effect of this microbiological growth was on the structural strength and integrity of the geotextile. Not only was it found that their mechanical characteristics were not significantly diminished in comparison with controls, but their strength was slightly increased in the more tightly woven fabrics tested. There are therefore opportunities to increase the efficiency of the water quality role of PPS by improving their performance; as a result, new types of geosynthetics are being developed specifically for permeable pavements, albeit with the focus on attenuation of peak flow (Nnadi et al., 2014).

3.4 WATER QUALITY AND SUSTAINABLE APPROACHES TO DRAINAGE

Whilst there are examples of sustainable approaches to drainage going back millennia (Charlesworth et al., 2017a), modern sustainable drainage has been in use globally for at least the last 35 years, for example, the USA, France, Sweden, New Zealand and Australia have all successfully integrated SuDS into their urban landscapes (Charlesworth et al., 2014). However, the design of SuDS is such that its focus, similar to conventional piped drainage, is on attenuation of the storm peak. The research has therefore focussed more on the water *quantity* reduction capabilities of SuDS systems and less on their abilities to improve water *quality* particularly their performance over the long term (Kirby, 2005). Studies utilising modelling techniques have similarly concentrated on reducing water amount (e.g., Lashford et al., 2014). There are studies at the laboratory and field scale from around the world which have provided evidence of their efficiency in dealing with typical urban and industrial pollutants carried in stormwater. Guidelines proposed by Wilson et al. (2004) and Woods Ballard et al. (2015) represent the most complete design and performance coverage of the numerous manuals published by CIRIA in the UK. The Melbourne Water (2015) web site sets out detailed guidelines for WSUD in Australia. In addition to these guidelines, there are books such as Butler and Davies (2004); Booth and Charlesworth (2014) and Charlesworth and Booth (2017) which include information on structural and sustainable approaches to address problems associated with urban pollutants in runoff. The United States Environmental Protection Agency has offered technological fact sheets about stormwater best management practices (BMPs) since the 1990s (USEPA, n.d.). This information is also available in different languages, for instance, in Spanish, which is one of the most complete manuals for SuDS published by CEDEX (Puertas Agudo et al., 2008; Rodríguez Sánchez and Díaz Martínez, 2009). National and international inventories and databases exist with information for researchers, local authorities and other institutions engaged in water cycle management. Probably the most significant one is the International Stormwater BMP Database (WERF et al., 2015) with more than 530 related studies from across the world. Other initiatives include Susdrain in the UK (www.susdrain.org/) and red SuDS in Spain (https://redsuds.es/mapa/), with more than 100 case studies each, classified both by location and device type.

The following sections therefore summarise some of the studies contributing supporting evidence of the efficacy and efficiency of SuDS devices and trains in improving the quality of stormwater in urban areas.

3.5 EVIDENCE OF THE EFFICIENCY AND EFFICACY OF SUDS IN IMPROVING THE WATER QUALITY OF URBAN STORM RUNOFF

This section is structured according to the *relevance* of the studies providing evidence of water quality improvements associated with the use of SUDS. There is far more evidence at the laboratory scale, and some examples of these are given, but there has been far less field monitoring of the individual devices, and even less of whole management trains.

3.5.1 Laboratory-Scale Studies

SuDS devices are either made up of hard infrastructure, for example, porous paving, or soft infrastructure, which utilises vegetation, for example, green roofs, or they can be made up of a combination of both whereby rain gardens can be installed into the urban streetscape (see Charlesworth et al., 2014 and Table 3.1). Experiments at the laboratory scale require the testing of physical models of the devices, and there has been much research undertaken of hard infrastructure, in particular PPS. This section begins with some examples of the testing of PPS in the laboratory under a variety of experimentally applied stresses.

Laboratory-scale studies enable a certain amount of experimental control to be exerted for the duration of the monitoring. This can include rainfall simulation to reproduce storm conditions, the application of contaminant loadings or the imitation of oil dripping from the sump of a motor vehicle; choice of simulated rainfall intensity and contaminants are therefore important to ensure repeatability (Nichols et al., 2015). All of these controls were exerted in a series of monitoring experiments carried out by Coventry University, UK, over a 3-year period on $1 m^3$ model rigs to test a variety of porous paving surfaces including porous asphalt, block pavers, porous concrete and a cut out from an in-use 10-year-old porous asphalt car park (see Figure 3.5; Charlesworth et al., 2017b). Background analyses were carried out for three months before the two test pollutant materials, polluted street dusts and clean/unused engine oil were added to the surfaces of the test rigs at monthly intervals in amounts equivalent to a month's worth of loading in a typical urban environment. Following the addition of the pollutants, the surfaces were artificially rained on, and samples of the effluent were collected and tested for Cd, Cu, Ni, Pb, Zn, hydrocarbons (oil and grease) and total suspended sediment (TSS).

Over the course of the monitoring programmes, more than 430 ml of oil (~1.7 g oil l^{-1} of rainfall) was applied to the surface of each of the models; these levels of oil were well in excess of those typically associated with storm water runoff which are generally $<<0.1$ g l^{-1} (Bond et al., 1999). In total, less than 0.5 ml of oil was detected in any of the effluents, indicating that close to

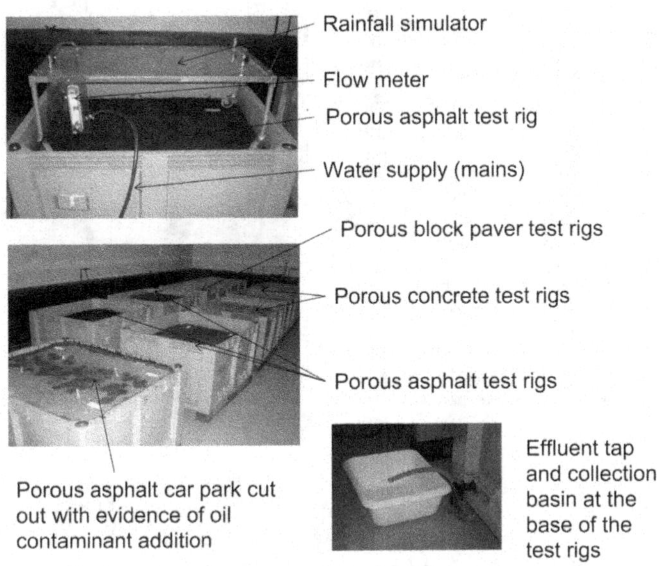

Figure 3.5 An example of laboratory monitoring of nine PPS test rigs at Coventry University, UK

99.9% of the added oil did not pass through the PPS structures. The results from all the test rigs indicated that effluent concentrations of heavy metals, TSS and hydrocarbons did not vary significantly from the values of those established for background. Concentrations of heavy metals were well within published environmental background values and below WHO guideline limits for potable water. Figure 3.6 shows the percentage retention of each pollutant based on the estimates of the total pollutant loading and the total pollutant output in the effluents. The estimates of heavy metals added were based on the measurements of their concentrations in the street dust, oil and artificial rainwater. In each case, more than 90% of the added contaminants appeared to have been retained within the PPS structures as shown in Figure 3.6. Oil has not been included in Figure 3.6 as 99.9% of the applied oil was retained and did not appear in the effluent. Cu and Zn were retained at greater than 98%, with Cd and Ni slightly lower at between 90% and 95%. Retention for TSS was up to 100%. Further details can be found in Charlesworth et al. (2017b). This study therefore suggested that the surface of a PPS appeared to have little influence on its ability to clean influent water and that the majority of the effluents were of potable water quality. In fact, in experiments to determine whether effluents from PPS could be used to irrigate food crops, Nnadi et al. (2015) demonstrated that such harvested stormwater could be used for irrigation even under nutrient leaching or oil dripping conditions.

Beneath the surface course, a PPS can comprise different structures made of different aggregates, from virgin stone to recycled material. Kazemi and Hill (2015) laboratory tested three

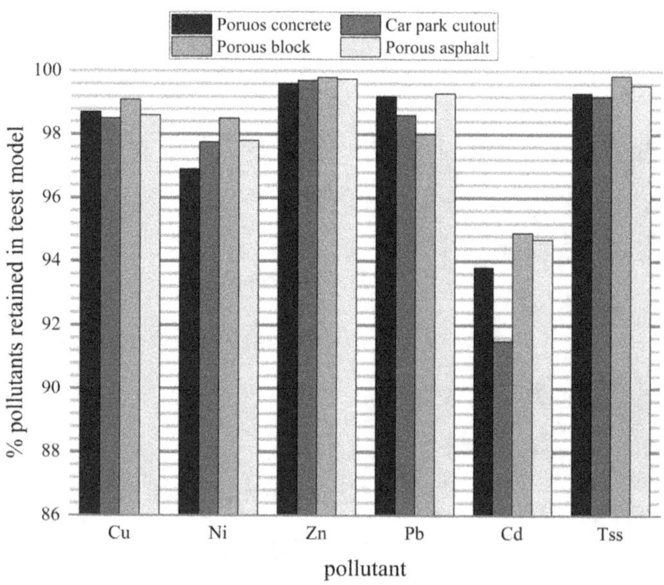

Figure 3.6 Percentage retention of pollutants by a variety of PPS test rigs monitored at the laboratory scale, Coventry University, UK. Where PPS models 1, 2 and 3 = porous asphalt; SC = car park cut out; 4 = porous concrete; PB = porous block pavers; TSS = total suspended sediment/solids

different types of basecourse aggregate, namely basalt, quartzite and dolomite. Their conclusions were that it is the initial quality of the urban runoff which determines its potential for reuse rather than the type of aggregate used in the PPS. Andrés-Valeri et al. (2013) investigated the potential use of basic oxygen furnace (BOF)-slag aggregate in PPS, finding the main drawback to be high pH and EC values of the effluent water. However, both parameters appeared to decline with time, giving an indication that in the longer term, they could achieve acceptable values, in particular, if water flow through the PPS was increased, thus reducing its residence time. Basic oxygen furnace-slag could therefore be suitable in environments that suffer from acid rain or where acid runoff is a problem, such as in some industrial or agricultural areas.

3.5.2 Field-Scale Evidence

Field trials from around the world have provided evidence of the potential benefits of SuDS in different climates, for example, Collins et al. (2008) in the USA, Pratt et al. (1995) in the UK, Lucke and Beecham (2011) in Australia, Pagotto et al. (2000) in France, Acioli et al. (2004) in Brazil, Jefferies et al. (2009) in Scotland and Perales-Momparler et al. (2014) in Valencia, Spain. The results from these various studies are slightly different and are dependent on the region's climatic conditions, yet important benefits for reducing runoff, improving water quality and providing amenity were demonstrated in all of them. Charlesworth et al. (2013) compared the use of SuDS at the building scale in both the UK and Spain with positive results in both cases, with adaptation required in terms of design and maintenance considering different climatic conditions. Acioli et al. (2004) constructed the first permeable reservoir pavements in Brazil for surface runoff control at source in a subtropical region. This experimental car park was 280 m^2 and the permeable surfaces tested were porous asphalt and reinforced grass.

There have been a variety of individual SuDS devices which have been studied at the field scale. Gomez-Ullate et al. (2011) investigated the main differences between several pervious surfaces and construction materials in permeable reservoir pavements in an experimental car park located in Santander, northern Spain (see Figure 3.7). This in-use car park had 45 sealed, or tanked, car park bays with six types of surface including porous asphalt, porous concrete, two types of interlocking concrete blocks and two types of reinforced grass. In terms of subsurface structure, two types of geotextile were used with a limestone aggregate, both in the base and in the sub-base layers (Andrés-Valeri et al., 2017).

After establishing background concentrations over a 6-month period of regular effluent sampling, an oil contaminant was added, and water quality parameters, including pH, total petroleum hydrocarbons (TPHs), Cu, Zn, Fe and Pb were determined.

Figure 3.7 Experimental car park, Las Llamas Park, Santander, Northern Spain

It was found that initial monitoring of the effluents before oil addition contained higher concentrations of the pollutants, due to their washing off from the construction materials but that these reduced as the experiment proceeded. However, it was found that porous asphalt and porous concrete were less efficient than the "open" pavements, such as those with block paver surface courses, which may reflect the use of bitumen binders, thus increasing the concentration of TPHs in the effluent from the porous surfaces. Results also emphasised the importance of including a geotextile in the PPS structure for water quality improvements regardless of the type of surface course used. In an extension to this study, Sañudo-Fontaneda et al. (2014a) monitored water quality from two other in-service PPS car parks in the north of Spain, with the same mix of surfaces as the study by Gomez-Ullate et al. (2011), and a variety of subsurface materials including BOF-slag as a recycled replacement for virgin aggregate. The results of these three studies showed that the limestone aggregate performed better overall than BOF-slag, particularly with respect to pH, EC and TPH. In terms of surface course, polymer modified porous concrete provided the best improvements overall in water quality, then interlocking concrete block pavement and thirdly porous asphalt.

Other devices monitored at the field scale include that by Lucke et al. (2014), who monitored four swales in Australia, finding that they successfully attenuated the storm peak, reduced runoff volumes and improved the quality of the stormwater. This was particularly the case in runoff with high concentrations of TSS and TP, with TSS removal of between 61.3% and 86.4%, of which 50% to 80% was generally removed within the first 10 m of the swale. By increasing the length of the swale to 30 m, a further 10% to 20% reduction in TSS concentrations could be expected. Two sustainable linear drainage systems, a swale and a filter drain were compared with a concrete ditch by Andrés-Valeri et al. (2014) in a monitoring study of over 25 months in which the devices received low levels of polluted runoff. Both SuDS devices performed better than the concrete ditch, with the filter drain performing better than the swale for all water quality parameters analysed, particularly electrical conductivity (EC), TSS and TPH.

In terms of treatment or management trains, some have been monitored for their water quality benefits; the studies by Heal et al. (2008) and Bastien et al. (2010) showed that treatment trains offer more advantages than individual SuDS devices. Probably the best-known study in the UK is the Hopwood Park Motorway Service Area, Bromsgrove, in the Midlands of England undertaken by Heal et al. (2008), with four SuDS management trains constructed in 1999 (Booth and McLaughlin, 2017). Interestingly, since the site was a UK environment agency (EA) demonstration site, the study included 5 years of monitoring dissolved NH_4-N, BOD, chloride, TSS, total Cu and total Zn carried out by the EA as well as deposited sediment concentrations of Cd, Cu, Pb, Zn, TPH and total PAHs. Whilst some of the samples early in the management train at Hopwood exceeded guideline values, by the time the water passed through the last of the ponds, all measured pollutants were well below, for example, for sediments in the train associated with the Lorry Park, TPH reduced by 5 times, and in the coach park, fuelling area and main access road management train, Zn reduced by 34 times.

The storm hydrograph is a plot of discharge or velocity of water against time, as is shown in Figure 3.1, but by plotting the concentration of the pollutants and sediments transported in the stormwater, polluto- and sedigraphs can be constructed, which show their change in concentration with time as the storm peak passes. Whilst there have been studies which produced pollutographs illustrating water quality improvements in SuDS devices and trains (e.g., Silva et al., 2010; Bertrand-Krajewski et al., 1998 and Figure 3.2), to the authors knowledge, there are very few which have used sedigraphs. One of these, Morgan et al. (2007), was able to show that stormwater ponds in urban areas that were designed to attenuate flow were more successful at removing sediment than those specifically designed with water quality improvements in mind; this is an important design consideration.

3.5.3 Modelling

As was the case with monitoring of SuDS devices, there is a wide literature utilising modelling to represent their ability to reduce the storm peak. Whilst there are less studies using modelling in terms of water quality, models can be used to represent the processes occurring in SuDS devices; their complexity makes modelling a necessity in providing understanding. However, there are as many different models as research groups studying the topic (Meyer et al., 2015). For instance, in the case of constructed wetlands, probably the most complex SuDS from a water treatment point of view, Meyer et al. (2015) classified existing models into those which attempted to represent the whole extent of processes occurring in the wetlands or mechanistic models, and more simplified models of use to engineers. Mechanistic models are focussed on the understanding of one or a group of processes with interactions that are present in the wetlands. The "simplified" models focus directly on performance, of direct value to engineers in the design of these systems.

Individual SuDS devices and treatment trains have been modelled in order to assess their performance and thus can provide decision support tools for use at local authority level. One example of this approach was by Jefferies et al. (2009), who developed the Stormwater Treatment Train Assessment Tool (STTAT) and scoring system for the Scottish Environment Protection Agency (SEPA). This system offered guidance on the selection of SuDS in an urban development considering the characteristics of the receiving water and hence the pollution risk to the development. In order to validate the model, 22 study sites were tested, including 16 treatment trains and 18 standalone SuDS of different types. The results allowed adjustment of the model such that it was judged to be robust when applied to a selection of circumstances.

By integrating a GIS approach with a semi-distributed stochastic model utilising an event or site mean concentration, Mitchell (2005) was able to address the problem of retrofitting SuDS to an urban area already in existence and provide a tool to support its planning of SuDS. Mapping the likelihood of pollution by diffuse sources, "hot spots" could be identified and loading maps produced of 18 key contaminants such as Cu. Used as a first pass during preliminary planning activities, the model that was produced was consistent with the EU Water Framework Directive (2000/60/EC) and an approach to pollution reduction which was applied to specified water quality objectives.

3.6 MAINTENANCE OF SUDS DEVICES

One of the main barriers to the use of SuDS, particularly in the UK and Spain, is the perceived inconvenience and expense of maintenance. To a certain point, the processes discussed in Section 3.2 lead to some of the problems which the present section discusses, introducing problems associated with the potential for seasonal release of pollutants in vegetation die back. Trapping pollutants in the device, particularly those associated with particulates, blocking or clogging, can reduce, or even in some cases, *avoid* water ingress, thus rendering its function of little use. This is particularly the case of PPS and there have been many laboratory studies devoted to this issue, including Davies et al. (2002) and Sañudo-Fontaneda et al. (2014b), which quantified the infiltration performance of concrete block paver, porous asphalt or concrete PPS. Nichols et al. (2015) concluded that the results obtained from accelerated laboratory testing depends more on the type and size of the sediments than on the intensity and duration of the simulated rain event.

In a comparison of 55 in-service PPS, 19 of which were constructed in the Netherlands and 36 in Australia, from 1 to 12 years in age. Boogaard et al. (2013) found that although their infiltration capacity appeared to decrease with age, nonetheless their performance was still generally acceptable, even after many years in service. This was also shown in laboratory experiments on permeable concrete pavers (González-Angullo et al., 2008), porous asphalt (Rodriguez-Hernandez et al., 2012) and porous concrete (Sañudo-Fontaneda et al., 2014b), whereby it was demonstrated that sufficient water was still able to infiltrate even when the surface appeared to be totally blocked.

In terms of the maintenance procedure for PPS, there have been very few studies of its effectiveness. In fact, Amato et al. (2010) concluded that no definitive quantitative estimates of street cleaning *in general* are possible, recommending more research and exchange of experiences in order to quantify the influence of the multiple variables involved: weather conditions, sweeping mechanism, particle size, timing, surface type, parked cars and so on. For PPS specifically, Suozzo and Dewoolkar (2012) tested the capacity of continuous cleaning operations (street sweeping and vacuum-truck cleaning) and spot-cleaning methods (pressure washing and vacuuming) to restore infiltration rates in pervious concrete pavements. They concluded that pressure washing is able to recover the infiltration capacity in severely clogged areas, but the combination of this method with vacuuming is recommended to remove particles from the pavement structure. In the case of porous asphalt, lack of maintenance means that when the surface does become clogged, and this after possibly 20 years of service, it is not possible to recover the original hydraulic conductivity even with high-pressure washing/vacuum cleaning, thus the recommendation would be to clean at intervals during the service life of the surface. For example, in the case of cementitious permeable pavement, Sansalone et al. (2008) recommended a 6-month cleaning frequency. Moreover, in cold regions, such as Sweden, regular cleaning of pervious surfaces must be undertaken combined with the avoidance of road salting during the winter (Al-Rubaei et al., 2013) in order to maintain effective infiltration.

The perceived problems with blocking or clogging of PPS have led to a focus on restoring infiltration rates, even though studies discussed earlier (e.g., Boogaard et al., 2013) have suggested that when the PPS appears to be extensively blocked, its performance is still more than adequate. Armitage et al. (2013) suggest that typically, maintenance every three months can include either vacuum sweeping and/or high-pressure jet-washing of the surface. Just employing sweeping alone can dramatically increase the permeability of the system. However, these activities have to be carried out carefully, or there is potential for previously trapped pollutants to be released into the environment.

Heal et al. (2008) concluded that the maintenance of SuDS is faster, cheaper and easier than that of conventional drainage in their study of the treatment trains at the Hopwood Park Motorway Service Area in England (see Section 3.5.2). Moreover, in order to design the SuDS train better, it was recommended that filter strips should be used as pre-treatment before water was passed into any ponds or wetlands in order to trap sediments early in the treatment train. Thus, it is possible to maximise the performance of ponds and wetlands with minimum maintenance cost since the removal of sediments in filter strips is easier than from other SuDS devices.

If maintained adequately, a SuDS system can perform for many decades, sometimes even for millennia (Charlesworth et al., 2017a). However, the accumulation of modern pollutants in the devices could potentially categorise them as hazardous waste at their "end-of-life," when their efficiency in terms of water quality improvement has declined, perhaps when being repaired, replaced or removed entirely. Nnadi et al. (2019) studied an 11-year-old block paver permeable car park which was being dismantled prior to the area being built on. They removed samples from layers of the structure of the PPS including the aggregate, pavers, dust and geotextile. They found that none of the sampled material from the site would be considered hazardous when compared against regulatory limits, and that it could be recycled and reused for other purposes. They do acknowledge that this is a single site, potentially only halfway through its lifetime, and that more research into end-of-life for SuDS devices is required.

3.7 INTEGRATING INTO THE CITY AS A WHOLE: WATER-SENSITIVE URBAN DESIGN

According to the World Bank, 56% of the global population amounting to 4.4 billion people live in urban areas. By 2050 it is predicted that the urban population will be more than double when almost 7 in 10 of the world's inhabitants will live in cities. Major water-related challenges faced by these increasingly populous cities across the world include water shortages (droughts), excess water (flooding) and water pollution; these will be intensified by the potential impacts of climate change and thus, as stated by Abbot et al. (2013), "the case for WSUD is obvious."

As was suggested in the Introduction (Section 3.1), water in a city should be considered an asset, a resource, rather than an inconvenience (Howe and Mitchell, 2012). Cities should be designed around water rather than water having to be constrained to fit around the constructed environment. Therefore, by integrating sustainable solutions, such as SuDS, to the problems of flooding, drought and water quality, WSUD offers the opportunity to create spaces that are water-centric, resilient and aesthetically pleasing. It was originally part of a water demand management strategy (Hedgcock and Mouritz, 1994) in which the natural environment formed the focus of planning, but has now evolved into a means of promoting place-making, facilitating a sense of place and local identity (Vernon and Tiwari, 2009). A general description of WSUD is that it takes the urban water cycle and incorporates water supply, management of wastewater, stormwater and groundwater, with urban design and environmental protection and represents a fundamental shift in the perception of the management of water in cities.

An in-depth treatise on WSUD is beyond the scope of this chapter, and there are a great many publications describing case studies of its application (e.g., Abbot et al., 2013; JSCWSC, 2009; Wong, 2006) and guidelines for its implementation (e.g., City of Melbourne, n.d.; Johnstone et al., 2012). To achieve transition to a water-sensitive city, there are six states suggested, beginning with water supply, then sewered, drained, waterway, water cycle and finally water sensitive (e.g., Brown et al., 2008; Wong and Brown, 2008, 2009). Wong and Brown (2009) asserted that there was rapid progress in WSUD in Australia and Singapore identifying what they termed "3 key pillars" underpinning the ability of a city to become water sensitive:

(i) the city should be able to access a range of sources of water provided by both centralised and decentralised infrastructure
(ii) ecosystem services provision for both the built and the natural environment
(iii) the political will and social acceptability for sustainability and water-sensitive cities

At the time Wong and Brown's (2009) article was published, there was not a single example they could give of a fully water-sensitive city, and whilst this is still true, there are examples of a few cities worldwide that have provided the lead on this approach. An example is the 3 ha Figtree Place, an urban redevelopment of 27 domestic units in Hamilton, an inner suburb of Newcastle, NSW, Australia (Coombes et al., 2000). Rainwater tanks and infiltration trenches were installed, and cleaned stormwater was directed to a central basin where it replenished the unconfined aquifer. The site was monitored for two years with samples taken from roof water, rainwater tanks, hot water systems and first flush pits; these were compared against the Australian Drinking Water Guidelines (1996). Results found that total water savings of about 60% were possible (45% reduction was achieved), and almost all stormwater was retained onsite. In terms of water quality, there was occasional exceedance of the guideline values for ammonia, pH, iron and lead, for rain collected from roofs, but when samples were taken from rainwater tanks and the hot water systems, these were found to comply with the potable guidelines, and it was thought that quality improved with storage due to flocculation, settlement and biodegradation within the tanks. This was confirmed by Coombes et al. (2005) in which the dominant process was found to be flocculation of the pollutants with the flocs then settling to the bottom of the tank or becoming attached to the tank walls. Biofilms were also found in the rainwater tanks which may be associated with the water quality improvements in the same way as they are in PPS (see Section 3.2).

By becoming a city catchment, there is potential to address water quality issues at the city scale by integrating SuDS devices into a WSUD. The smaller scale has shown the benefits, and Charlesworth (2010) has shown how SuDS devices and trains can mitigate and adapt to climate change, thus a water-sensitive city can provide multiple benefits which could potentially have global impacts.

3.8 CONCLUSIONS

This chapter has shown how the use of structural measures, such as SuDS, as individual devices or designed into management or treatment trains, can address problems associated with urban stormwater pollution, basically pollutants that are part of suspended solids or that can be easily biodegraded. As NbS, they are multiple-benefit and flexible in that there is equal emphasis on their ability to reduce the storm peak and provide amenity and biodiversity opportunities. The approaches taken can also assist in mitigating and adapting to climate change. They can form an integral part of WSUD, reducing water demand and providing liveable cities in which water is a vital part, and actually planned into the constructed environment. This is a change in philosophy whereby water is kept in the city, not rapidly ejected from it and herein lies the first problem, illustrated by the singular lack of an example of a city that has achieved water-sensitive status thus far.

A second problem is that much of this constructed environment already exists, and thus measures will need to be retrofitted, which makes designing and planning difficult, and at times impossible in the space available, in the structural strength of the buildings themselves, or in the layout of busy, highly trafficked roads. The design of SuDS and its integration into the urban environment needs to be site specific, whereby an assessment is made of any particular pollutants which are of concern; for example, there is now less focus on Pb, but more so on elements such as Zn and Cu. If emerging pollutants are also considered, and the usual "cocktail" of urban pollutants in urban runoff is to be addressed, then the full SuDS management train would need to be implemented. As was discussed in Section 3.2, further considerations include whether the pollutant is particulate-associated, and therefore if TSS is a problem, in which case settling in, for example, a detention basin would be effective. More advanced treatment may be needed for recalcitrant pollutants and "forever chemicals," for example, herbicides, phthalates, and PFASs (Kumar et al., 2023), and thus there is need for focus on research to improve the structure of PPS and the performance of geotextiles incorporated in them, as well as other SuDS devices and NbS.

Currently, SuDS are still very much designed along traditional lines, that is, for storm attenuation purposes first and foremost. Predictions of climate change include the potential for increased storminess, increased runoff velocity and volume and decreased rainfall and drought conditions, justifying the approach to a certain extent. However, associated with these impacts, the first flush of high concentrations of pollutants may well increase, erosion of polluted urban soils will be promoted and these contaminants will be transported in the stormwater runoff. There will therefore be a need for more storage space in cities for excess stormwater and for concomitant treatment of the transported pollutants which SuDS can provide. This ability to remove stormwater from combined sewer systems can prolong the life of the infrastructure, obviating the necessity for upgrading and the ensuing inconvenience and cost.

As to the future for SuDS, the social implications of their use are important, and Wong and Brown (2009)'s third key pillar for the attainment of water sensitivity highlighted social acceptability as a requirement (see Section 3.6). Unfortunately, there is little research on this particular topic, but Ardler Village in Dundee, Scotland (UK) is an example of such work: one of the first villages to be retrofitted with SuDS in the late 1990s, it included several swales, two ponds and one basin. A public perception study was carried out with a public participatory GIS, and Jose et al. (2015) concluded that SuDS can provide many cultural and social benefits to people who live near them. The generation of new urban habitats with wildlife was very much appreciated by those living in proximity with a sustainably drained area, with vegetated SuDS providing the best in terms of aesthetics: open spaces, a sense of well-being and educational and recreational opportunities.

Finally, as is shown in Wong and Brown (2009)'s third WSUD pillar, which contains the requirement for political will to encourage its implementation (Section 3.6), local and regional policies will need to be developed in relation to sustainable stormwater management across the world as they are fundamental in ensuring the uptake of SuDS, as well as encouraging the wider adoption of WSUD.

References

Abbot, J., Davies, P., Simkins, P., Morgan, C., Levin, D. and Robinson, P. (2013) *Creating Water Sensitive Places – Scoping the Potential for Water Sensitive Urban Design in the UK*. CIRIA, C724, pp. 129, London, UK.

Acioli, L. A., Agra, S. G., Goldenfum, J. A. and Silveira, A. (2004) Experimental Study of Permeable Reservoir Pavements for Surface Runoff Control at Source in a Subtropical Region. 5th International Conference NOVATECH – Sustainable Techniques and Strategies in Urban Water Management, 5 International Conference. GRAIE, Lyon, France.

Allen, D., Olive, V., Arthur, S. and Haynes, H. (2015) Urban sediment transport through an established vegetated swale: Long term treatment efficiencies and deposition. *Water*, 7, 1046–1067.

Al Mamun, A., Shams, S. and Nuruzzama, M. (2022) Review on uncertainty of the first-flush phenomenon in diffuse pollution control. *Applied Water Science*, 10(53). https://doi.org/10.1007/s13201-019-1127-1

Al-Rubaei, A. M., Stenglein, A. L., Viklander, M., Blecken, G. T. (2013) Long-term hydraulic performance of porous asphalt pavements in Northern Sweden. *Journal of Irrigation and Drainage Engineering*, 139(6), 499–505.

Amato, F., Querol, X., Johansson, C., Nagl, C., Alastuey, A. (2010) A review on the effectiveness of street sweeping, washing and dust suppressants as urban PM control methods. *Science of the Total Environment*, 408(16), 3070–3084.

Andrés-Valeri, V. C., Castro-Fresno, D., Sañudo-Fontaneda, L. A., Coupe, S. and Rodriguez-Hernandez, J. (2013) *Water Quality Assessment in Pervious Pavements with BOF-slag Sub-base in a Parking Area in the North of Spain*. 8th International Conference NOVATECH – Planning and technologies for sustainable urban water management. GRAIE, Lyon, France.

Andrés-Valeri, V. C., Castro-Fresno, D., Sañudo-Fontaneda, L. A., Coupe, S. and Rodriguez-Hernandez, J. (2014) Comparative analysis of the outflow water quality of two sustainable linear drainage systems. *Water Science and Technology*, 70(8), 1341–1347.

Andrés-Valeri, V. C. S., Perales-Momparler, L. A. Sañudo-Fontaneda, I. Andrés-Doménech, D. Castro-Fresno and Escuder-Bueno, I. (2017) Sustainable drainage systems in Spain. In: Charlesworth, S. M. and Booth, C. (eds) *Sustainable Surface Water Management; a Handbook for SuDS*. Wiley Blackwell Publishing.

Armitage, N., Vice, M., Fisher-Jeffes, L., Winter, K., Spiegel A. and Dunstan, J. (2013) Alternative Technology for Stormwater Management: The South African Guidelines for Sustainable Drainage Systems. WRC Report No. TT 558/13. Available at: www.wrc.org.za/

Bastien, N., Arthur, S., Wallis, S. and Scholz, M. (2010) The best management of SuDS treatment trains: A holistic approach. *Water Science and Technology*, 61(1), 263–272.

Bayon, J.R., Jato-Espino, D., Blanco-Fernandez, E. and Castro-Fresno, D. (2015) Behaviour of geotextiles designed for pervious pavements as a support for biofilm development. *Geotextiles and Geomembranes*, 43(2), 139–147.

Bertrand-Krajewski, J.-L., Chebbo, G. and Saget, A. (1998) Distribution of pollutant mass vs volume in stormwater discharges and the first flush phenomenon. *Water Research*, 32(8), 2341–2356.

Berwick, N. (2017) Sustainable drainage systems: Operation and maintenance. In: Charlesworth, S. M. and Booth, C. (eds) *Sustainable Surface Water Management; a Handbook for SuDS*. Wiley Blackwell publishing.

Bolognesi, A., Casadio, A., Ciccarello, A., Maglionic, M. and Artina, S. (2008) Experimental Study of Roadside Gully Pots Efficiency in Trapping Solids Washed off During Rainfall Events. 11th International Conference on Urban Drainage, Edinburgh, Scotland, UK.

Bond, P. C. and Pratt, C. J. (1998) *Developments in Permeable/Porous Pavements Observations on Mineral Oil Bio-Degradation*. Standing Conference on Stormwater Source Control. Coventry University.

Bond, P. C., Pratt, C. J. and Newman, A. P. (1999) A Review of Stormwater Quantity and Quality Performance of Permeable Pavements in the UK. Proc. 8th Int. Conf. on Urban Storm Drainage, Sydney, Australia, 248–255.

Boogaard, F., Lucke, T. and Beecham, S. (2013) Effect of age of permeable pavements on their infiltration function. *Clean – Soil, Air, Water*, 2014, 42(2), 146–152.

Booth, C. A. and Charlesworth, S. M. (eds) (2014) *Water Resources in the Built Environment: Management Issues and Solutions*. Chichester: Wiley-Blackwell.

Booth, C. A. and McLaughlin, A.-M. (2017) Contemporary landscapes and buildings of motorway service areas. In: Charlesworth, S. M. and Booth, C. (eds) *Sustainable Surface Water Management; a Handbook for SuDS*. Wiley Blackwell Publishing.

Botturi, A., Ozbayram, E. G., Tondera, K., Gilbert, N. I., Rouault, P., Caradot, N., Gutierrez, O., Daneshgar, S., Frison, N., Akyol, C., Foglia, A., Eusebi A. L. and Fatone, F. (2021) Combined sewer overflows: A critical review on best practice and innovative solutions to mitigate impacts on environment and human health. *Critical Reviews in Environmental Science and Technology*, 51(15), 1585–1618. DOI: 10.1080/10643389.2020.1757957.

Brown, R., Keath, N. and Wong, T. (2008) *Transitioning to Water Sensitive Cities: Historical, Current and Future Transition States*. 11th International Conference on Urban Drainage, Edinburgh, Scotland, UK.

Butler, D. and Davies, J. W. (2004) *Urban Drainage*. 2nd edition. London: Spon Press.

Charlesworth, S. (2010) A review of the adaptation and mitigation of Global Climate Change using Sustainable Drainage in cities. *Journal of Water and Climate Change*, 1(3), 165–180.

Charlesworth, S. M. and Booth, C. (eds) (2017) *Sustainable Surface Water Management; a Handbook for SuDS*. Wiley Blackwell Publishing.

Charlesworth, S. M., Sañudo–Fontaneda, L. and Mays, L. W. (2017a) Back to the future? History and contemporary application of sustainable drainage techniques. In: Charlesworth, S. M. and Booth, C. (eds) *Sustainable Surface Water Management; a Handbook for SuDS*. Wiley Blackwell Publishing.

Charlesworth, S. M., Beddow, J. and Nnadi, E. (2017b) The fate of pollutants in porous asphalt pavements, laboratory experiments to investigate their potential to impact environmental health. *International Journal of Environmental Research and Public Health*, 14, 666; doi:10.3390/ijerph14060666.

Charlesworth, S. M., Bennett, J. and Waite, A. (2016) An evaluation of the use of individual grass species in retaining polluted soil and dust particulates in vegetated sustainable drainage devices. *Environmental Geochemistry and Health*, 38(4), 973–985. DOI 10.1007/s10653-016-9791-7.

Charlesworth, S. M., Lashford, C. and Mbanaso, F. (2014) Hard Sustainable Drainage Infrastructure in the Urban Environment. Review of Current Knowledge, Foundation for Water Research.

Charlesworth, S. M., Perales-Momparler, S., Lashford, C. and Warwick, F. (2013) The sustainable management of surface water at the building scale: Preliminary results of case studies in the UK and Spain. *Journal of Water Supply: Research and Technology – AQUA*, 62(8), 534–544.

Charlesworth, S., De Miguel, E. and Ordóñez, A. (2011) A review of the distribution of particulate trace elements in urban terrestrial environments and its application to considerations of risk. *Environmental Geochemistry and Health*, 33, 103–123.

Charlesworth, S. M., Everett, M., McCarthy, R., Ordóñez, A. and De Miguel, E. (2003) A comparative study of heavy metal concentration and distribution in deposited street dusts in a large and a small urban area: Birmingham and Coventry, West Midlands, UK. *Environment International*, 29(5), 563–573.

Charlesworth, S. M. and Lees, J. A (1999) Particulate-associated heavy metals in the urban environment: Their transport from source to deposit, Coventry, UK. *Chemosphere*, 39(5), 833–848.

Charlesworth, S. M. and Lees, J. A. (2001) The application of some mineral magnetic measurements and heavy metal analysis for characterising fine sediments in an urban catchment, Coventry, UK. *Journal of Applied Geophysics*, 48, 113–125.

Chocat, B., Ashley, R., Marsalek, J., Matos, M. R., Rauch, W., Schilling, W. and Urbonas, B. (2007) Towards the sustainable management of urban storm water. *Indoor and Built Environment*, 16(3), 273–285.

City of Melbourne (n.d.) City of Melbourne WSUD Guidelines – applying the model WSUD guidelines: An initiative of the Inner Melbourne Action Plan. Available at: http://tinyurl.com/prehr97.

Collins, K. A., Hunt, W. F. and Hathaway, J. M. (2008) Evaluation of various types of permeable pavement with respect to water quality improvement and flood control. Proceedings of the 2nd National Low Impact Development Conference, Wilmington, United States, 275–289.

Colwell, S. (2000) Characterization of Performance Predictors and Evaluation of Mowing Practices in Biofiltration Swales. Unpubl. MSc Thesis. University of Washington. 104pp.

Coombes, P. J., Argue, J. R. and Kuczera, G. (2000) Figtree place: A case study in water sensitive urban development. *Urban Water*, 4(1), 335–343.

Coombes, P. J., Dunstan, R. H., Spinks, A., Evan, C. A. and Harrison, T. (2005) An overview of a decade of research into the quality of rainwater supplies collected from roofs. *Proceedings of 12th International Conference on Rain Water Catchment Systems*, New Delhi, India.

Coupe, S. J., Smith, H. G., Newman, A. P. and Puehmeier, T. (2003) Biodegradation and microbial diversity within permeable pavements. *European Journal of Protistology*, 39(4), 495–498.

Davies, J. W., Pratt, C. J. and Scott, M. A. (2002) Laboratory study of permeable pavement systems to support hydraulic modelling. Proceedings of the 9th International Conference on Urban Drainage (9ICUD) Portland, Oregon, September 8–13.

Deletic, A. (1998) The First Flush Load of Urban Surface Runoff. *Water Research*, 32(8), 2462–2470.

Deletic, A. (2005) Sediment transport in urban runoff over grassed areas. *Journal of Hydrology*, 301, 108–122.

Dickie, S., McKay, G., Ions, L. and Shaffer, P. (2010) Planning for SuDs – making it happen. CIRIA C687. Construction Industry Research and Information Association, London.

Ellis, J. B., Revitt, D. M. and Lundy, L. (2013) Behaviour of selected emerging pollutants in stormwater best management practices (BMPs). NOVATECH 2013.

Escarameia, M., Todd, A. J. and Watts, G. R. A. (2006) Pollutant removal ability of grassed surface water channels and swales: Literature review and identification of potential monitoring sites. Published Project Report PPR169. TRL Limited. Available at: http://tinyurl.com/nrf9dlz

Gomez-Ullate, E., Novo, A. V., Bayon, J. R., Hernandez, J. R. and Castro-Fresno, D. (2011) Design and construction of an experimental pervious paved parking area to harvest reusable rainwater. *Water Science and Technology*, 64(9), 1942–1950.

González-Angullo, N., Castro-Fresno, D., Rodriguez-Hernandez, J. and Davies, J. W. (2008) Runoff infiltration to permeable paving in clogged conditions. *Urban Water Journal*, 5(2), 117–124.

Heal, K. V., Bray, R., Willingale, S. A. J., Briers, M., Napier, F., Jefferies, C. and Fogg, P. (2008) Medium-term performance and maintenance of SUDS: A casestudy of Hopwood Park Motorway Service Area, UK.11th International Conference on Urban Drainage, Edinburgh, Scotland, UK.

Hedgcock, D. and Mouritz, M. (1994) Water sensitive residential design: A vision for a browner city? Proc. 1994 National Greening Australia Conference. A vision for a greener city: The role of vegetation in urban environments.

Howe, C. and Mitchell, C. (Eds) (2012) Water Sensitive Cities. IWA

Humphrey, J., Rowett, C., Tyers, J., Gregson, M. and Comber, S. (2023) Are sustainable drainage systems (SuDS) effective at retaining dissolved trace elements? *Environmental Technology*, 44(10), 1450–1463, DOI:10.1080/09593330.2021.2004454

Hung, C.-L., James, J. L. A. and Carbone, G. J. (2018) Impacts of urbanization on stormflow magnitudes in small catchments in the Sandhills of South Carolina, USA. *Anthropocene*, 23, 17–28. https://doi.org/10.1016/j.ancene.2018.08.001

Jefferies, C., Duffy, A., Berwick, N., McLean, N. and Hemingway, A. (2009) Sustainable Urban Drainage Systems (SUDS) treatment train assessment tool. *Water Science and Technology*, 60(5), 1233–1240.

Jimenez, B. (2008) Wastewater risks in the urban water cycle. in *Urban Water Security: Managing Risks*. Jimenez, B. and Rose, J. Ed., Taylor and Francis Group, pp. 151–200 ISBN: 978-0-415-48567-8.

Johnstone, P., Adamowicz, R., de Haan, F. J., Ferguson, B. and Wong, T. (2012) *Liveability and the Water Sensitive City – Science-Policy Partnership for Water Sensitive Cities*. Melbourne, Australia: Cooperative Research Centre for Water Sensitive Cities. Available at: http://tinyurl.com/nmuol5y

Jose, R., Wade, R. and Jefferies, C. (2015) Smart SUDS: Recognising the multiple-benefit potential of sustainable surface water management systems. *Water Science and Technology*, 71(2), 245–251.

JSCWSC (Joint Steering Committee for Water Sensitive Cities) (2009) Evaluating Options for Water Sensitive Urban Design – A National Guide. 79 pp. Available at: http://tinyurl.com/natsozq.

Kazemi, F. and Hill, K. (2015) Effect of permeable pavement basecourse aggregates on stormwater quality for irrigation reuse. *Ecological Engineering*, 77, 189–195.

Kim, L.-H., Kayhanian, M., Lau, S.-L. and Stenstrom, M. K. (2003) A new modeling approach for estimating first flush metal mass loading. Diffuse Pollution Conference, Dublin, 2003, 4B SUDS 4-38.

Kirby, J. T. (2005) SuDs – Innovation or tried and tested practice? *Municipal Engineer*, 158, 115–122.

Krishnakumar, S., Singh, D. S. H., Godson, P. S. and Thanga, S. G. (2022) Emerging pollutants: Impact on environment, management, and challenges. *Environmental Science and Pollution Research*, 29, 72309–7231. https://doi.org/10.1007/s11356-022-22859-3

Kumar, M., Sridharan S., Sawarkar, A. D., Shakeel, A., Anerao, P., Mannina, G., Sharma, P. and Pandey, A. (2023) Current research trends on emerging contaminants pharmaceutical and personal care products (PPCPs): A comprehensive review. *Science of the Total Environment*, 859, 160031.

Lapointe, M., Rochman, C. M. and Tufenkj, N. (2022) Sustainable strategies to treat urban runoff needed. *Nature Sustainability*, 5, 366–369.

Lashford, C., Charlesworth, S. M., Warwick, F., Blackett, M. (2014) Deconstructing the sustainable drainage management train in terms of water quantity – preliminary results for Coventry, UK. *CLEAN – Soil, Air, Water*, 42(2), 187–192.

Lashford, C., Lavers, T., Reaney, S., Charlesworth, S., Burgess-Gamble, L. and Dale, J. (2022) GI, NBS, NFM, SuDS and more – a review of the terminology and application of sustainable flood management techniques in the UK. *Water*, 14, 1204. https://doi.org/10.3390/w14081204

Lashford, C., Rubinato, M., Cai, Y., Hou, J., Charlesworth, S., Coupe, S., Tait, S. (2019) Urban flood management: A comparative analysis of practice in UK and China and lessons to be learnt. *Sustainability*, 11, 213; doi:10.3390/su11010213

Lee, J. H. and Bang, K. W. (2000) Characterization of urban stormwater runoff. *Water Research*, 34(6), 1773–1780.

Loganathan, P., Vigneswaran, S. and Kandasamy, J. (2013) Road-deposited sediment pollutants: A critical review of their characteristics, source apportionment, and management. *Critical Reviews in Environmental Science and Technology*, 43(13), 1315–1348, DOI: 10.1080/10643389.2011.644222

Lucke, T. and Beecham, S. (2011) Field investigation of clogging in a permeable pavement system. *Building Research and Information*, 39(6), 603–615.

Lucke, T., Mohamed, M. A. K. and Tindale, N. (2014) Pollutant removal and Hydraulic reduction performance of field grassed swales during runoff simulation experiments. *Water (Switzerland)*, 6(7), 1887–1904.

Lundy, L. (2017) Urban water and sediment quality. In: Charlesworth, S. M. and Booth, C. (eds) *Sustainable Surface Water Management; a Handbook for SuDS*. Wiley Blackwell Publishing.

Melbourne Water (2015) Stormwater management. Water Sensitive Urban Design (WSUD). Available at: http://tinyurl.com/o9avrst

Meyer, D., Chazarenc, F., Claveau-Mallet, D., Dittmer, U., Forquet, N., Molle, P., Morvannou, A., Pálfy, T., Petitjean, A., Rizzo, A., Samsó Campà, R., Scholz, M., Soric, A. and Langergraber, G. (2015) Modelling constructed wetlands: Scopes and aims – a comparative review. *Ecological Engineering*, 80, 205–213.

Mitchell, G. (2005) Mapping hazard from urban non-point pollution: A screening model to support sustainable urban drainage planning. *Journal of Environmental Management*, 74, 1–9.

Morgan, C. T., Heal, K. V., Wallis, S. G. and Lunn, R. J. (2007) Assessing the effects of design and climate change on sediment removal in urban stormwater ponds. Water Quality and Sediment Behaviour of the Future: Predictions for the 21st Century. Proceedings of Symposium HS2005 at IUGG2007, Perugia, July 2007. IAHS Publ. 314.

Newman, A. P., Pratt, C. J., Coupe, S. and Cresswell, N. (2002) Oil bio-degradation within permeable pavements by microbial communities. *Water Science and Technology*, 45(7), 51–56.

Nichols, P. W. B., White, R. and Lucke, T. (2015) Do sediment type and test durations affect results of laboratory-based, accelerated testing studies of permeable pavement clogging? *Science of the Total Environment*, 511, 786–791.

Nnadi, E. O., Coupe, S. J., Newman, A. P. and Mbanaso, F. U. (2015) Stormwater harvesting for irrigation purposes: An investigation of chemical quality of water recycled in pervious pavement system. *Journal of Environmental Management*, 147, 246–256.

Nnadi, E. O., Coupe, S. J., Sañudo-Fontaneda, L. A. and Rodriguez-Hernandez, J. (2014) An evaluation of enhanced geotextile layer in permeable pavement to improve stormwater infiltration and attenuation. *International Journal of Pavement Engineering*, 15(10), 925–932.

Nnadi, E., Newman, A., Coupe, S. and Mbanaso, F. (2019) An evaluation of stormwater treatment efficiency of Sustainable Drainage Systems (SuDS). In: Nnadi, E. O. (Ed.) *Stormwater: Sources, Monitoring and Management*. Nova Science Publishers, Inc.

Pagotto, C., Legret, M. and Le Cloirec, P. (2000) Comparison of the hydraulic behaviour and the quality of highway runoff water according to the type of pavement. *Water Research*, 34(18), 4446–4454.

Paul, P. and Tota-Maharaj, K. (2015) Laboratory studies on granular filters and their relationship to geotextiles for stormwater pollutant reduction. *Water*, 7, 1595–1609.

Perales-Momparler, S., Hernández-Crespo, C., Vallés-Morán, F., Martín, M., Andrés-Doménech, I., Andreu Álvarez, J. and Jefferies, C. (2014) SuDs efficiency during the start-up period under mediterranean climatic conditions. *Clean – Soil, Air, Water*, 42(2), 178–186.

Pratt, C. J., Mantle, J. D. G. and Scholfield, P. A. (1995) UK research into the performance of permeable pavement, reservoir structures in controlling stormwater discharge quantity and quality. *Water Science and Technology*, 32(1), 63–69.

Pratt, C. J., Newman, A. P. and Bond, P. C. (1999) Mineral oil bio-degradation within a permeable pavement long term observations. *Water Science and Technology*, 39(2), 103–109.

Puertas Agudo, J., Suárez López, J. and Anta Álvarez, J. (2008) *Gestión de las aguas pluviales: iImplicaciones en el diseño de los sistemas de saneamiento y drenaje urbano*. Madrid: Ministerio de Fomento, CEDEX, Centro de Estudios Hidrográficos. Monografías M-98.

RESTORE (2013) Rivers by Design: Rethinking development and river restoration. Environment Agency. Available at: https://assets.publishing.service.gov.uk/government/uploads/system/uploads/attachment_data/file/297315/LIT8146_7024a9.pdf

Rodriguez-Hernandez, J., Castro-Fresno, D., Fernández-Barrera, A. H. and Vega-Zamanillo, A. (2012) Characterization of infiltration capacity of permeable pavements with porous asphalt surface using cantabrian fixed infiltrometer. *Journal of Hydrologic Engineering*, 17(5), 597–603.

Rodríguez Sánchez, J. J. and Díaz Martínez, A. (2009) *Guía técnica de diseño y gestión de balsas y otros dispositivos de retención de contaminantes en carreteras*. Madrid: Ministerio de Fomento, CEDEX, Centro de Estudios de Técnicas Aplicadas. Manuales y recomendaciones R-18.

Sansalone, J., Kuang, X., Ranieri, V. (2008) Permeable pavement as a hydraulic and filtration interface for urban drainage. *Journal of Irrigation and Drainage Engineering*, 134(5), pp. 666–674.

Sañudo-Fontaneda, L. A., Charlesworth, S. M., Castro-Fresno, D., Andrés-Valeri, V. C. A. and Rodriguez-Hernandez, J. (2014a) Water quality and quantity assessment of pervious pavements performance in experimental car park areas. *Water Science and Technology*, 69(7), 1526–1533.

Sañudo-Fontaneda, L. A., Coupe, S. J., Charlesworth, S. M. and Rowlands, E. G. (2018) Exploring the effects of geotextiles in the performance of highway filter drains for sustainable and resilient road drainage. *Geotextiles and Geomembranes*, 46, 559–585.

Sañudo-Fontaneda, L. A., Rodriguez-Hernandez, J., Calzada Pérez, M. A. and Castro-Fresno, D. (2014b) Infiltration behaviour of polymer-modified porous concrete and porous asphalt surfaces used in SUDS techniques. *Clean – Solid, Air, Water*, 42(2), 139–145.

Scholes, L., Revitt, M., Gasperi, J. and Donner, E. (2008) Priority pollutant behaviour in stormwater Best Management Practices (BMPs). Source Control Options for Reducing Emissions of Priority Pollutants (ScorePP). Sixth Framework Programme, Sub-Priority 1.1.6.3, Global Change and Ecosystems. Project no. 037036.

Semadeni-Davies, A., Hernebring, C., Svensson, G. and Gustafsson, L.-G. (2008) The impacts of climate change and urbanization on drainage in Helsingborg, Sweden: Suburban stormwater. *Journal of Hydrology*, 350, 114–125.

Silva, A., Nascimento, N., Seidl, M. and Vieira, L. (2010) SWITCH in Belo Horizonte, Brazil: Infiltration and detention systems for more sustainable stormwater control in Belo Horizonte. *Reviews in Environmental Science and Biotechnology*, 9, 7–13.

Shuster, W. D., Bonta, J., Thurston, H., Warnemuende, E. and Smith, D. R. (2005) Impacts of impervious surface on watershed hydrology: A review. *Urban Water Journal*, 2(4), 263–275.

Song, C. (2022) Application of nature-based measures in China's sponge city initiative: Current trends and perspectives. Nature Based Solutions, 100010. https://doi.org/10.1016/j.nbsj.2022.100010

Spicer, G. E., Lynch, D. E., Newman, A. P. and Coupe, S. J. (2006) The development of geotextiles incorporating slow release phosphate beads for the maintenance of oil degrading bacteria in permeable pavements. *Water Science and Technology*, 54(6–7), 273–280.

Suozzo, M. J. and Dewoolkar, M. M. (2012) Long-term field monitoring and evaluation of maintenance practices for pervious concrete pavement in Vermont. *Transportation Research Record*, 2292, 94–103.

UN (2023) *Blueprint for Acceleration: Sustainable Development Goal 6 Synthesis Report on Water and Sanitation*. New York: United Nations.

USEPA (United States Environmental Protection Agency) (2009) Stormwater wet pond and wetland management guidebook. EPA Contract 68-C-99-253. 80 pp.

US EPA (nd) Storm Water Technology Fact Sheets. Available at: http://water.epa.gov/scitech/wastetech/mtbfact.cfm

Vernon, B. and Tiwari, R. (2009) Place-making through water sensitive urban design. *Sustainability*, 1, 789–814.

Warwick, F. (2017) Surface water strategy, policy and legislation. In: Charlesworth, S. M. and Booth, C. (eds) *Sustainable Surface Water Management; a Handbook for SuDS*. Wiley Blackwell Publishing.

Water Environment Research Foundation (WERF), Federal Highway Administration (FHWA), American Public Works Association (APWA), Environmental and Water Resources Institute (EWRI) of ASCE (2015) International Stormwater BMP Database. Available at: www.bmpdatabase.org/index.htm

Water, U. K. (2008) Lessons learnt from Summer Floods 2007: Phase 2 report–Long-term Issues. Water UK's Review Group on Flooding, London.

Watkins, S. and Charlesworth, S. M. (2014) Sustainable drainage systems – features and designs. In: Booth, C. and Charlesworth S. M. (eds) *Water Resources in the Built Environment –Management Issues and Solutions*. Chichester: Wiley-Blackwell.

Weis, J. S. and Weis, P. (2004) Metal uptake, transport and release by wetland plants: Implications for phytoremediation and restoration. *Environment International*, 30, 685–700.

Wilson, S., Bray, R. and Cooper, P. (2004) *Sustainable Drainage Systems. Hydraulic, Structural and Water Quality Advice*. CIRIA: London. C609.

Wong, T. (2006) An overview of water sensitive urban design practices in Australia. *Water Practice & Technology*, 1(1). IWA Publishing. doi: 10.2166/WPT.2006018

Wong, T. H. F., Fletcher, T. D., Duncan, H. P. and Jenkins, G. A. (2006) Modelling urban stormwater treatment – a unified approach. *Ecological Engineering*, 27, 58–70.

Wong, T. and Brown, R. (2008) Transitioning to Water Sensitive Cities: Ensuring Resilience Through a New Hydro-Social Contract. 11th International Conference on Urban Drainage, Edinburgh, Scotland, UK.

Wong, T. and Brown, R. (2009) The water sensitive city: Principles for practice. *Water Science and Technology*, 60(3), 673–682.

Woods Ballard, B., Kellagher, R., Martin, P., Jefferies, C., Bray, R. and Shaffer, P. (2007) *The SUDS Manual*. London: CIRIA (C697).

Woods Ballard, B., Wilson, S., Udale–Clarke, H., Illman, S., Ashley, R. and Kellagher, R. (2015) *The SuDs Manual*. CIRIA. London.

WWAP (United Nations World Water Assessment Programme)/UN-Water. (2018) *The United Nations World Water Development Report 2018: Nature-Based Solutions for Water*. Paris, UNESCO.

4 Impacts of Urban Drainage Management on Water Quality Following Extreme Rainfall

Antonio Lastra De La Rubia

4.1 INTRODUCTION

With a few exceptions, the sewer system emerged to empty the storm water collected in the cities and to transport this water out of town. Wherever possible, the sewer systems emptied into a river or into the sea. When the urban drinking water supply became popular, it also became necessary to remove the sewage produced by the new sanitary uses. And this was done through the same open channel used for the storm water. This situation resulted in several hygiene-related problems both in the cities and in the environment (only in the extraordinary Minoic civilisation and in the Indo Valley a closed pipe-based sewer system was built to remove the sewage from the cities (De Feo, et al., 2014)).

Consequently, there was a pressing need to treat this wastewater in order to preserve the water bodies in a sufficiently healthy state given that these water bodies were the source of the drinking water supply itself and of the agricultural field use too. This area (the treatment of the urban and industrial wastewater) has advanced greatly in the sanitation network in recent years.

However, once the problem of the removal and treatment of the wastewater had been solved (in the technical sense), it became apparent that the most critical issue was caused by flooding. The problem was due to the notable increase of the impervious surfaces in the cities. The solution adopted to deal with this problem was again, the construction of big sewers (combined or not) to remove the increasing amount of water. It was a quantity-based solution.

This quantity-based solution was rendered insufficient by the population rise and the migratory movement from rural areas to the cities that produced more and more impervious surfaces and thus more water to remove. At this point, the combined system needed to release water in the event of rain to avoid flooding and the separated system released the whole amount of runoff water to the water bodies without treatment. Either way the environmental impact was gradually increasing.

The main problem was that both solutions were made under the premise that the storm water wasn't contaminated at all and, for that reason, it was possible to release this water directly into the river without treatment (separated system) or with a dilution factor (combined system). Nevertheless, storm water in an urban area is heavily polluted, and this pollution is even bigger in flood time. When this occurs, the water bodies are vulnerable with the problem being that both aquatic life and human health are negatively affected. This is because these water bodies (river, lakes, aquifers and sea) are the main source of our drinking water. During a flood, the pollutants carried by the runoff are a serious threat because not only are there more pollutants but also the urban drainage network has more problems to manage this water.

However, a proper management of the urban drainage network in flood time is able to drastically reduce the pollutants released into the water bodies. The degree of resilience of a city depends, directly, on how they preserve their environment in a flood situation.

In this chapter, a brief review will be carried out concerning the main factors that a city must consider to reduce the impact in the event of flooding. These factors should be considered in all the management phases, from the project phase (weir project, storm water tank project, development of sustainable drainage systems (SuDS) etc.) to the maintenance and cleaning operations and even the operations during the flood.

4.2 URBAN DRAINAGE MAINTENANCE

The fundamental premise necessary to address is that during a flood, the storm water is contaminated. As shown in Figure 4.1, the water collected in an urban area contains pollutants from several sources:

- Air pollution precipitation
- Street dirt washed away including car parks, parking lots and sidewalks (oil, petrol, pesticides etc.)
- Roof and other building components washed away
- General garbage disposal
- Organic residues generated by both animals and humans

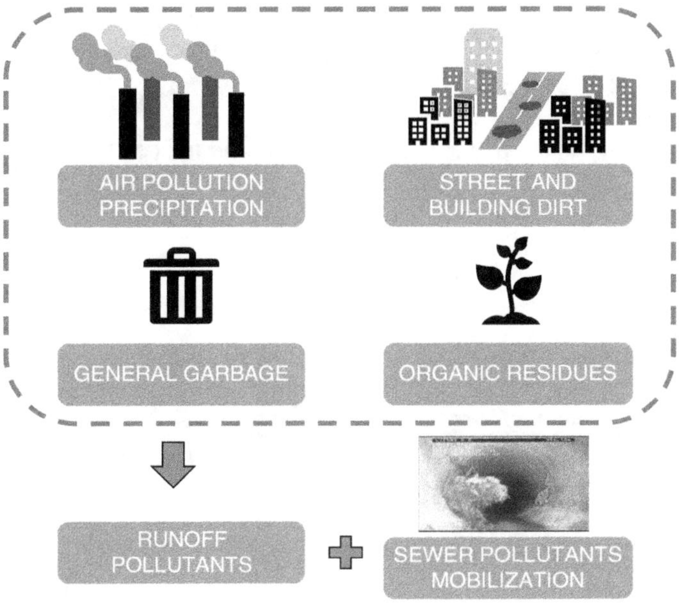

Figure 4.1 Urban pollutants sources

- Movement of pollutants attached to the inner surface of the sewer pipes
- Movement of pollutants attached to the soil

The last point implies that the water drives the dirt layer inside the sewer pipes when the flow and the water velocity grow above a breaking point. This problem appears mostly in the combined systems, but it can also appear in the separated systems, because even when theoretically there are not any sewage connections, there are in fact always a few. These kinds of sewers are huge, and the wastewater flow is low; therefore, this water becomes stagnant, and a decantation process begins. During rainfall, this wastewater ends in the environment. However, it is possible to say that in a combined system, the resuspension of sewer sediments is at the origin of 70% to 80% of the suspended matter and of 10% to 70% of *E. coli* – for example in a case study in Paris (Passerat et al., 2010) – but the remaining suspended matter (*E. coli* and other pollutants) comes from the surface runoff, and those pollutants are also flowing in the discharge of the separated system.

The contamination due to the street and road washing is especially important in terms of microbial risk. It is possible to assume that the runoff water in a flood has about 20–30 times more *E. coli* and enterococci than the allowed limit for bathing water (Sterk et al., 2008), and the urban drainage manager has to operate the network efficiently to prevent these pollutants from ending up in a water body.

Another major issue is the state of cleanliness of the other drain elements in the urban drainage networks such as the surface water drains, road channels and other kinds of inlets as not only might this garbage decrease the drain capacity, but also because all the pollutants contained within will join the storm water on its way to the sewer system. An example is the presence of decomposing leaves in the surface water drains which are placed there by the street sweepers.

That's why the first and most basic step to mitigate the contamination of the water bodies during an urban flood event is the preventive cleaning of the network especially before the wet season (Figure 4.2). However, this is not enough; it is also interesting to have a short-time rain forecast that could allow for performing an extra cleaning task whenever a severe rain event is expected in order to make sure that all the urban drainage elements will perform as expected. It is important to emphasise that the cleaning tasks (in normal or exceptional situations) are carried out not only to avoid smell and blockage problems but also to prevent the dirt located in the sewer network surface elements from ending up in a water body (Figure 4.3).

The state of cleanliness of the sewer system is especially important to avoid exposing the environment to the microorganisms that cause contamination during an urban flood event. It is known that floods originated by the combined system carrying a larger number and amount of pathogens than the floods originated only by surface runoff (De Man et al., 2013). A special case of the management of sewers in terms of the quantity of solids retained and the need to properly remove and dispose them of is observed in arid and semiarid regions, where rainy season is well defined, yet limited, allowing for a heavy accumulation of such solids in the systems, as exemplified in Jimenez et al. (2004).

The pollutographs during an urban flood event can be quite different depending on the catchment characteristics, the rainfall graph and other circumstances. Therefore, the most important task is to establish the pollutograph at points where there exist discharges to a water body. This pollutograph must be related to the shape and recurrence factor of the rainfall.

Citizens also share responsibility in the existence of pollutants found in the runoff and in the sewer in the case of extreme rainfall (Figure 4.4).

The state of cleanliness of the streets and roads is crucial to reduce the load of pollutants carried by this runoff. Moreover, it is important to prevent local urban floods, which is crucial for the environmental health. Discharging into the toilet the things such as oils, wet wipes, paints and other matters, which should be discharged into the garbage, puts dangerous material in the sewer that is going to be released during rainfall and could finish up in the environment. It is also important to keep the streets and roads clean, notably in low-income regions of some countries. The communication campaigns aiming at making the people aware of this potentially dangerous behaviour is generally followed by the reduction of the water contamination during urban floods to a great extent.

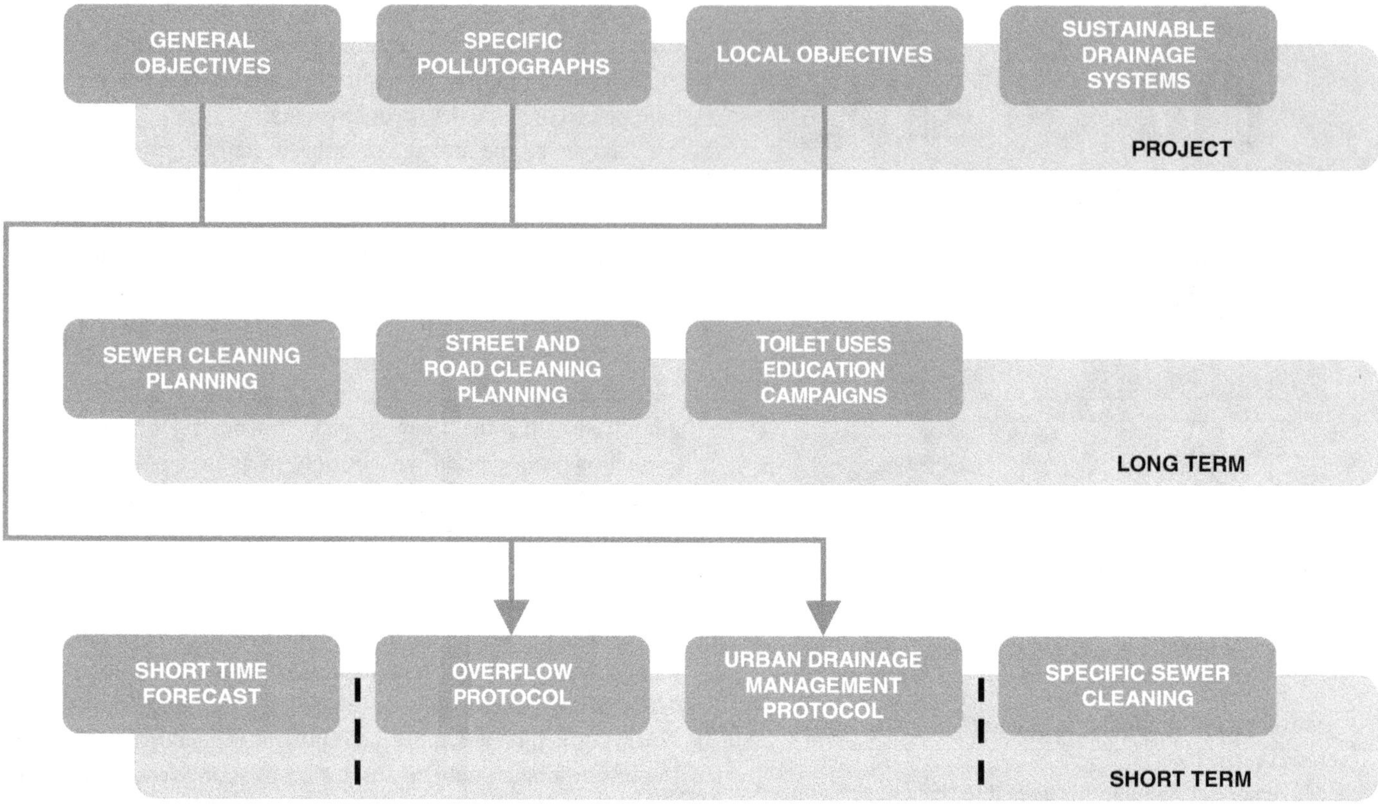

Figure 4.2 Urban drainage management

Figure 4.3 During a flood, these wet wipes could be released in the environment

Figure 4.4 The street dirt is a main problem for the urban drainage system

4.3 SHORT-TIME RAIN FORECAST

A short-time rain forecast system is interesting not only to know when it is going to rain but also to anticipate the kind of rainfall that is going to happen. It is very important to note that in an urban flood event, the pollution directly depends on the rainfall hyetograph. In this respect, the hyetograph repartition over the time is more important than the total volume of rainfall. This is easy to understand because, in the same way that a sewer pipe has a maximum gravity flow that can pass through it, the released pollutants stick to the inner walls of the sewer and this will lead to a minimum water flow velocity. It is possible to say that the important thing during an urban flood is not only the volume of water but also the duration of the peak of the rain event. The same phenomenon occurs with the release of pollutants.

In the dry season, the urban drainage manager needs to be able to rely on enough operational flow transfer capacity during the heaviest rainfalls. A common mistake is to fill the storm water tanks with every rainfall event without information about upcoming rainfall. This situation may mean that the tanks are full when the heaviest rainfall comes. Therefore, a good weather forecast is necessary for enabling an efficient drainage management in order to avoid not only flooding but also the release of pollutants into the water bodies.

4.4 DILUTION FACTOR

Another important issue to be considered in the whole drainage network life cycle is the expected objective that the infrastructure was designed for. One of the most common methodologies used to design and manage sewer infrastructure (overflow weirs, storm water tanks, sewage pump) is the dilution factor. A flood is a great opportunity to verify if thinking about dilution is a good idea or not.

The dilution factor was born when rainwater was thought to be clean. With this concept, it was accepted that a mix between rainwater and sewage had less pollutants than the sewage itself, so the key was to find the ratio between the rainwater flow and the normal sewage flow to have a mixed volume of enough clean water to be released into the environment without concern. An important part of urban drainage infrastructure has been designed with this concept all around the world.

However, current knowledge says that the dilution concept is not true in most of the urban areas because the maximum value of pollutants is even higher in the storm water than in the sewage itself during an extreme rainfall event. However, the dilution factor could be a good idea to compare two different combined sewer overflows (CSOs) at the same point (Gutierrez Andrés and Gallin, 2009).

In Figure 4.6 it is possible to see an example for a city characterized by a Mediterranean climate with continental influences (Madrid, Spain). The study of a rainfall event focuses on the concentration of pollutants at two particular points of the system: downstream in the combined sewer network; and before a storm water tank. The observed dilution factor value is 21 (between the mix volume and the peak flow of sewage in dry time). The values for biochemical oxygen demand (BOD), chemical oxygen demand (COD), suspended solids and ammoniacal nitrogen are higher than the same values in the sewage on a normal dry day (a graphic of the pollutants at the same point without storm water is reflected in the Figure 4.5). These are commonly observed parameter values concluding that the dilution factor does not minimize the pollution concentration, and if we rely on the dilution factor, we could have a contamination problem during a flood.

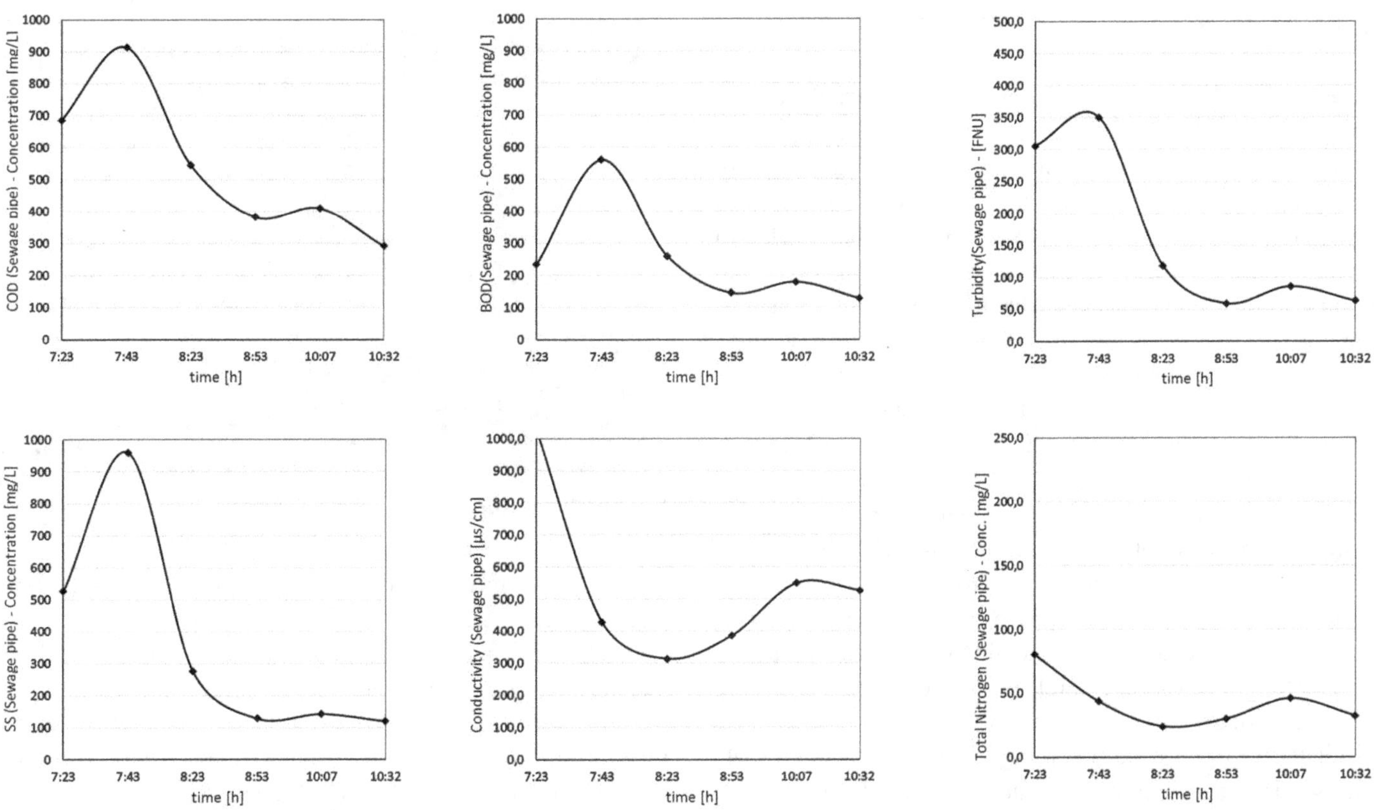

Figure 4.5 Pollutant concentration during an extreme rain event before a storm water tank (Combined system, Madrid, Spain)

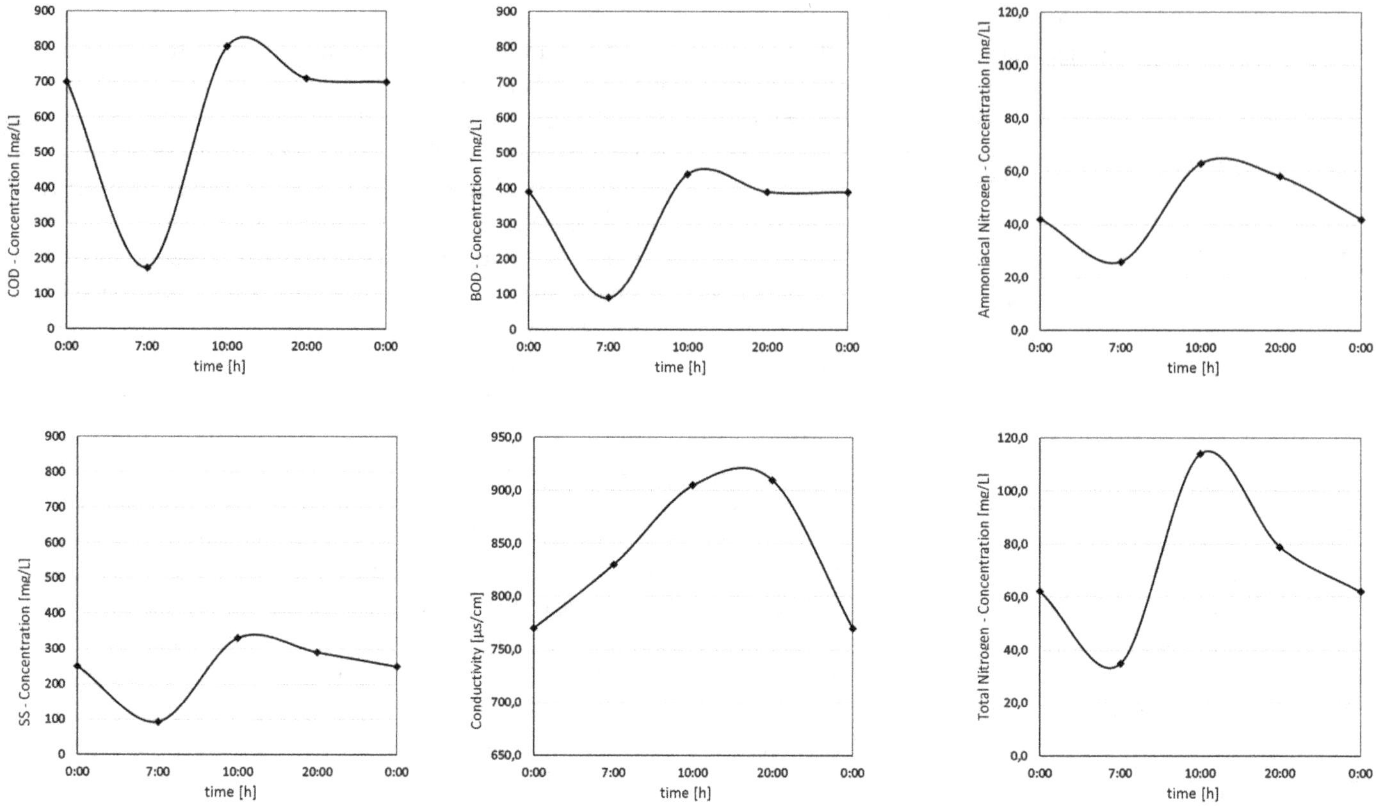

Figure 4.6 Common pollutant concentration on a dry day (Combined system, Madrid, Spain)

4.5 RETURN PERIOD

It might be assumed that in subsequent rain events, the pollutant measure could be lower because roads, streets, pipes and air have been cleaned by the first rainfalls. But that may not be true. If the previous rainfalls have had low-intensity peaks, the rainfall with a high-intensity peak (the one that is able to flood) is going to be highly contaminated. Furthermore, the first rain event after a dry season is able to carry a high amount of pollutants. The conclusion is that in order to predict how much pollution is going to be generated by a rainfall event, it is necessary to know, not only the rainfall characteristics such as intensity and duration, but also the rain patterns of the previous days (Launay et al., 2013).

In the project phase, the use of the "return period" concept is common. This concept is under discussion because it is useful only to define rainfall, but it is not useful to define the rain period in which that rain occurs so, maybe, it would be better to talk about a new "return period" that would include the probability of a previous rainfall event prior to the one that we are considering.

Another problem associated with the use of the "return period" factor is that it is often misunderstood because the non-experts could understand that it is the period in which there is not going to be another rain with the same value (Figure 4.7), and this is obviously incorrect. It is better to work with the "average recurrence interval" (Parzybok et al., 2011), which is "the average number of years between two similar events of rainfall for a given duration at a specific location."

Figure 4.7 Unexpected storm water overflow

Good management practice is to show the results of the short-time forecast not with rain deep values but with "return period" or "average recurrence interval." Because it is easy and in real time, and because it gives a better understanding of the characteristics of the upcoming rain, this is a powerful tool that allows

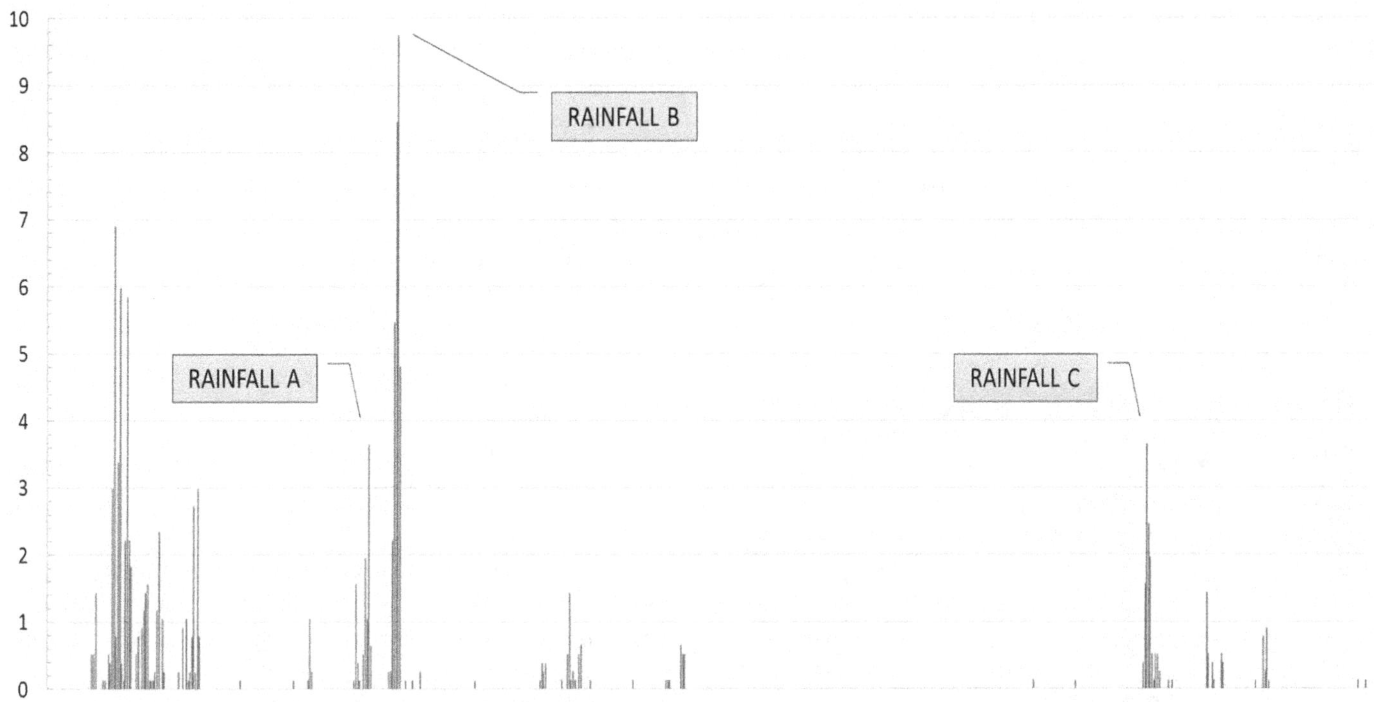

Figure 4.8 Hourly data from several consecutive rainfalls (Madrid, Spain)

the infrastructure manager (Zhou et al., 2015) to decide what to do to mitigate the effect of the upcoming rain and the water contamination that goes with it.

Figure 4.8 shows the hourly rain data of several consecutive days in Madrid. The rainfall C is similar to rainfall A but has more pollutants because it occurs after a dry period. However, the rainfall B is more contaminated than the rainfalls A and C because, despite the previous rainfalls, the peak intensity is higher than the other two rainfalls. So, both premises seem to be true; firstly, a wet season implies that the pollutants in consecutive rainfalls are decreasing and, secondly that, an extreme rainfall with a high-intensity peak (and an associated flood) may transport large amounts of pollutants even in the middle of a wet season.

Another ratio that is generally used to design and operate a release point is the spill frequency and/or volume. But there are controversial opinions about the use of this value (Lau et al., 2002) because it does not work the same for all the weather seasons. However, it is an easy tool to use when it is not possible to have a proper characterisation of pollutants in a drainage overflow weir.

The conclusion is that the quality of storm water depends on the rainfall characteristics, the catchment characteristics, the previous rainfall events, seasonal characteristics (both in the rainfall and in the catchment) (Liu et al., 2013), state of cleanliness (both in the sewer and in the catchment) and, in a few one-off exceptions, the dilution ratio is not the best data to use for the design nor for the management of an urban drainage network.

4.6 FIRST FLUSH

A structural remedy to fight against the excess of runoff in an urban area is the construction of storm water tanks. This infrastructure is usually located underground and, at first, it was designed to store rainwater during an event in order to avoid flooding in urban areas. They were located, for this reason, in the middle of the urban catchment to laminate the drainage network hydrograph. This lamination is necessary to absorb the peak of the rain event when the generated flow exceeds the sewer pipe maximum flow capacity.

In recent years, the storm water tanks are being used not only to avoid flooding but also to limit the impact of an overflow to the environment. The storm water tank has to store the gap between the flow of water (from a combined system or from a separated one) that the sewer system or the WWTP is able to manage and the flow of water that the infrastructure is receiving.

The design of the storm water tank (Figure 4.9) usually focusses on quantity of water, considering what is called the "first flush" which is a controversial issue. The first flush is the first runoff in a rain event. Theoretically, this first water is more contaminated than the rain that it is coming after, so the standard way to design the tank is to fix a theoretical rain event for a return period and the gap of time during which the tank will absorb water in these conditions.

Several methods exist to estimate the first flush. The traditional one says that in a first flush, 70% to 80% of total pollutants mass is transported in the first 25% to 30% runoff volume. Methods that are more modern calculate mean concentration of pollutants for each increment of runoff (Bach et al., 2010). In places where there is a huge difference between dry and wet seasons, the first rain after a dry period is also called "first flush." In a flooding event, it is expected that the real rainfall exceeds the theoretical one so the tank stores only a few minutes of rain and the tank management is really important to face the release of pollutants in the overflow weir.

First flush is a concept that does not correspond to the reality everywhere in the same way. There are currently many researches focussing on places where the "first flush" event occur (Huang et al., 2007; Kim et al., 2005), and others on places where this phenomenon does not happen (Goonetilleke et al., 2005; Suarez and Puertas, 2005). There are even huge differences between the behaviour of the different pollutants. For example, there are urban areas where the opposite of a first flush was observed with *E. coli* (McCarthy, 2008).

Figure 4.9 Storm water tank

The main factors affecting first flush are the same as in the dilution discussion (Shamseldin, 2011):

- Rainfall characteristics (shape, duration, intensity etc.).
- Seasonal characteristics: the behaviour is different in dry season than in wet season.
- Catchment size.
- Catchment typical characteristics: land uses, impervious area and so on.

In Madrid (Spain), the Council built many storm water tanks to protect the river from the pollution in rainfall events. Some of these tanks are among the biggest storm water tanks in Europe (one of them has a capacity of 400,000 m^3 and another one 250,000 m^3). Because of this, the company that is responsible for the management of the sewer system (Canal de Isabel II) has great expertise with regards to the kind of pollutants associated with the runoff in a large Mediterranean city.

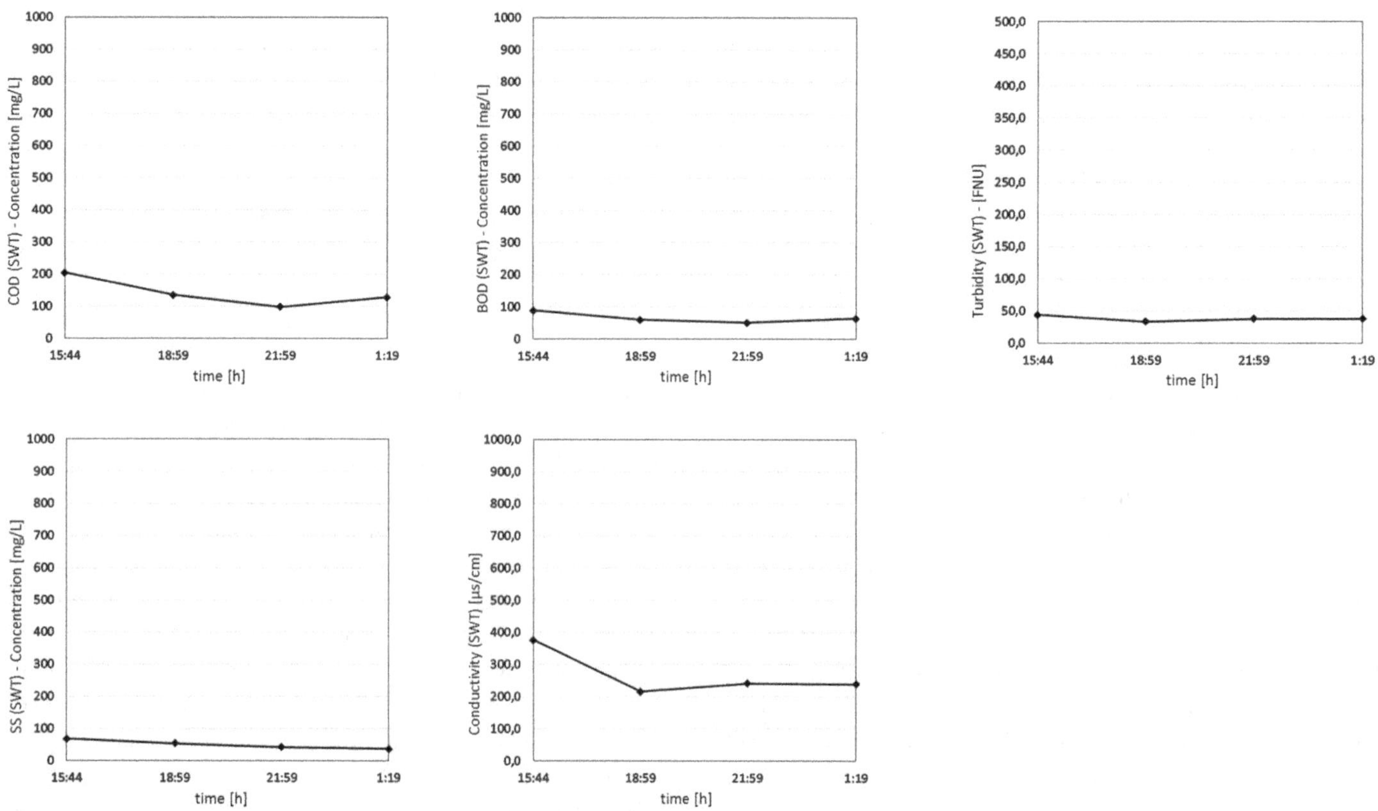

Figure 4.10 Pollutant concentration during an extreme rain event into a storm water tank (a few hours after the rainfall)

The first conclusion in the studies done in the combined sewer system in Madrid is that the first flush did not happen, at least in the way that first flush is explained in the bibliography. What really happened was that the pollutants concentration remained above the medium concentration paths in sewage for three, four and even more hours and they decreased when the rain was decreasing too. And that situation occurs even in rainfall with a low return period (2 years). In Figure 4.10 it is possible to see the different pollutants' concentration as normal behaviour. Some of these pollutants graphs even show a secondary peak two hours after the first one.

However, in the same studies it is possible to see that the behaviour of the pollutants is different in each catchment and in each tank even when all of them are in the same city and when the land uses are similar between the different catchments. Therefore, the second conclusion is that only a parametric study in each tank is useful to know what is happening about the load of pollutants generated by a rainfall (normal or extreme).

4.7 OVERFLOW WEIR LAYOUT

A decision that is necessary to take during both the design and the operational phases is about where to release the excess of water when, during a flooding, the urban drainage cannot transport all the water or the storm water tank is full (or to keep space for the next rain).

Figures 4.11–4.14 show the most common storm water tank layout. When the tank is managed focussing only on water quantity (to avoid flooding), all the different layouts are working, but when it comes to focus on avoiding the water contamination, the location of the overflow weir becomes an important issue.

In a study made on Madrid storm water tanks, it is possible to see that a few hours after the rain, the combined system water that has been stored in the tanks has significantly less pollutants than in the beginning (see Figure 4.10) and less than the sewage flowing in that moment through the sewer.

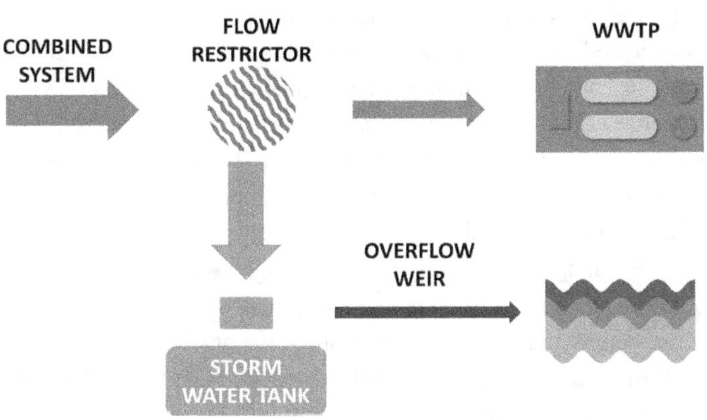

Figure 4.11 Storm water tank off-line. Overflow weir upstream the tank

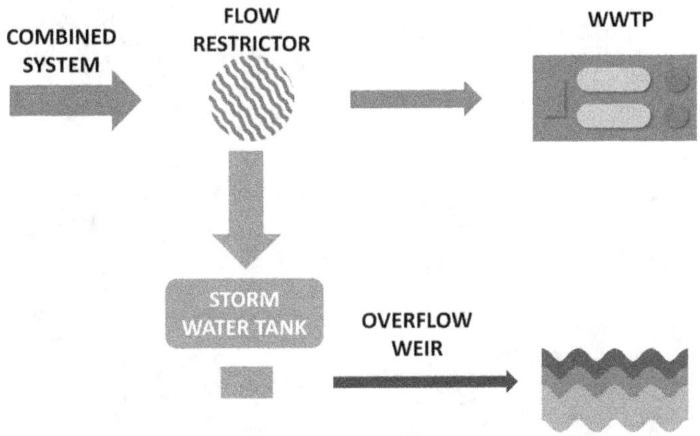

Figure 4.12 Storm water tank off-line. Overflow weir downstream the tank

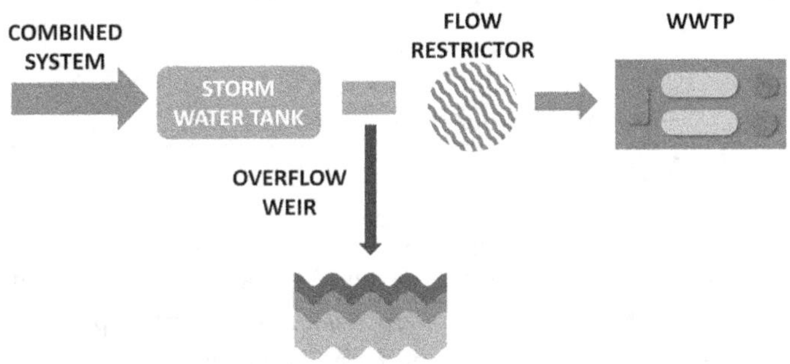

Figure 4.13 Storm water tank on-line. The overflow in excess is directly discharged first into the storm water tank. This way the pollutant peak attenuation benefits both the WWTP and water body receptor

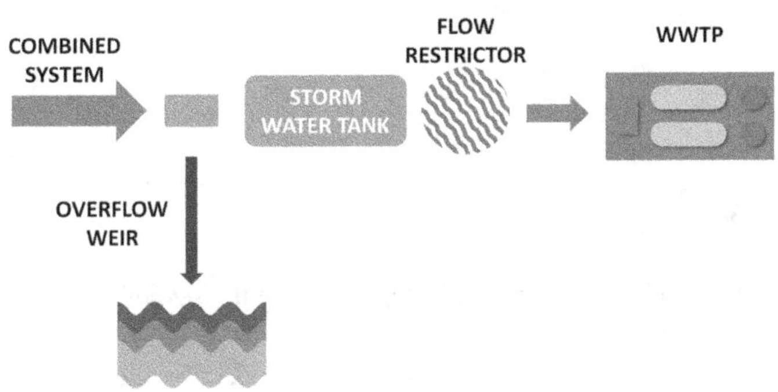

Figure 4.14 Storm water tank on-line. Overflow weir upstream the tank

With this information, it is possible to say that in the studied situations, the tank works like a decantation tank so it is one of the best practices to release the excess of water in an overflow weir downstream from the tank. In other words, the incoming water has to go through the tank and release the excess at the end of it. With this operation, the impact in the water bodies is lower than in the other tank layout.

Figure 4.15 Storm water treatment before a CSO

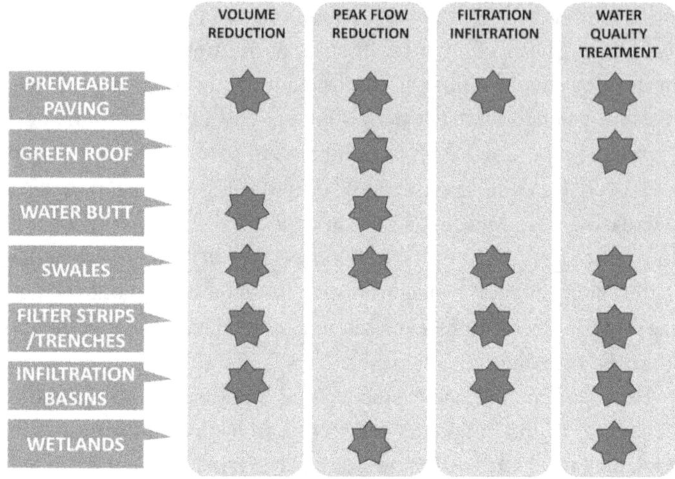

Figure 4.16 Sustainable drainage systems

4.8 STORM WATER TANK MAINTENANCE

If we choose to release the water downstream from the tank, it is mandatory to be sure that the tank is clean enough when an extreme rainfall event is expected. This is an important practice to minimise the mobilisation of the pollutants into the tank and so to reduce the release of the pollutants into the environment.

Then an important issue is the disposal of a good cleaning system. A system which has to be capable of giving a quick response if a rainfall is expected so a manual cleaning has to be discarded. The different types of cleaning system (Figure 4.15) are:

- Scraper systems
- Spray and jet systems
- Vacuum flushing systems
- Tipping buckets
- Flushing gates

The decision will be made depending on the size of the tank, the water evacuation system, the disposal of a reliable power supply (some systems don't need power to operate) and the existence of a water source to fill the system if this is the case (e.g., tipping buckets).

4.9 SUSTAINABLE DRAINAGE SYSTEMS

The main function of the SuDS is to collect, store and clean runoff before it drains into the sewer system. The aim of these techniques is to restore the natural water cycle to the land which has been modified because of urban development. In a first approach, these techniques try to mimic the natural condition of the runoff, meaning that the runoff is partially stored in these areas and the excess of water delays its arrival to the drainage system, so it is possible to deal with more volume of water with the same sewer infrastructure. Following what it is said in the CIRIA SuDS Manual (Woods Ballard et al., 2015): "This is achieved by harvesting, infiltrating, slowing, storing, conveying and treating runoff on site." The SuDS have shown themselves to be a great solution to prevent or minimise floods and the water pollutions associated with them.

There are several kinds of SuDS, but for a pollutant control the most important are (Figure 4.16):

The key for the development of a good SuDS is to establish from the very beginning who is going to pay for its implementation (the land owner, planning authorities, the urban drainage manager, garden authorities etc.), how and who is going to maintain it and, finally, who is responsible if something goes wrong. Without clarity about these aspects, the efficiency of the SuDS could be compromised.

The IPCC (Intergovernmental Panel on Climate Change) in their last assessment cycle, Jimenez et al. (2014) warns that if gas emissions continue to increase at the current rate, the average temperature could rise above dangerous levels. In the next century, it is expected that the consequences of this anthropogenic-induced climate change will modify the frequency and the severity of extreme rainfall events. With these modifications, the flood probability is growing year by year. And with the increasing severity of the rainfall, the runoff is going to be more polluted than currently (Easterling et al., 2000).

The management of the urban drainage system has to be the solution to mitigate the problem associated with more floods and more contaminated floods. The improvement of the storm water tank management and the development of sustainable drainage solutions are the greatest tools to fight against the effects of the climate change and enhance the resilience of the cities.

4.10 CONCLUSIONS AND RECOMMENDATIONS

In an urban flooding, the runoff is heavily contaminated, and the release of the excess of water to the water bodies is dangerous for them. This water is so contaminated that it can affect people even before it has drained into the sewer system. Improving the city resilience is a task that has to be managed not only by the urban drainage manager but also by the city council, the citizens and all the planning authorities.

A good urban drainage management is the key to minimising the effects of the bad quality water. In this chapter, we have noted that the most important things to do to avoid the contamination of the environment are as follows:

In the long term:

The first thing is to analyse the different pollutants of the sewer water correlated with the different kinds of rains and the climate seasons in each network point in which there is an overflow spill or a storm water tank. These pollutographs are not only to create a successful project in the best way but also to manage the network with the best results. With little investment, the improvement could be really significant. The best practice is to work, directly with the local pollutographs and not with the concepts such as dilution or first flush.

Another task to do is to create a realistic sewer cleaning programming thinking about the rainy season to be sure that, when the extreme rain comes, the sewer is clean enough. The cleaning programming has to include all the drainage infrastructures and the associated elements.

It is necessary to involve all the citizens and the administrations to support and enforce the avoidance of dangerous discharges in the sewer. While designing a street and road cleaning programme, it is important to keep the beginning of the wet season in the mind.

In the short term:

A performant short-time forecast is a good idea for the anticipation of extreme rain events and the related floods. The announcement of an upcoming extreme rain event could be used to perform an extra cleaning of the most important urban drainage elements, and thus to prepare the infrastructure (e.g., releasing the water stored in the storm water tank from the former rains to have enough available volume to absorb the water coming). It is a good idea not to show the rain forecast with rain deep values but with "return period" or "average recurrence interval" (Zhou et al., 2015) for a better understanding.

During the rain event:

It is important to have an operating protocol based in the results of the pollutograms in each important point of the network to have all the information to manage properly these infrastructures. It is necessary, for example, to know where it is best to release the excess of water when the sewer and the storm tanks are full because, as it has been said in this chapter, there are huge differences (talking about water quality) between the different options. During a flood event it is not possible to take decisions on the go. A good operating protocol is the key to a good management process.

Other considerations:

With the climate change threat and a future with more extreme rainfall, it is mandatory to change the traditional way in which the urban drainage planning has been done. The development of the sustainable drainage techniques to slow, store, infiltrate and treat the runoff is not only a good way to give a bit of nature to the cities but also the best way to improve the resilience of an urban drainage network.

References

Bach, P. M., McCarthy, D. T. and Deletic, A. (2010) Redefining the storm water first flush phenomenon. *Water Research Journal* 44(8), 2487–2498. https://doi.org/10.1016/j.watres.2010.01.022.

De Feo, G., Antoniou, G., Fardin, H. F., El-Gohary, F., Zheng, X. Y., Reklaityte, I., Butler, D., Yannopoulos, S. and Angelakis, A. N. (2014) The historical development of sewers worldwide. *Sustainability* 6, 3936–3974. https://doi.org/10.3390/su6063936.

De Man, H., Leenen, I., Van Knapen, F. and Roda Husman, A. M. (2013) *Potential health risk associated with exposure to microorganisms in urban Flood.* In proceedings NOVATECH 2013, Lyon. France.

Easterling, D. R., Meehl, G. A., Parmesan, C., Changnon, S. A., Karl, T. R. and Mearns, L. O. (2000) *Climate extremes: Observations, modeling, and impacts*. Science, Washington, USA.

Goonetilleke, A., Thomas, E., Ginn, S. and Gilbert, D. (2005) Understanding the role of land use in urban stormwater quality management. *Journal of Environmental Management* 74(1), 31–42.

Gutierrez Andrés, J. and Gallin, A. (2009) *New simple methodology to estimate degree of dilution in CSO discharges from sewer systems*. Proceedings of WaPUG Spring Meeting, UK.

Huang, J., Du, P., Ao, Ch., Lei, M., Zhao, D., Ho, M. and Wang, Z. (2007) Characterization of surface runoff from a subtropics urban catchment. *Journal of Environmental Sciences* 19(2), 148–152.

Jimenez, B., Mendez, J., Barrios, J., Salgado, G. and Sheinbaum, C. (2004) Characterization and evaluation of potential reuse options for wastewater sludge and combined sewer systems in Mexico. *Water Science and Technology* 49(10), 171–178.

Jiménez Cisneros, B. E., Oki, T., Arnell, N. W., Benito, G., Cogley, J. G., Döll, P., Jiang, T. and Mwakalila, S. S. (2014) Freshwater resources. Chapter 3 In: Climate change impacts, adaptation, and vulnerability. Part A, Edition 1st. Global and sectoral aspects. Contribution of working Group II to the Fifth Assessment Report of the Intergovernmental Panel on Climate Change. Cambridge University Press, Cambridge, UK and New York, NY, 229–269 pp.

Kim, L. H., Kayhanian, M., Lau, S. L. and Stenstrom, M. K. (2005) A new modelling approach for estimating first flush metal mass loading. *Water Science Technology* 51(3–4), 159–167.

Lau, J., Butler, D. and Schütze, M. (2002) Is combined sewer overflow spill frequency/volume a good indicator of receiving water quality impact? *Urban Water* 4(2), 181–189.

Launay, M., Dittmer, U., Steinmetz, H. and Roda Kuch, B. (2013) *Temporal variations of organic pollutants levels during storm events in an urban receiving water body*. University of Stuttgart, Germany.

Liu, A., Egodawatta, P., Guan, Y. and Goonetilleke, A. (2013) Influence of rainfall and catchment characteristics on urban stormwater quality. *Science of the Total Environment* 444, 255–262.

McCarthy, D. T. (2008) A traditional first flush assessment of *E. coli* in urban stormwater runoff. 11th International Conference on Urban Drainage, Edinburgh, Scotland, UK.

Parzybok, T., Clarke, B. and Hultstrand, D. M. (2011) *Average recurrence interval of extreme rainfall in real-time*. Earthzine. [Available online at http://earthzine.org/2011/04/19/average-recurrence-interval-of-extreme-rainfall-in-real-time/.]

Passerat, J., Ouattara, N. K., MoucheL, J. M., Rocher, V. and Servais, P. (2010) Impact of an intense combined sewer overflow event on the microbiological water quality of the Seine River. *Water Research* 45(2), 893–903.

Shamseldin, A. Y. (2011) *Storm water first flush analysis in the Auckland Region*. Prepared by Auckland UniServices Ltd for Auckland Regional Council. Auckland Council Technical Report 2011/007.

Sterk, G., Ten Veldhuis, J. A. E., Clemens, F. L. H. R. and Berends, B. R. (2008) *Risk assessment for urban pluvial flooding*. 11th International Conference on Urban Drainage, Edinburgh, Scotland, UK.

Suarez, J. and Puertas, J. (2005) Determination of COD, BOD and suspended solids loads during combined sewer overflow events in some combined catchment in Spain. *Ecological Engineer* 24(3), 199–217.

Woods Ballard, B., Wilson, S., Udale-Clarke, H., Illman, S., Scott, T., Ashley, R. and Kellagher, R. (2015) *The SUDS manual*. CIRIA, UK.

Zhou, Y., Lau, W. K. and Huffman, G. J (2015) Mapping TRMM TMPA into average recurrence interval for monitoring extreme precipitation events. *Journal of Applied Meteorological and Climatology*, 54(5), 979–995.

5 Online Forecast Model for Urban Flooding and Induced Water Contamination

Simon Berkhahn, Thomas Graf, Insa Neuweiler, Aaron Peche, Robert Sämann, Lothar Fuchs, Stefan Krämer, and Julian Wahl

5.1 INTRODUCTION TO FORECAST MODELS IN URBAN HYDROLOGY

Due to high population density and extended impervious surfaces, urban areas have a high vulnerability. As urbanisation will continue to increase, vulnerability will also increase in the future. One class of risks for urban areas is related to water and flooding. Pluvial floods occur when exceptional rainfall (>25–30 mm/hour) that happens usually over fast timespans (up to 3 hours) leads to a failure of the drainage system and to severe flooding in a city with inundation of streets and basements. Pluvial floods often occur with little warning in areas not obviously prone to flooding. The risk/threat of pluvial floods lies in its high temporal dynamics: events can occur within 30 minutes to a few hours. The potential damage of pluvial flood events is high since early warning is extremely challenging due to the fast onset and fast response time of urban areas.

In the last decades, Europe has suffered from severe pluvial flood events that produced a large number of casualties and significant economic losses in several cities and countries. In Germany, a prominent example is the pluvial flood in Dortmund in July 2008, where local rainfall rates of 200 mm during 3 hours lead to large damage to infrastructure and buildings (Grünewald, 2009). Another example is a storm event with extreme rainfall rates in Deidesheim in 2004, which caused a damage of 3.5 Mio Euro (www.umweltbundesamt.de/en/topics/climate-energy/climate-change-adaptation/adaptation-tools/project-catalog/urbas-urban-flash-floods-prediction-management-of). The URBAS-Database of pluvial urban flooding in Germany lists 15 events in the city of Hannover, which is the area of interest of the present chapter.

In urban areas, climate change and urbanisation are significant factors that increase the risk of pluvial flooding in urban areas. Therefore, pluvial flooding events are expected to happen more often in the future. According to the German Insurance Association (GDV), Germany must adapt to extreme storm events during the next decades. This includes an increasing number of heavy rainfall events and extreme floods raising the mean annual damage level from about 500 Mio EUR to 1,000 Mio EUR (GDV, 2011).

To ease the consequences of flood events, real-time forecasting and warning systems are useful tools in order to provide for better preparation activities ahead of the flood event (e.g., lowering the water level in a reservoir or a pond, activation of pumps, evacuation of residents, traffic regulation, and road closure) and to enhance the preparatory lead time. These measures aim at reducing the loss of life and property in vulnerable (e.g., urban) areas. Existing warning systems were mainly implemented in large catchments where fluvial (river) flooding occurs. Therefore, while early warning systems have been put into practice to address fluvial flood events (e.g., Buchholz et al., 2013; www.hvz.baden-wuerttemberg.de/), these systems are still missing for pluvial (rainfall) flood events especially in urban areas.

Early warning systems for urban areas have to be based on a prediction of a pluvial flood with a lead time of at least 30 minutes. It includes rain forecast, prediction of flooded areas, and prediction of transport paths of contaminants. For contaminant transport into the subsurface, the lead time is not that critical; however, a lead time of one to few hours is desirable.

Vulnerability of urban areas can manifest itself in a cascading event where accidentally released hazardous substances can potentially lead to deterioration of urban water quality during a pluvial flood. As pluvial floods may cause flooding of areas where it is unexpected, substances that are, for example, temporarily stored (and considered safe under normal circumstances) could be released, or contaminated sites could be unexpectedly flooded. This leads to mobilisation and release of contaminants into the surface water and a fast transport over long distances. By infiltration or by leakage through the sewer system, contaminants can be transported into urban groundwater, where they may cause a long-term risk to local water resources.

Increasing vulnerability of urban areas calls for the implementation of early warning systems, because traditional solutions to the urban drainage problems are based on the concept of providing sufficient capacity for a given rainfall event (design event). However, this traditional approach often neglects the dynamics of the system and the lack of flexibility of its operation. For example, a pumping facility is activated based on a predefined water level at a wet well, independently of the situation elsewhere in the system. This could lead to the situation where the capacity of the system is exceeded locally, whereas elsewhere in the system, storage capacity may still be available. Therefore, the system will only be fully utilised when it is loaded with the design load, while for all other loadings, the system will perform suboptimally, in which case a certain amount of storage capacity will never be used. In such situations, it is always beneficial to investigate the benefits of applying real-time control before upgrading the system.

Although sewer systems in Germany are usually of high standard, pluvial flood may have an impact on storage capacity of drainage systems. Real-time operation of drainage systems is used since several years but has mainly focussed on the reduction of combined sewer overflow, the filling of retention basins, or the inflow to treatment plants. In large urban catchments with a long time of concentration, it is possible to simulate the flows and water levels based on predicted or measured rainfall and react on this predicted situation within the sewer system (Fuchs and Beeneken, 2005).

A structure of real-time control systems that aims at increasing the efficiency of the existing drainage system can have a different level of complexity. Generally, there are three different levels of control: local, regional, and global control. Any control system is comprised of the following elements: the sensor to monitor the systems state and variables, the flow or water level regulator to control or maintain the process, a controller to control the flow/water level regulator, and a communication system to link the sensors, controllers, and regulators together in the system. It is mostly used together with a SCADA system (supervisory control and data acquisition system) which is positioned on top of a real-time control system in order to control a process that is external to the SCADA system. Furthermore, in order to send and receive the data remotely, the entire system must be telemetered. Using the resulting telemetry, the systems engineer or operator can obtain the sensor measurement information remotely in real time.

An efficient control strategy can substantially reduce (or optimise) the amount and frequency of CSO (combined sewer overflow) spills, flooding from a critical manhole or pump running time. In addition, it may lead to substantial cost savings as the number and storage capacities of the reservoirs required to keep spilling (from either an overflow structure or a manhole) below a certain limit depends on the efficiency of the applied control strategy. In real-time flood forecasting, the scope is different. In that case, the focus is not on a single point like a CSO or a basin but on all critical points of the entire catchment and to simulate how parts of a given catchment are inundated.

Real-time pluvial flood forecasting is further complicated by the fact that real-time observations of the system are seldom available. Furthermore, it is often fast response and short highly intensive rain storms that cause pluvial floods in urban areas. Calibration in real time is therefore typically unachievable and instead one will have to rely on the best possible rainfall forecast in order to predict the risk of local flooding in real time.

To predict the consequences of a pluvial flooding in urban areas, numerical-mathematical models are becoming important tools for engineers in urban drainage. To forecast a pluvial flood, a first step is to forecast the rainfall, which can be done either by numerical weather forecast or by nowcasting based on radar measurements of the rainfall. Radar rainfall estimates suffer from different sources of errors that must be considered in an accurate quality control processing of the observations. This processing focusses on the correction of non-meteorological echoes, correction for underestimation because of beam blockages, conversion from reflectivity to instant rainfall (i.e., Z–R relationship), and calibration. After the quality control, short-term radar-based nowcasting could be calculated, which allows forecasting the rainfall with a high degree of accuracy.

The prediction of pluvial flooding consists of modelling the water fluxes in the city, including 1D pipe flow in the sewer systems, 2D overland flow, and 3D subsurface flow. However, physically based models of the hydraulic system in an urban area are complex. They require fine spatial grids with millions of grid cells. The sewer system and overland flow and subsurface flow systems need to be hydraulically coupled. Also, these models require detailed information about the surface coming from a digital terrain model (DTM) with a high resolution and additional information to be integrated into the DTM like curb height. These models are mostly used to reanalyse previous floods or to produce inundation maps based on a design storm with a certain return period. Contrarily to the reanalysis of previous floods, these approaches might also be used in real-time operation of drainage systems.

Numerical models including all flow processes in an urban environment are today not standard. They require several hours of CPU time to predict a flow scenario, which is too slow to take measures to prevent damage. For this reason, either neural networks or other simplified approaches (e.g., using databases generated from scenario simulations) are needed that can be used for a reasonable forecast. Only few studies exist, where such attempts have been made (Sommer, 2011; Sen, 2013). For example, Sen (2013) formulated an approach to use real-time rainfall monitoring in order to forecast flood inundation for the city of Kolkata. That approach includes a fully coupled hydrodynamic urban flow model, but did not consider contaminant transport. Given the similarity of ideas, the approach to generate a coupled flow model will be the base for our attempt to generate a coupled urban flow model, which will then be used for forecasting of pluvial flood events.

5.2 HYDRAULIC INTERACTIONS AND CONTAMINANT TRANSPORT IN URBAN FLOOD MODELS

Almost all urban floods are caused by rainfall events of short duration and high intensity. Flood water flows rapidly along preferential flow paths and may overwhelm the capacity of local drainage systems (Schirmer et al., 2013). Urbanisation has led to a rapid increase of impermeable surface areas, which increases surface runoff and leads to enhanced local flooding because of excess rainfall (Aronica et al., 2012). In the event of exceptional rainfall, the subsequent urban flood will create a series of hydraulic interactions as shown in Figure 5.1: (1) Inflow/outflow of rainwater in/on the pipe network. (2) Infiltration/exfiltration of rainwater/subsurface water. (3) Leakage of pipe water into the subsurface/drainage of water from the subsurface into pipes. According to Lerner (1986, 2002), leakage loss can be up to 50% depending on the city and the country. To estimate the relative importance of each hydraulic interaction that should be included in a groundwater model, flow velocities and storage in each domain can be compared. Water flow velocities are likely to increase in the order 3D subsurface (smallest) – 2D land surface – 1D pipe network (highest). Conversely, water storage decreases in the same order: 3D subsurface (highest) – 2D land surface – 1D pipe network (smallest).

Flow in the subsurface is also relevant if hazardous substances are released during and after a pluvial flood event. The risk of contamination of aquifers used for drinking water needs to be considered, and information on subsurface flow is needed to estimate the transport paths.

5.2.1 The Role of Groundwater for Flood Prediction Modelling

During a flood event induced by exceptional rainfall, both the infiltration into the subsurface and the flow of rainwater into the pipe network are sinks for stormwater. Although the groundwater may not be used for water production, urban aquifers are a potential storage location for stormwater, reducing surface runoff from sealed areas (Göbel et al., 2004). However, even if the storage in the subsurface is high, subsurface flow velocities are very small compared to flow velocities of surface water. As a consequence, subsurface storage during a very quick and intense rainfall event is limited. Subsurface storage can be further reduced by pipe water leaking into the subsurface. At some locations, it might even happen that the subsurface storage is exceeded such that subsurface water exfiltrates from the subsurface onto the surface, thereby exacerbating potential flood damage. For these reasons, it is important that the numerical groundwater model account for the hydraulic coupling between the land surface and the subsurface and also the leakage from sewer pipes into the subsurface (and back). A more detailed review of possible hydraulic interactions in urban environments is given by Schirmer et al. (2013).

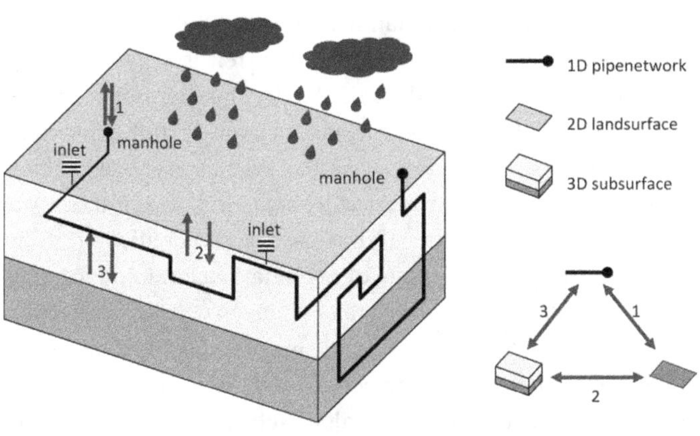

Figure 5.1 Conceptual model and dimensionality of flow domains including hydraulic interaction

Numerous studies on urban water flow exist where different levels of physical complexity to represent hydraulic interactions have been adopted. In a review paper of modelling studies for the city of Dresden, Grischek et al. (1996) concluded that groundwater resources are more affected by natural river fluctuations (discharge) and by precipitation than by urban impacts. The impact of urban floods, however, was not taken into account in the reviewed modelling studies. Grischek et al. (1996) also noted that urbanisation causes significant groundwater pollution (by, for example, chlorinated hydrocarbons, nitrate, and sulphate). Finally, the approach of developing and using a coupled flow model to predict contaminant transport in the subsurface, as proposed here, is a worthwhile endeavour.

Hunter et al. (2008) gave an overview of other hydraulic models to simulate urban flooding in the absence of hydraulic coupling to groundwater. The vast majority of urban flood models treat the land surface as 2D simulation domain, in which either the full shallow-water equation or the diffusive wave approximation is being solved. Simulating urban flash floods for the cities of Nîmes (France) and Sumacárcel (Spain), Abderrezzak et al. (2009) noted that flow is strongly 3D in character but did not consider hydraulic coupling to the 3D subsurface where groundwater quality and quantity are likely affected by the floods.

The importance of 3D modelling and the significance of groundwater on urban water management are discussed by Mansell and Wang (2010) who compared the overall water balances of Glasgow and Beijing. Glasgow uses remote rural surface water catchments, while Beijing relies more on groundwater. The overall water balance of the two cities is modelled using a numerical tank model representing the processes of infiltration, evaporation, runoff, and storage for both paved and unpaved surfaces. The results of the modelling activity highlight the importance of accounting for groundwater recharge by stormwater runoff.

Rauch et al. (2002) described the main concepts required for describing the dynamics of drainage in an entire urban area. Several types of receiving waterbodies appear in reality (rivers, lakes, sea, groundwater, etc.) such that, according to Rauch et al. (2002), urban flow modelling efforts should also focus on groundwater to guarantee the long-term goal of "good water quality." In doing so, the effects of different urban water management strategies (e.g., abstraction for industrial and domestic water supplies and stormwater infiltration) on the quantitative and qualitative states of the resource groundwater would be predictable on a long-term perspective. Similarly, Basnayaka et al. (2013) undertook a modelling study to investigate the impact of shallow groundwater, and they argued that shallow groundwater plays a main role in urban flood processes.

According to Sommer (2011), including a groundwater model in an urban flood model is important because groundwater levels will significantly rise due to three reasons: (i) direct infiltration of surface water into the aquifer during a flood, (ii) inflow of surface stormwater into the sewer system and outflow from the sewer system into the groundwater, and (iii) increased percolation and inflow from the groundwater recharge areas after an exceptional rainfall event. Following Sommer (2011), rising urban groundwater tables represent the following three threats: (i) instability of buildings by buoyancy effects (Beyer, 2003), (ii) increased infiltration of potentially contaminated groundwater into the sewage system (Karpf and Krebs, 2004), and (iii) recontamination of soils and sediments by polluted groundwater or a remobilisation of pollutants and endangering of drinking water (Marre et al., 2005).

To address the important effect of groundwater during urban floods, Sommer (2011) studied the interactions between flood and groundwater for the city of Dresden during floods in August 2002 and in spring 2006. For that study, Sommer (2011) constructed a coupled model including all three flow domains shown in Figure 5.1. However, Sommer (2011) did not apply their fully coupled flow model to cases of contaminant transport as is discussed in this chapter.

5.2.2 Contaminant Transport during Pluvial Flood Events

Due to high water levels during pluvial flooding, the risk of mobilisation of potentially dangerous substances increases (Egodawatta et al., 2007). Contaminants of interest in urban areas can be divided into accumulated and local spill contaminants. In the first case, substances are accumulated on streets and in the soil. During a flood event, they are washed into the stormwater. For example, during dry weather periods, dust of particulate matter from traffic and industrial fume accumulates on streets and facades. Also, heavy metals from the abrasion of tires or released polycyclic aromatic hydrocarbons (PAH) from combustion in motors of cars accumulate on street surfaces (van Metre et al., 2000; Vaze and Chiew, 2002; Zhang et al., 2015, 2017). During a heavy rain event, those substances are washed off. This creates a high concentration in the drainage system. Substances like heavy metals are transported within this first flush and are not removed in a treatment plant but are transported directly to receiving waters via the drainage network (Sansalone and Buchberger, 1997; Li et al., 2007; Semadeni-Davies et al., 2008). Also, sediments with heavy loads of adsorbed contaminants are mobilised during flood events and reach the sewer system (Wildi et al., 2004).

Another problem is the release of potentially dangerous microbes into the stormwater. Water that spills from manholes of sewer systems could, for example, release bacterial pollutants such as *Campylobacter coli* which cause health risk (Sales-Ortells and Medema, 2015; Nguyen et al., 2017).

Local spills happen by accidents that occur during a pluvial flood event. Chemicals such as oil, gasoline, pesticides, or fertiliser that are stored in basements, garages, and depots can be mobilised in case of high water levels and contaminate large areas. Surveys showed that up to 20% of damaged households during a pluvial flood were contaminated by chemicals or sewer outflow (Rözer et al., 2016; Spekkers et al., 2017). Furthermore, these chemicals can infiltrate into the subsurface and threaten the quality of drinking water wells. Although such contamination events might be rare, the potential damage can be very high if substances are toxic. During a pluvial flood, high precipitation may lead to exceeding of the capacity of urban drainage systems, resulting in flow interaction between drainage pipe network and the surface via manholes and street inlets, such that the transport paths of substances after a spill are not necessarily clear.

Predicting transport of contaminants during a flash flood is important to plan countermeasures for the impact of contaminant release and spreading. Numerical models are used for this purpose. For the case of distributed sources, contaminants will mostly spread over the drainage system of the city. The critical question is, how much contaminant mass will be transported to the treatment plants and how much will be released into receiving water or into the soil (Wildi et al., 2004). This question is related to the budgets of water and substances over the full domain. In case of the release of a very hazardous substance at one location (a local spill), the contaminant will spread in a part of the sewer system and on the surface. In this case, a fast prediction of the transport paths and the points of potential release into receiving waters and into the subsurface are of interest. A fast prediction is essential. Such real-time contaminant spreading forecasting models may help civil protection and disaster management authorities to deal with the hazardous substances. The modelling of transport paths is directly related to the prediction of the flow fields in a model. Temporal connections between pipes due to overland flow as well as due to possibly reversed flow directions create complex dynamic flow patterns, which influence the potential areas of concern. The prediction of preferential paths is therefore not straightforward.

5.3 MODELLING FLOW AND CONTAMINANT TRANSPORT IN URBAN ENVIRONMENTS FOR PLUVIAL FLOOD FORECASTING

When modelling flow and contaminant transport during a pluvial flood event, different modelling concepts are required for the different parts. In this section, we introduce models as examples that could be combined for a forecast system. The input for such a forecast system is discussed in Section 5.4, and a test example is shown in Section 5.5.

5.3.1 Pipe Flow and Surface Flow: HYSTEM-EXTRAN 2D

Water flow in a network of urban drainage pipes as well as flow on the surface is simulated with HYSTEM-EXTRAN 2D [HE2D] (itwh, 2014). HYSTEM-EXTRAN 2D is a dual drainage model consisting of a fully distributed rainfall runoff model and a coupled 1D/2D hydrodynamic model for the sewer network flow and the overland flow, respectively. Infiltration losses are computed based on a modified Horton infiltration model.

HE2D conceptually describes pipe flow as open-channel flow where pressurised pipe flow is transformed into free-surface flow using either a Preissmann slot (Preissmann, 1961) or an iterative procedure. HYSTEM-EXTRAN 2D simulates open-channel pipe flow using the complete 1D Saint-Venant equations (Saint-Venant, 1871) discretised with a finite volume (FV) scheme. The Saint-Venant equations are given by the continuity equation (5.1) and the momentum equation (5.2):

$$\frac{\partial A}{\partial t} + \frac{\partial (v \cdot A)}{\partial x} = 0 \tag{5.1}$$

$$\frac{1}{g}\frac{\partial v}{\partial t} + \frac{v}{g}\frac{\partial v}{\partial x} + \frac{\partial H_{\mathrm{PW}}}{\partial x} = I_s - I_r \tag{5.2}$$

where A [L^2] is cross-sectional surface area, t [T] is time, v [L T^{-1}] is flow velocity, x [L] is pipe length, g [L T^{-2}] is gravitational acceleration, H_{PW} [L] is pipe water level, I_s [–] is bottom slope, and I_r [–] is friction slope. I_r is parameterised using the Darcy–Weißbach equation in combination with the Prandtl–Colebrook friction law (itwh, 2014). A common assumption is to use circular concrete pipes with a surface roughness of $\epsilon = 1.5$ mm (itwh, 2014).

Surface water flow is computed with a diffusive wave approximation of the 2D depth-averaged shallow water equations (SWE) by using an explicit cell-centred FV method. The diffusive wave approximation assumes the influence of inertia on the flow to be insignificant by neglecting the temporal and spatial derivatives of the flow velocities of the equation for the conservation of momentum. Hence the moment equation simplifies to:

$$g\nabla h = g(-\nabla z - I_E) \tag{5.3}$$

where h [L] is water depth, z is surface elevation [L], and I_E is the energy gradient.

By inserting the Manning–Strickler flow formula for approximating the energy gradient, an analytical expression for computing the velocity vector is obtained:

$$v = -\frac{h^{2/3}}{n}\sqrt{\left|\nabla(h+z)\right|} \cdot \frac{\nabla(h+z)}{\left|\nabla(h+z)\right|} \tag{5.4}$$

By inserting the velocity vector v into the equation for the conservation of mass, the equation for the diffusive wave approximation is obtained, including only the water depth h as unknown variable:

$$\frac{\partial h}{\partial t} + \nabla \cdot (v \cdot h) = f_s \tag{5.5}$$

5.3.2 Subsurface Flow: OpenGeoSys

Subsurficial water flow in fully and partly saturated porous media is simulated with OpenGeoSys [OGS] (Kolditz et al., 2012). OpenGeoSys describes flow in the subsurface using the Richards' equation (Richards, 1931), formulated in a form that includes water pressure as primary variable:

$$\phi\rho\frac{\partial S}{\partial p}\frac{\partial p}{\partial t} - \nabla \cdot \left[\rho\frac{\kappa\kappa_r}{\mu}(\nabla p - \rho g)\right] = \frac{\rho Q_{\mathrm{leak}}}{V_{\mathrm{unit}}} \tag{5.6}$$

where ϕ [–] is porosity, ρ [M L^{-3}] is fluid density, S [–] is water saturation, p [M L^{-1} T^{-2}] is water pressure, ∇ [L^{-1}] is nabla operator, κ_r [–] is relative permeability, and κ [L^2] is intrinsic permeability, which is connected to saturated hydraulic conductivity K_s [L T^{-1}] in the form $K_s = \kappa\rho g / \mu$, where μ [M L^{-1} T^{-1}] is fluid dynamic viscosity. Furthermore, V_{unit} [L^3] is unit volume and Q_{leak} [L^3 T^{-1}] is leakage flow.

Typically, the soil retention curve in the form of a capillary pressure saturation relationship is parametrised using van Genuchten's closed form (van Genuchten, 1980):

$$-p = \frac{\rho g}{\alpha}\left[S_e^{-\frac{1}{m}} - 1\right]^{\frac{1}{n}} \qquad m = 1 - \frac{1}{n} \tag{5.7}$$

where α [L^{-1}] is air entry value, m [–] and n [–] are pore size distribution parameters, and the effective saturation S_e [–] is given by

$$S_e = \frac{S - S_r}{S_m - S_r} \tag{5.8}$$

where S_r [−] is residual saturation and S_m [−] is maximum saturation. That parameterisation leads to the following formulation for relative permeability (van Genuchten, 1980)

$$\kappa_r = S_e^{1/2} \cdot \left[1-\left(1-S_e^{1/m}\right)^m\right]^2 \tag{5.9}$$

5.3.3 Coupled Model OpenGeoSys: Hystem Extran

In order to simulate urban flow processes, the surface flow and pipe flow simulator HE2D was coupled to the groundwater flow simulator OGS, both of which are C-based codes. The conceptual model and data transfer are depicted in Figure 5.2. The coupling is based on exchange fluxes between the soil and the potentially defect subsurface pipes.

The coupling enables a bidirectional data transfer between both software packages. It was realised using the FIFO method Named Pipes (Bustamante, 2007), which is considered a cost-effective and easy-to-implement in-memory push-migration solution (Laszewski and Nauduri, 2011). This inter-process communication enables a more cost-effective data transfer compared with, for example, file-based coupled models because reading from or writing into files is not required. In the coupled model, HE2D acts as a server and OGS as a client.

For simulations on a scale larger than a few pipe defects (such as an urban scale), the non-iterative coupling scheme is based on a timestep-wise updating of source terms. HYSTEM-EXTRAN 2D calculates a pipe water level H_{PW} [L] at a given timestep t and the value is transferred to OGS using the server pipe. In OGS, H_{PW} and the water pressure in the soil located directly below the colmation layer p [M L^{-1} T^{-1}] are used in order to determine the leakage flow Q_{leak} [L^3 T^{-1}] using a transfer function called leakage function (LF). At timestep $t+\Delta t$, Q_{leak} is used as a source term in the groundwater model and transferred to HE2D using the client pipe where it is used as a source term in the pipe flow model.

The LF describes the leakage flux q_{leak} [L T^{-1}] as a function of H_{PW} and p in the form $q_{\mathrm{leak}} = f\left(H_{\mathrm{PW}}, p\right)$. Multiplication of q_{leak} with the defect area A_{defect} [L²] gives the leakage flow Q_{leak}. The LF was derived from a series of transient simulations with the coupled model on a fully discretised single-pipe-defect domain. It represents leakage through the colmation layer and enables to upscale the leakage flow process. A fine spatial discretisation of the colmation layer in the groundwater model can be omitted. This results in reduced computation times. Results from single-defect domain simulations and the LF for a test area are visualised in Figure 5.3. This LF represents leakage in a setup with a sandy aquifer, a sandy trench backfill and a clayey-silty colmation layer. More details on the LF, the coupled model, and a successful model validation are described by Peche et al. (2017).

5.3.4 Modelling of Contaminant Transport after a Spill Event

Models that describe the transport of contaminants build up on flow models that describe the development of water heights and flow velocities in pipes and on streets. Such simulation models usually describe the drainage network as a set of pipes and manholes. Methods to solve the equations that describe the transport of contaminants in the flow fields can be separated into Eulerian and Lagrangian methods. Eulerian methods split the area of transport (pipes and surface) into finite, discrete volumes or elements by superposing a grid over the domain. The transport problem between neighbouring volumes and elements is solved in a fixed

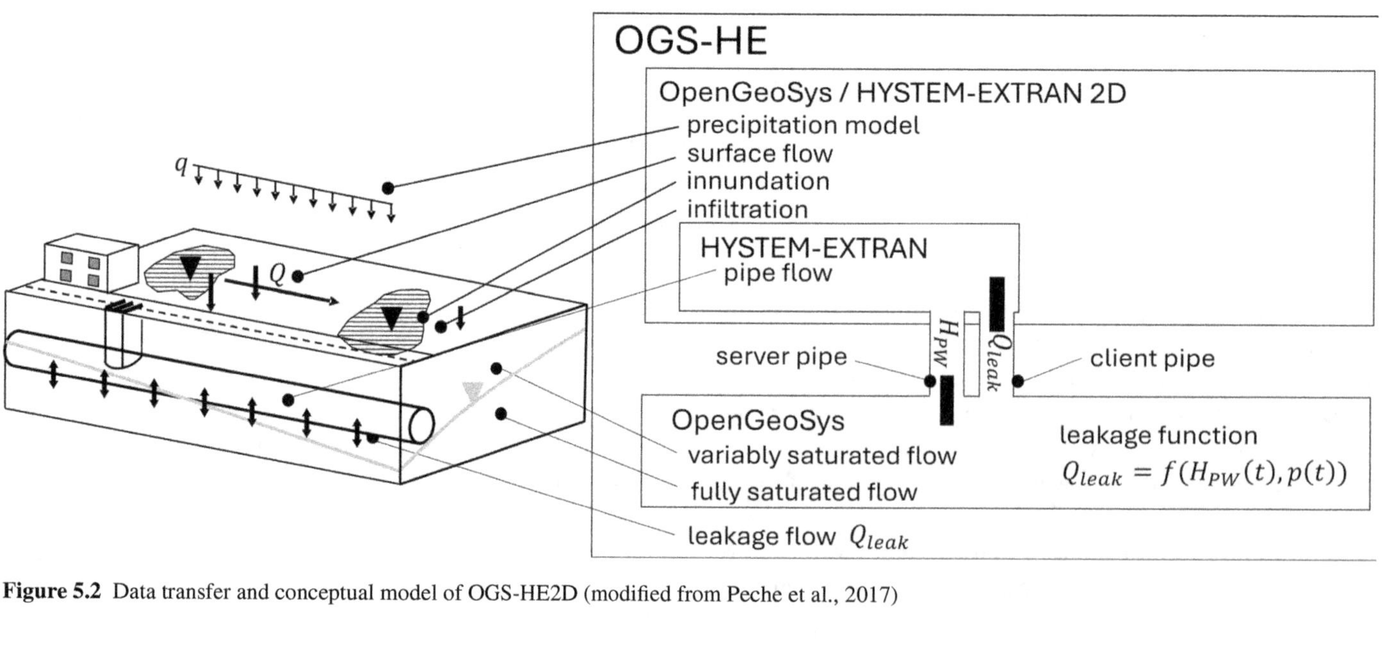

Figure 5.2 Data transfer and conceptual model of OGS-HE2D (modified from Peche et al., 2017)

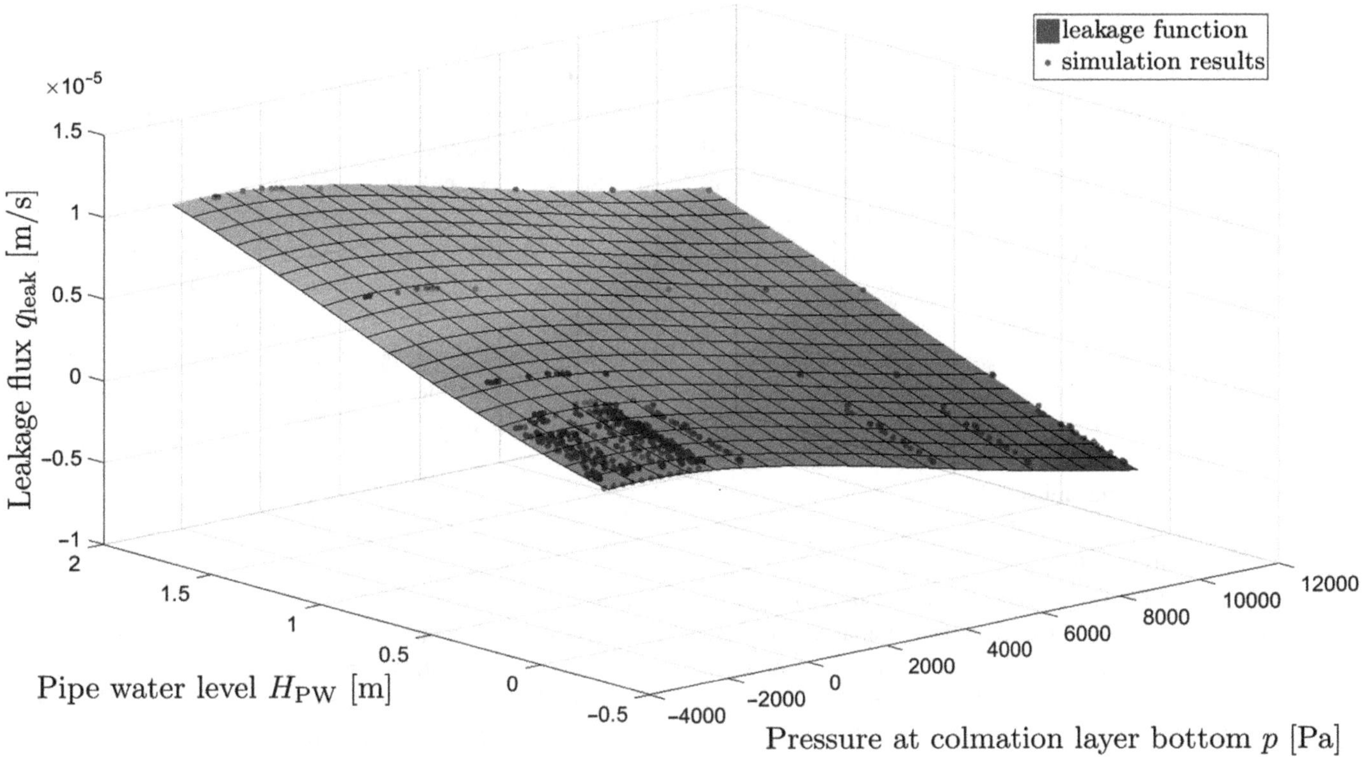

Figure 5.3 Leakage function for the test area (modified from Peche et al., 2017)

in space (Eulerian) framework. The finite difference (FD) method is commonly used to solve transport problems in water fluxes that are associated to flood and rainfall events. This method has the advantage that the organisation of fixed elements and the interaction between elements is easy to implement, but it is complex to guarantee mass conservation (Vojinovic and Tutulic, 2009). Finite Volumes are capable of conservation of mass and do not have problems with high flow velocities (Versteeg and Malalasekera, 2007). Standard open source or commercial codes implement this method with the purpose to quantify and to predict the mass of sewage load at treatment plants (Cambezet et al., 2008). It requires low computational effort if it is possible to discretise the domain with low resolution. In an urban area, where domain sizes are large, this is the case if constant fluxes or total loads should be predicted. However, the coarse resolution can result in large errors when contaminant transport after a spill event, where distinct travel paths are needed, have to be predicted. The reason is that within one element of an FV method, all variables (e.g., water level, velocity, and solute concentration) are treated as a mean value over the domain. If, for example, a pipe and its representing element have a length of several tens of metres, no propagation of a concentration front can be seen. The great length of elements combined with the calculation of mean values lead to flattening of concentration peaks, which is called smearing or "numerical diffusion" (Dehghan, 2004; Pathirana et al., 2011). To avoid this, the grid has to be set up with a finer resolution. Increasing the number of elements increases the number of calculations and slows down the computation of results, which could make a fast forecast of transport paths after a spill impossible (Salamon et al., 2006).

If high resolution needs to be considered, Lagrangian methods can be beneficial compared to Eulerian methods. They are also called discrete particle methods, because the moving elements within a domain are seen as particles that may react with each other (Niu and Phanikumar, 2015). An element represents a defined volume of matter and moves according to physically based equations. The transport is described in a moving framework, as the position of each single particles is followed. The modelling of physical particle interaction is computational expensive because nearest neighbours can only be found inside the crowd by calculating the distance between all particles (Liang et al., 2015). Less effort can be achieved by neglecting particle interaction. A particle is then moved in one-time step according to the local flow field, followed by a movement due to mixing (diffusion or dispersion). This movement is modelled with the use of random walk approaches, where the mixing of particles is described as random jumps (Banton et al., 1997; Visser, 1997). Particle-based methods are described as object-oriented code which can easily be parallelised to run on multiple CPUs or graphic cards (GPUs) (Liang, 2015). This makes the method very efficient. The number of particles to represent a solute mass is adjustable. More particles generate more trustful forecasting results, because the generation of random numbers is only a tool to emulate physical behaviour on the micro scale (Yeung and Pope, 1989; Salamon et al., 2006). Particle-based transport models are popular to describe transport of solute in subsurface flow on length scales from pore scale (micrometre) to kilometre scale problems (Blunt, 2013). In particular, transport processes in complex fracture networks are often described with particle-based methods (Delay et al., 2005).

To calculate transport paths of spilled contaminants during a strong rainfall event, where the flow field changes with time, Eulerian and Lagrangian methods can be combined to take advantage of each method. The basic assumption that has to be made here is that the contaminant does not influence the flow behaviour. Rain volume flow can, for example, be calculated with a Eulerian FV method, which generates the dynamic velocity field in each part of the flooded domain. The contaminant is set up as a cloud of particles and moves in the previously calculated velocity field. Eulerian–Lagrangian coupling techniques in urban hydrology have been used to quantify sedimentation of particulate contaminants in sewer or drainage water, for example, in sedimentation tanks (Yan et al., 2014).

The application of Eulerian–Lagrangian models to city-scale transport models can be useful to create a real-time contaminant forecasting after a spill event, which can predict possibly affected areas before they actually become contaminated. When a spill event is reported, information about substance, injection time, and location is known and can be used to start a particle cloud simulation. As it is not likely that the exact transport behaviours of the substance, such as sorption coefficients, settling, or chemical reactions, are known, the aim is to predict transport paths assuming that the contaminant does not react at all. This could lead to under-prediction of transport times or to over-prediction of affected areas and should thus be considered a conservative estimate. This estimate is uncertain due to the uncertainty of the flow field.

To run the transport simulation, the dynamic flow field has to be calculated first, based on actual rain gauges and precipitation forecast. As the calculation of the flow is computationally expensive and might take a long time, alternatively, a database of pre-calculated flow fields could be used. An ensemble database could, for example, be created based on the rainfall intensity signal. The flow field is chosen, for which the input rain fits best the actual weather conditions. Based on the selected flow field, the particle-based transport model predicts the possible flow path of the spilled substance. A detailed tracking of concentration peaks and dynamic maps of contaminated area is possible within a few minutes on a desktop computer.

Also, a group of flow fields could be chosen to carry out an ensemble of transport paths. An ensemble approach might be beneficial, as uncertainties and variability in input parameters cause the uncertainty of prediction of travel paths. The selection of a velocity field from stored results contains errors because no exact match of a rain event can be selected from a database. Actual and scenario rain gauges will always differ in intensity and duration. A second error is due to collecting the report of a spill event. Start and duration of a spill are usually not exactly known and delay between a spill, detection, and report creates variations in time. A delay can cause a mismatch of the flow field, as it is variable in time during flash flood events. Spreading patterns of spilled contaminants can differ substantially when the spill location or time is varied. The outlined concept of particle-based transport simulation after a spill during a pluvial flood event is illustrated with an example in Section 5.3.5.

5.3.5 Artificial Neuronal Network to Predict Inundated Areas

The lead time for the prediction of urban floods is small in comparison to, for example, river floods. One reason for this is the short live time of storms. Henonin et al. (2013) discussed several different approaches to simulate flood forecasts. According to Henonin et al. (2013), the time from the precipitation forecast until the arrival of the rain event is normally too short for simulations with physically based 2D flood models, which calls for simplifications in flood forecast models. Duncan et al. (2013) used artificial neural networks (ANN) for flood prediction at single points in the sewer network. Artificial neural networks can be seen as a well-known tool for flood prediction. This section aims at closing the gap between accuracy and computation time in 2D forecasts of urban floods.

With the help of ANN, a data-driven model for 2D flood prediction in urban areas was build up. The model uses time series of rain events as input and generates spatially distributed maps of maximum water levels as output. The ANN is trained with simulation results of HE2D.

Since the middle of the 20th century, ANN are developed. The methodology of ANN roots in modelling the human brain. The aim was to develop an information processing model that is similar to the structure of the brain. The concept of ANN is based on graphs consisting of edges and nodes. The neurobiology roots are also shown in the terminology of ANN. The edges are called synapses, and the nodes are known as neurons. The neurons are distributing the incoming information to the next neurons. This happens via the synapses. Every synapse has a weight which regulates the priority of the information for the next neuron.

ANN are more or less complex combinations of the introduced simple elements, neurons, and synapses. To understand the behaviour of those components in the network, it is useful to have a look at perceptrons. The perceptron was first described by Rosenblatt (1958). The perceptron can be seen as a precursor of the ANNs used nowadays.

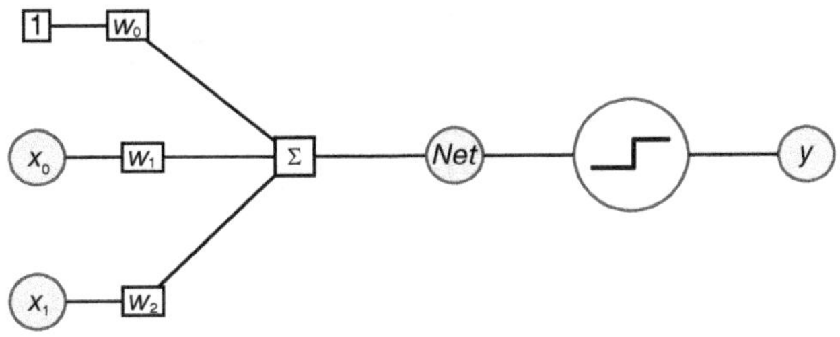

Figure 5.4 Structure of a simple perceptron

As it is shown in Figure 5.4, the perceptron consists of one neuron with multiple weighted inputs. The inputs for the classic perceptron are binary values. The neuron sums up the input values x_i weighted with the synapses w_i:

$$net = \sum_{i=1}^{n} x_i w_i \tag{5.10}$$

The result of this sum is called net input and is processed with a so-called activation function:

$$y = \phi(net - \Theta) \tag{5.11}$$

where y is the output of the neuron. In case of flood prediction, y represents the water level. The threshold Θ was introduced by McCulloch and Pitts (1943) to represent the behaviour of the natural neuron. The idea of the activation function is that a natural neuron is sending information depending on how strongly it is activated. The perceptron as the simplest type of ANN uses the sign function as activation function. Therefore, the perceptron can linearly classify separable cases.

The perceptron can be trained by adjusting the weights of the synapses and the threshold. To simplify the training, the threshold is replaced by a bias neuron. This bias neuron has a constant value of 1. The variation of the threshold is realised by adjusting the synapse which connects the bias with the neuron. Before starting the training of the perceptron, the weights have to be initialised with random values. The training starts with presentation of a random sample from the training data. The calculated output y is compared with the desired output $\hat{y}$:

$$\delta = \hat{y} - y \tag{5.12}$$

The mismatch δ is used to adjust the weight in dependency of the corresponding input. In the present study, the desired output $\hat{y}$ is the water level calculated with HE2D. The rule for adaptation is:

$$w_i^{t+1} = w_i^t + \eta \cdot \delta \cdot x_i \tag{5.13}$$

where t is the iteration step of the training and η is the learning rate, which is used to improve the convergence of the training (Priddy and Keller, 2005). The two steps of sample presentation and weight adaptation is repeated until the error (here the RMSE) for the entire training set reaches the chosen limit.

To solve more complex problems, the multilayer perceptron (MLP) was developed. As shown in Figure 5.5, the MLP consists

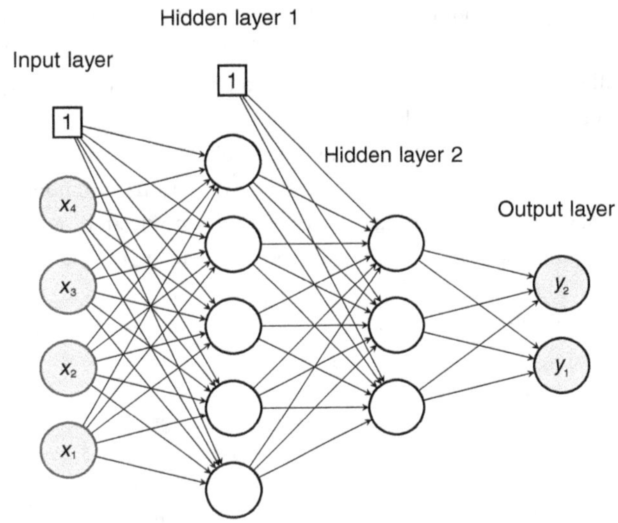

Figure 5.5 Structure of a multilayer perceptron (MLP)

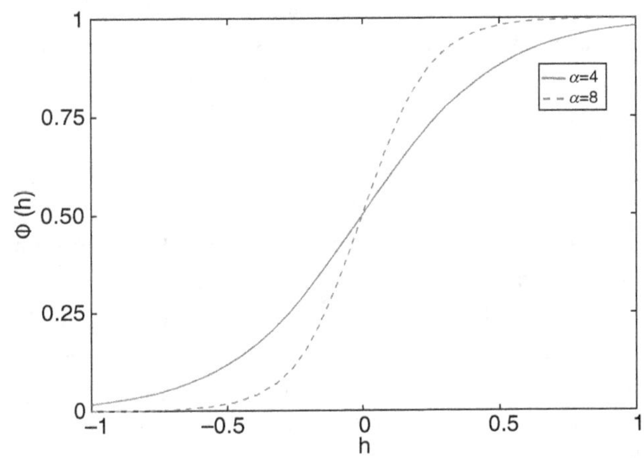

Figure 5.6 Sigmoid activation function with two different slopes

of input neurons, one or more hidden layers, and an output layer. The hidden neurons are equal to the neuron of the perceptron. Only the indices in the formulation have to be extended. The net input to the neuron j can be written as:

$$net_j = \sum_{i=1}^{n} x_i w_{ij} \tag{5.14}$$

where w_{ij} stands for the synapse from neuron i to neuron j. Sigmoid functions like:

$$\phi(h) = \frac{1}{e^{-\alpha h} + 1} \tag{5.15}$$

are often used as activation function for neurons in MLP (Figure 5.6). The function variable h is the net input net_j. The value α controls the slope of the curve. If the sum of weighted inputs is very small (Figure 5.6a) the output of the neuron will be near to the minimum of the activation function ϕ. On the other hand, a high value of the sum will generate an output near to the maximum. The value range of the perceptron output depends on the activation function. In the case of a sigmoid function, the value range is from 0 to 1. For that reason, the training data has to be scaled in the same range. In case of flood prediction, the smallest desired water level in the training set is scaled to 0 and the highest value is scaled to 1.

The training of an MLP is more complicated than the classic perceptron. That is easy to understand when thinking about the calculation of δ for a hidden layer. There is no desired output $\hat{y}$ for a hidden neuron. This issue is handled by using the backpropagation algorithm.

This method calculates the influence on the error for every synapse:

$$\delta_j = \frac{\partial E}{\partial o_j} \cdot \frac{\partial o_j}{\partial net_j} = \begin{cases} (o_j - \hat{o}_j) \cdot o_j \cdot (1 - o_j) & \text{if } j \text{ is output neuron} \\ \sum_{l=1}^{n} \delta_l \cdot w_{jl} \cdot o_j \cdot (1 - o_j) & \text{if } j \text{ is hidden neuron} \end{cases} \quad (5.16)$$

where E is the error and o_j is the output of neuron j. The $\hat{o}_j$ symbolises the desired output. By starting the calculation of δ_j at the output layer and proceeding backwards to the input layer, every δ_l has already been calculated. For every synapse, the adjustment of the weight is calculated by:

$$\Delta w_{ij} = -\eta \cdot \frac{\partial E}{\partial w_{ij}} = -\eta \cdot \delta_j \cdot o_i \quad (5.17)$$

where o_i is the output of the previous neuron. This algorithm is the basis for most training algorithms of MLP.

When using MLP for 2D flood prediction, one has to think about the input and output data. The input data is a time series of precipitation with five-minute time steps. The rain for the prediction model is partially uniform. This leads to an input vector with 24 neurons for a 2-hour rain event. For the output vector, the number is much higher. The calculation grid of HE2D has over one million cells for a medium-sized city. Considering a fully coupled MLP, this would lead to a number of synapses which are far too large for fast training and prediction.

With a regular rectangular grid with edge length of six metres, the number of cells can be reduced by 80% to 90%. The resolution with this approach is still sufficient to detect the hotspots and the spatial distribution of the flood. Test scenarios with this reduced resolution with about 120,000 water level cells (output neurons) have shown limitations in memory requirements.

To solve this problem and to be prepared for extension of the study area, the ANN-grid approach was presented by Berkhahn et al. (2017). The study area is divided in $n \times n$ grid cells. For every cell, an independent ANN is build up and trained. The input vector for those ANNs stays the same, but the size of output variables is reduced depending on the grid size. With this approach one can regulate the memory requirements. Every single ANN has to represent a relatively small area. The disadvantage of the approach are unphysically sharp edges at the borders of the grid cells. One reason for this issue is the loss of the interconnection between the grid cells. Another reason is the overfitting effect.

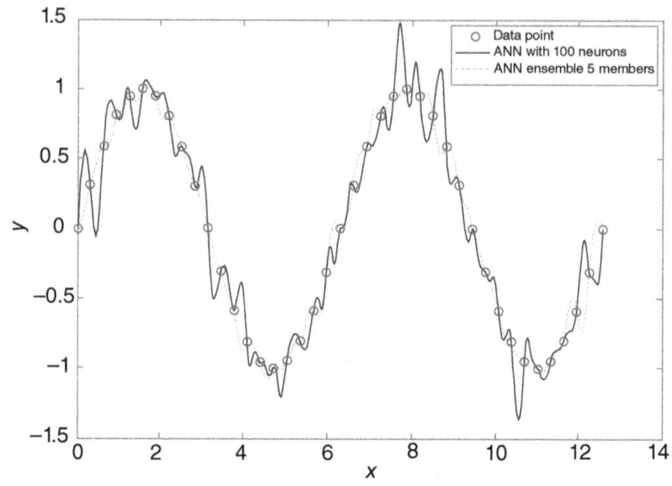

Figure 5.7 Example for overfitting and the ensemble approach

The overfitting effect occurs when an ANN has too many layers or too many neurons per layer. Another reason is that a training process can have too many iterations. The dark solid line in Figure 5.7 shows the overfitting effect for a synthetic test case. All data points are matched but the ANN gives oscillating results between the data points. In the same figure, the light dashed line shows the result of an ensemble of ANN. This ensemble approach was presented by Hansen et al. (1992). The ANN is trained several times with different initial weights. The results of every ANN are averaged, and the oscillations are smoothed out. With this method, the overfitting problem is solved. The sharp edges at the borders of the grid cells are eliminated or negligibly small with the ensemble results.

5.4 REQUIRED INPUT FOR ONLINE FLOOD FORECASTING

5.4.1 Radar Data Correction

Urban flood management is driving a growing need to predict the performance of urban drainage systems with a lead time of a few hours. The key for the runoff predictions are rainfall observations and predictions highly resolved in space and time. In contrast to rain gauges, weather radars have an unprecedented potential to provide this type of information; operational weather radar systems in Europe provide areal rainfall information at spatial scales between ≤1 km² and at time resolutions ≤5 minutes. In general, these data meet the requirements for urban drainage (Schilling, 1991; Berne et al., 2004). In contrast to rain gauges, as basis for storm tracking and extrapolation methods, radar data have the potential to forecast rainfall structures for real-time control in urban drainage systems (Krämer et al., 2007; Achleitner et al., 2009) or for flood management.

However, weather radar data at X and C-band frequencies are affected by the following physical effects:

- Clutter echoes caused by signal scattering, for example, on wet surfaces, wind turbines, anomalous propagation of the radar signal due to thermal layers, signal shading, or insects.
- Wet radome attenuation in case rainfall structures cross the radar site.
- Rain-induced signal attenuation along the radar signal path.
- Conversion of radar reflectivity (Z) in to the target figure rainfall intensity (R).

As a result of these effects, C-band radar rainfall data are uncertain and they systematically underestimate the "true" rainfall at a magnitude of 50%. The underestimation is mainly due to the effect of wet radome attenuation and rain-induced signal attenuation (Figure 5.8), which have a strong nonlinear influence.

For the use in urban drainage or flood models, the radar data have to be corrected. A common approach is the adjustment of observed radar raw data by using a network of ground gauges to compensate these phenomena. This approach is problematic for several reasons:

1. The variability of microphysical rainfall processes and the resulting areal rainfall distribution at the microscale (≤2 km²), which is comparable to the radar grid resolution, are high. Hence, a single gauge observation is not representative for the corresponding radar measurement of the raster element aloft (Pedersen et al., 2010; Schellart et al., 2012).
2. Convective rainfall structures, especially when initiated by local convection, which is most relevant for flood management in urban areas, have spatial scales at the magnitude of 8–16 km² (Gires et al., 2012). To capture these small-scale rainfall structures, a network of gauges with a higher density is required. For example, the German Weather Service operates a network of about 1.300 telemetering gauges, the average density for the total area of Germany is 1 gauge/275 km².
3. To obtain robust radar gauge adjustment factors, it is necessary to integrate radar and gauge observations over a certain time span of at least 30–60 minutes. This interval is required to consider time shifts between radar observations in the atmosphere and gauge measurements at the ground. Further, random effects, which are induced by the different spatiotemporal scales of the radar sampling and gauge measurements, and storm characteristics are reduced (Zawadzki, 1975; Krajewski and Smith, 2002). Transmission, processing, and integration of data result in a time-lag of the adjusted radar data which reduces the lead time of quantitative reliable radar forecasts.
4. In the forecasting context, any adjustment and correction have to be applied for the entire radar image (PPI) since errors in radar observations persist in the forecast rainfall structures in space and time.

The alternative is the correction of the physical effects on radar measurements, which is based on the radar data itself. Therefore, a methodology was developed for single polarised radar data which consist of the following steps:

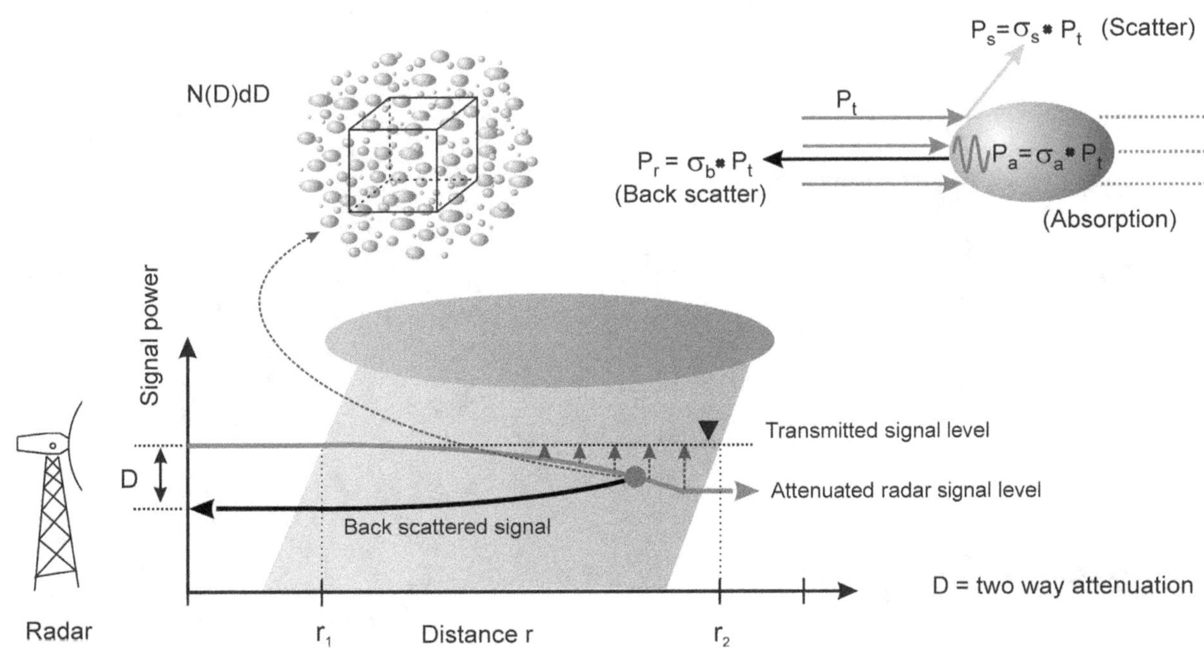

Figure 5.8 Attenuation of radar signal due to intervening rainfall (Krämer, 2008)

1. Clutter correction: radar- and catchment-specific algorithms are applied.
2. Wet-radome attenuation correction: the attenuation correction is calculated on the radar reflectivity field at the radar site at a distance of 4 km outside the blind range of the radar. The conversion between rainfall rate and water film thickness on the radome surface is estimated from in-situ measurements by Manz et al. (2001) and Kurri and Huskoonen (2008).
3. Attenuation correction of the radar signal along the signal path: the attenuation correction is based on the algorithm proposed by Hitschfeld and Bordan (1954). To avoid the known instability of the correction, variable attenuation coefficients are applied (Rahimi et al., 2005; Krämer and Verworn, 2009; Jacobi and Heistermann, 2015).
4. Conversion of radar reflectivity (Z) into rainfall intensity (R): following Steiner et al. (1998), R–Z relations that are characteristic for different rainfall processes (stratiform, convective rainfall) are assigned to the corrected radar reflectivity fields.
5. Spatiotemporal interpolation of 5-minute radar resolution to 1-min resolution using optical flow algorithms (Lucas and Kanade, 1981).
6. Transformation of radar data from polar into Cartesian coordinates.

These corrections are performed individually for each radar data set. To meet the requirements for real-time application of radar data in the field of urban drainage, all the corrections are performed within 1 minute. This is less than the update interval of radar data ($dt = 5$ min).

Figure 5.9 demonstrates the effect of radome and radar signal attenuation for a convective rainfall event in 2017 for the C-band radar Hannover-Langenhagen operated by the German Weather Service DWD (WMO-Code 10339). Figure 5.9a shows the uncorrected radar reflectivity data for a single-time stamp (PPI). Rainfall structures in light grey (>40 dBZ ≈ 10 mm/hr) indicate convective processes. Figure 5.9b shows the same time stamp but for corrected radar reflectivity. Areal differentiated corrections can be identified at locations of extreme rainfall.

In Figure 5.10, cumulated rainfall is depicted for the same event. The black triangles indicated rain gauges in the city of Hanover. Figure 5.10 shows rainfall for uncorrected radar data (a) and for corrected data (b). In Figures 5.10c and 5.10d, time series of gauge and radar observations, respectively, are given corresponding to the locations close to the catchment Oberricklingen marked by the circles in Figure 5.10b. Obviously, the uncorrected radar data underestimate the true rainfall. In contrast, the results for the corrected data indicate a fair agreement with the gauges. The attenuation correction leads to spatial differentiated increase in rainfall.

5.4.2 Radar Nowcasting

Rainfall forecasting based on radar data is often termed as radar nowcasting since the information content decreases rapidly from the "true" observation image up to the level of complete loss of information within 2–3 hours. The skill to predict rainfall depends on the several aspects:

- Prediction skill of the model: For radar nowcasting, there are models which use different types of input data and algorithms. A common approach is the radar tracking: relevant rainfall structures are identified in subsequent radar images (PPI), storm movement vectors are computed using

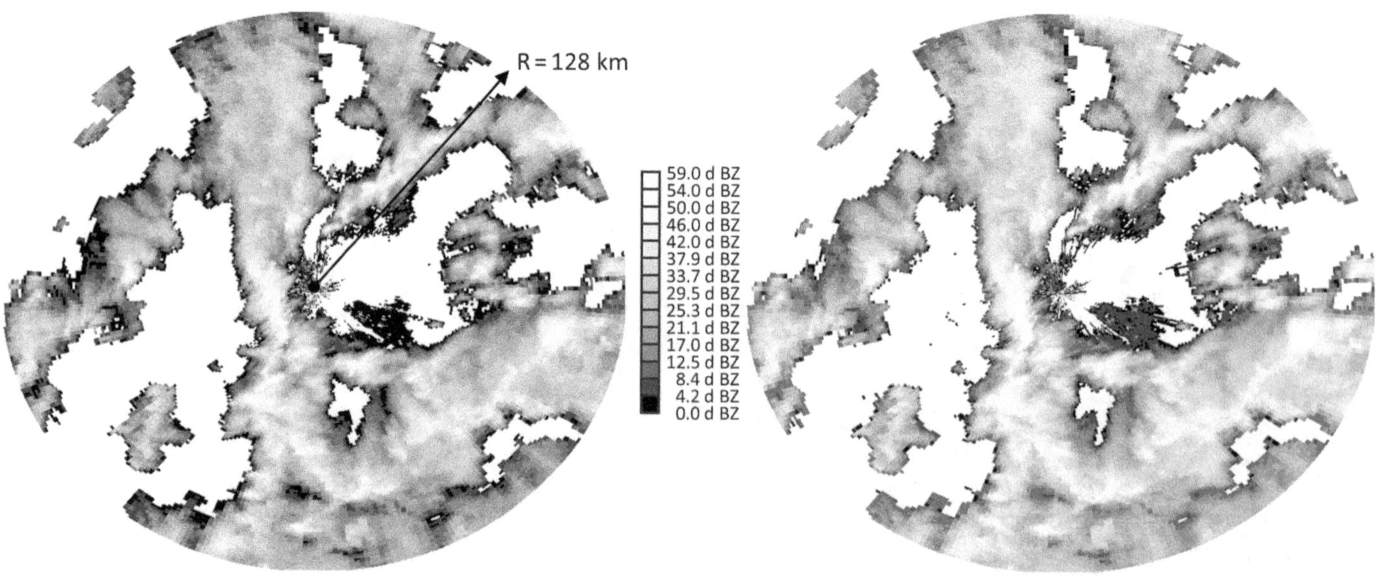

Figure 5.9 PPI, Radar Hannover Langenhagen ((a) uncorrected data; (b) corrected data for radome and signal attenuation)

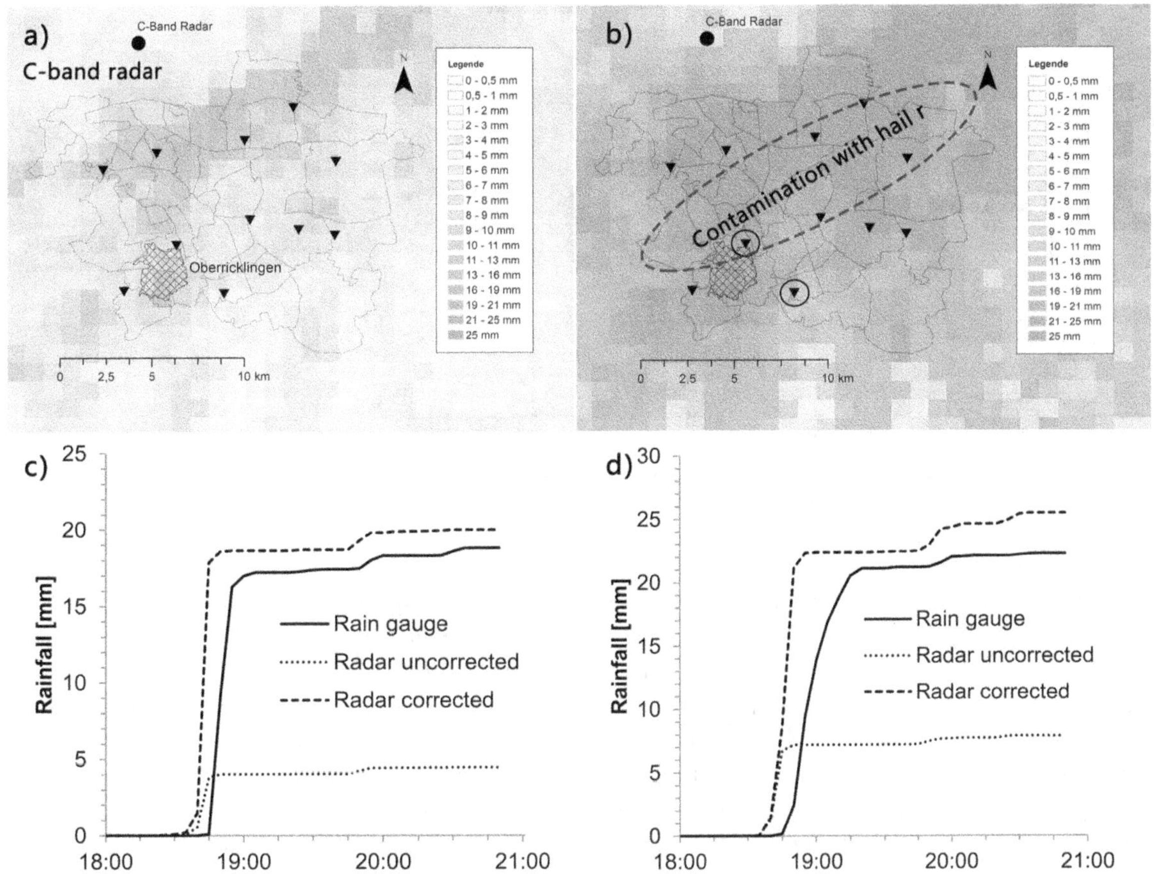

Figure 5.10 Cumulated rainfall for a convective event, (a) uncorrected radar rainfall, (b) corrected radar rainfall, (c) and (d) comparison of uncorrected and corrected radar rainfall with gauges (marked by circles in b)

correlations techniques, and the rainfall structures are advected in space with time. To meet operational requirements of urban drainage applications, the HyRaTrac model was developed (Krämer et al., 2007; Schellart et al., 2014). A central requirement is the computational performance since radar correction, nowcasting, and rainfall runoff simulations have to be performed within update times of radar data, for example, <5 minutes.

- Type of rainfall process: The prediction skill depends on the type of rainfall process and varies between several minutes in case of convective downbursts and several hours for frontal systems at the mesoscale (Han et al., 1998). Verworn and Krämer (2008) analysed the predictability of 42 rainfall events of a one-year-continuum relevant for combined sewer overflows. They found that 70% of the events were predictable until up to 30 minutes lead times within reliable bounds; 50% of the events showed reasonable lead times up to 60 minutes.
- Size of the target catchment: Large target areas (catchments) are more effective in filtering out small-scale variations and uncertainties in predicted precipitation fields than smaller target areas (Sempere-Torres et al., 2005; Rossa et al., 2011).
- Quality of radar data: The quality and the quantitative reliability of radar observations and derived data are essential for the quality of rainfall forecasts and subsequent rainfall runoff simulations and predictions (Ligouri et al., 2012).

Figure 5.11 demonstrates the effect of radar data correction on the predictability of the convective rainfall event shown in Figure 5.10 for the Oberricklingen catchment (size ≈ 6 km²) using the HyRaTrac model. The uncorrected data (Figure 5.11a) significantly underestimate the rainfall for the first peak for all lead times; with corrected data the prediction skill at lead time 20 minutes (Figure 5.11b) is much better than the observation for the uncorrected data when compared with the gauges. In addition, it can be recognised that the radar measurement in the atmosphere is 5–10 minutes before the gauge recordings at the ground.

5.4.3 Digital Terrain Model

An essential input to 2D hydrodynamic flood models is the surface terrain data describing the topography of the area subject to a flood hazard analysis. The quality in terms of accuracy and resolution of terrain data fundamentally determines the predictive

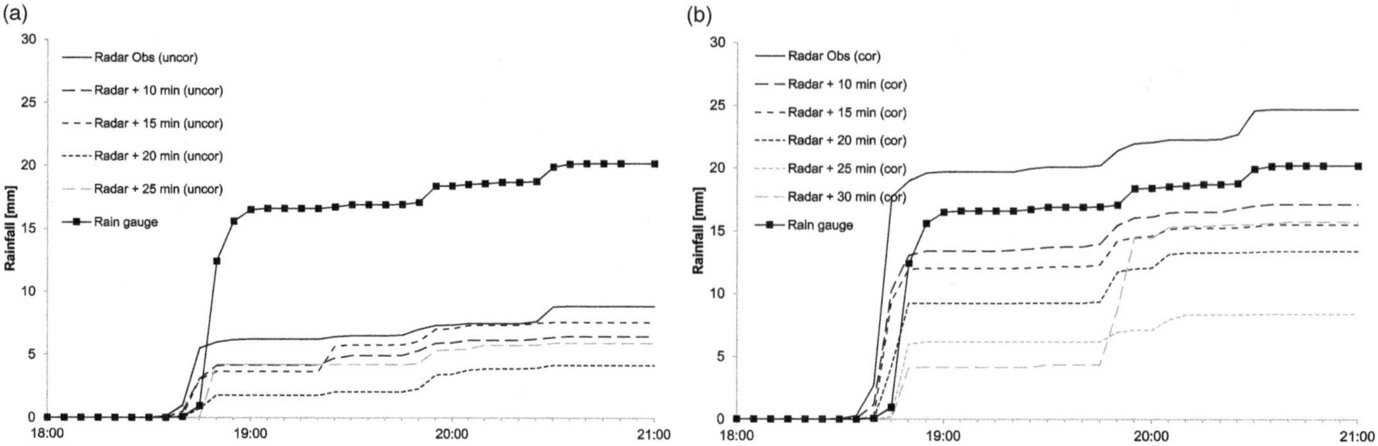

Figure 5.11 Forecast skill at different lead times using the HyRaTrac model for a convective rainfall event, (a): uncorrected radar data and (b): corrected data

capacity of flood models. Only if primary hydraulically relevant features of the surface are correctly represented in the data, reliable predictions of flow paths and inundation extends can be obtained (Casas et al., 2006; Leitão et al., 2009). Terrain data are commonly stored in the form of DTMs, which provide a continuous representation of the bare surface of the Earth, while excluding above-ground features such as vegetation, buildings and bridges (Mandlburger, 2015). Digital terrain models provide the basis for the setup of 2D hydraulic models and allow for a topographic analysis of preferential flow paths and for assessing the flood mitigation measures.

DTMs commonly have been provided in the form of raster-based datasets which are generated by interpolating preprocessed irregularly distanced terrain points into a structured grid. More recently triangulated irregular network (TIN)-based DTMs have become more widely available. These offer several advantages with regard to flood modelling in comparison to raster-based formats (Bates et al., 1996; Abdullah et al., 2013). These include a direct storage and thus a preservation of the exact location of spot heights and the possibility of incorporating breaklines for accurately representing line-shaped geometries and sharp relief changes. That is, local peaks or depressions can be preserved and elements such as curbs, walls, and river banks can be represented as continuous linear features. Another advantage of TINs is that the point density can be varied according to the topographic variability of the surface. Therefore, the original bulk of terrain points can be reduced to a subset of topographically critical points, leading to lower data sizes. Moreover, TINs can be readily visualised in three dimensions which can be helpful for an error identification based on visual inspection.

The accuracy of a DTM is determined by the techniques and methods employed for acquiring and treating the data. The quality of the terrain data is influenced by the accuracy of the survey device and the density and distribution of measurement points. Data for a DTM can be collected via ground surveys, photogrammetric techniques, or LiDAR surveys. Airborne LiDAR surveys, also referred to as airborne laser scanning (ALS), have evolved and are the most important method for a cost-efficient collection of high-resolution terrain data (Mandlburger, 2015). Countries such as Germany, Switzerland, and the Netherlands have conducted nationwide ALS surveys (Alho et al., 2009). Airborne laser scanning typically results in measurement spacings in the range of 0.1 m to 5 m, with a 0.3-m accuracy regarding location and a vertical accuracy of 0.15 m. Advances during the last years in ALS have increased the vertical accuracy to 0.05 m (Abdullah et al., 2013). More recently, vehicle-based laser scanning, also referred to as mobile mapping, has been used to obtain even higher point densities and accuracies (Jaakkola et al., 2008).

Before using the data from LiDAR surveys within a DTM, they require processing involving thinning, filtering, and interpolation. Thinning is used to reduce large amounts of points to more manageable volumes, by removing points without topographic significance. Filtering algorithms are applied for excluding noise, man-made objects, and vegetation.

For the purpose of flood modelling, DTMs often need to be modified so as to ensure a realistic representation of flow paths. This includes mainly the implementation of flow-relevant features such as buildings and walls, but concerns also stream with a potential contribution to rainfall-induced floods. Additionally, there might be artefacts from the filtering process within the DTM, which need to be removed (Abdullah et al., 2013). Particularly for larger domains, the areas which need to be corrected might only become apparent through an analysis of the 2D model's simulation results. Flow-relevant structures are either incorporated into the DTM via geoprocessing operations or directly implemented into the computational grid of the 2D model. In

2D models, a common approach is to represent buildings as gaps within the computational domain with a no-flow boundary condition along their outline (itwh, 2014).

5.5 CASE STUDY IN OBERRICKLINGEN (HANNOVER, GERMANY)

5.5.1 Site Description

We show the simulation of urban flood prediction of a small subcatchment in the city of Hanover in northern Germany (Figure 5.12). The geographical area has an extension of 6 km². A small hill is located on the west side. A receiving river is located at the east side of the domain. The surface slope is relatively flat with a value of 0.6% for surfaces and pipes. The pipe network is separated in sewer and drainage networks. Only the drainage network is part of the prediction model. It has many internal links to handle overload volumes and direct them to four outfalls into the river in the east. The drainage network of the subcatchment has a length of 53 km and is represented with 1220 elements. Length of pipes varies between 1 and 130 m. The study area can be characterised as a residential area with a relatively low percentage of paved surfaces, but it also includes industrial sites in the north with a higher degree of imperviousness. Green areas cover more than 50% of the study area.

The subsurface of the test area consists of sandy gravel material with clay lenses in the eastern part and fine sandy material in the western part. The hill on the western side acts as a groundwater divide, and groundwater flow is from west towards the river

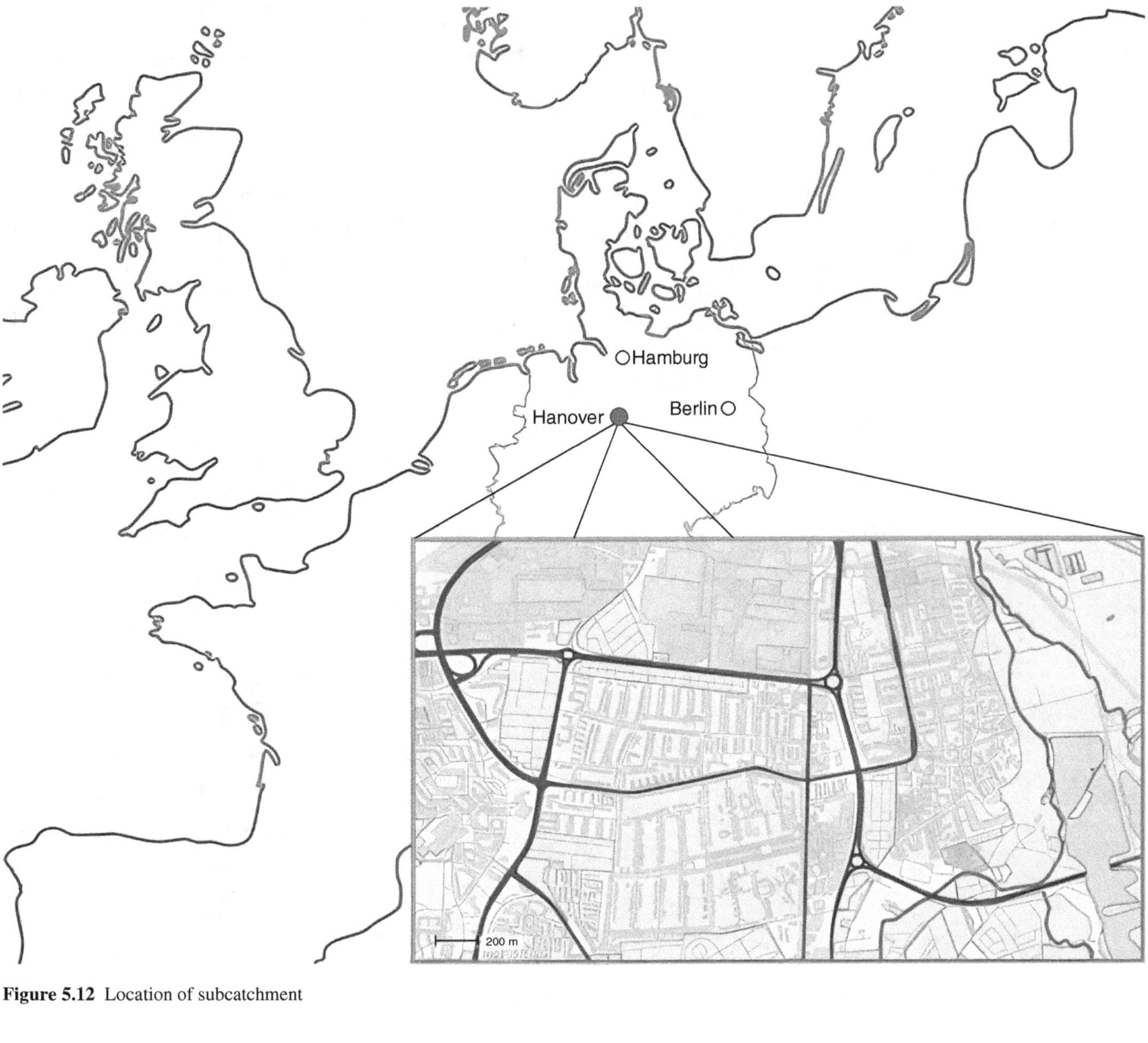

Figure 5.12 Location of subcatchment

in the east. The river is in an effluent connection with groundwater (flow from the groundwater into the river). With regard to the pipe network, the pipes located in the eastern part of the test area are predominantly below the groundwater surface, while the pipes located in the western part are above the groundwater surface.

An artificial Euler type-2 design storm with a return period of 50 years is used to demonstrate the complexity of the flow pattern. Spatially constant rain is applied over a time span of 1 hour in 5-minute resolution with a total volume of 29.2 mm/m². The maximum peak of 17 mm occurs 15 minutes after the onset of the rain. Spill-outs from drainage to surface already happen at return periods of 5 years in this area, but ponds are small and transport paths on the surface are not crucial.

5.5.2 Water Flow

5.5.2.1 SURFACE FLOW

The 2D surface flow model covers an area of 5.3 km² discretised into an unstructured grid of approximately 1.2 million triangular elements. The mean triangle size is 4 m². Areas with greater topographical variability are generally represented by a finer resolution, as the element size was generated in dependence of the surface terrain gradient. The mesh is constructed based on a merged DTM including terrain data from car-based laser scanning with a resolution of 0.1 m for street areas and ALS with a resolution of 0.5 m for the other areas (Figure 5.13). The DTM was modified in order to ensure a hydraulic connectivity of flood-relevant structures such as underpasses.

Buildings are cutout of the computational domain and are represented by internal, no-flow boundary conditions. The effective

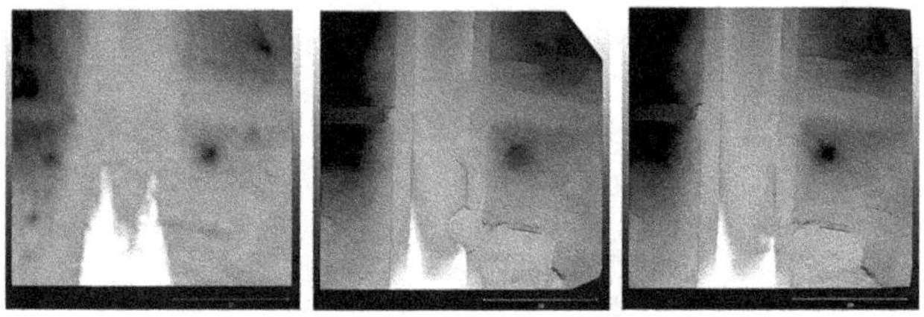

Figure 5.13 0.5-m resolution DTM from ALS, 0.1-m resolution DTM from car-based laser scanning and merged DTM (from left to right)

Figure 5.14 Surface water depths and degree of affectedness of buildings for a 1-hour rainfall of a 50-year return period

rainfall amounts at building areas are directly conveyed to the sewer network via a linear reservoir model. The percentage of roof area connected to the sewer network, which was supplied by the local drainage department, is accounted for by a constant runoff coefficient. For the other areas (e.g., streets and green areas), the surface flow model receives net rainfall inputs from the rainfall runoff model which accounts for the land use of the site. Surface areas are classified according to their hydrological characteristics in paved and unpaved areas and according to their roughness in terms of a spatially varying Manning–Strickler roughness coefficient. The surface model is coupled to the sewer network model via storm inlets and manholes. For inlets, the constant maximum inflow capacity of 10 l/s is assumed.

In order to reduce the required simulation time, building polygons, which are incorporated into the computational grid, are generalised to avoid unnecessarily small triangular elements. The Douglas–Peucker algorithm was applied to remove insignificant vertices while preserving the general geometry of the polygons. Furthermore, building polygons adjacent to each other were merged and small gaps within polygons were filled.

For a 1-hour design storm with a return period of 50 years, the simulation takes approximately 1 hour on a standard computer. The corresponding flood depths of this event are displayed in Figure 5.14. The affectedness of buildings is determined by intersecting the water depths with the buildings. These results supply the basis for a quantitative empirical loss model for determining damage potentials.

The physically based model is applied to generate the training data set for the fast flood forecasting model consisting of ANN.

5.5.2.2 FAST WATER LEVEL PREDICTION

Berkhahn et al. (2017) presented the first results of fast prediction of inundated areas using an ANN for synthetic rain events. The results have shown a good representation of the hotspots and overall a small error in the test area. The computation time for flood simulation in a medium-sized city was under one minute. This means that the calculation time is within the necessary range for a real-time prediction.

5.5.2.3 PIPE FLOW AND SUBSURFACE FLOW COUPLING

In this case study, we show exemplary results from a series of pipe leakage simulations with the coupled model. The pipe network domain and subsurface mesh are visualised in Figure 5.15.

Initial condition of all simulations was the subsurface pressure distribution of a calibrated groundwater model of Oberricklingen and a pipe water level of zero in all pipes. A hypothetical statistical standard defect of the size $A_{defect} = 2.25 \times 10^{-4}$ m² (originally calculated for the city of Dresden by Karpf (2012)) was distributed over each metre of pipe in the network. Pipe leakage was simulated totally for 12

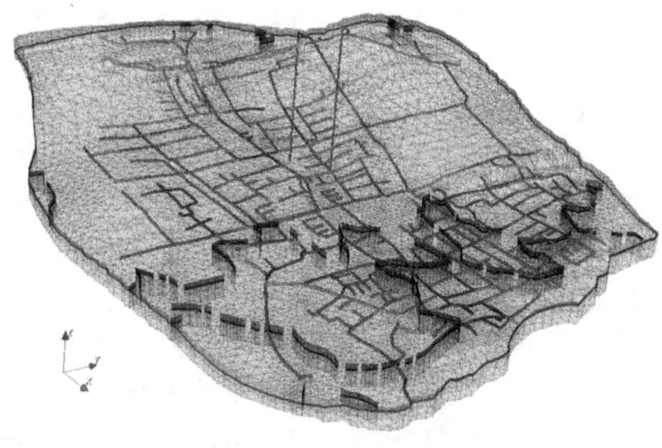

Figure 5.15 Pipe network domain and subsurface mesh

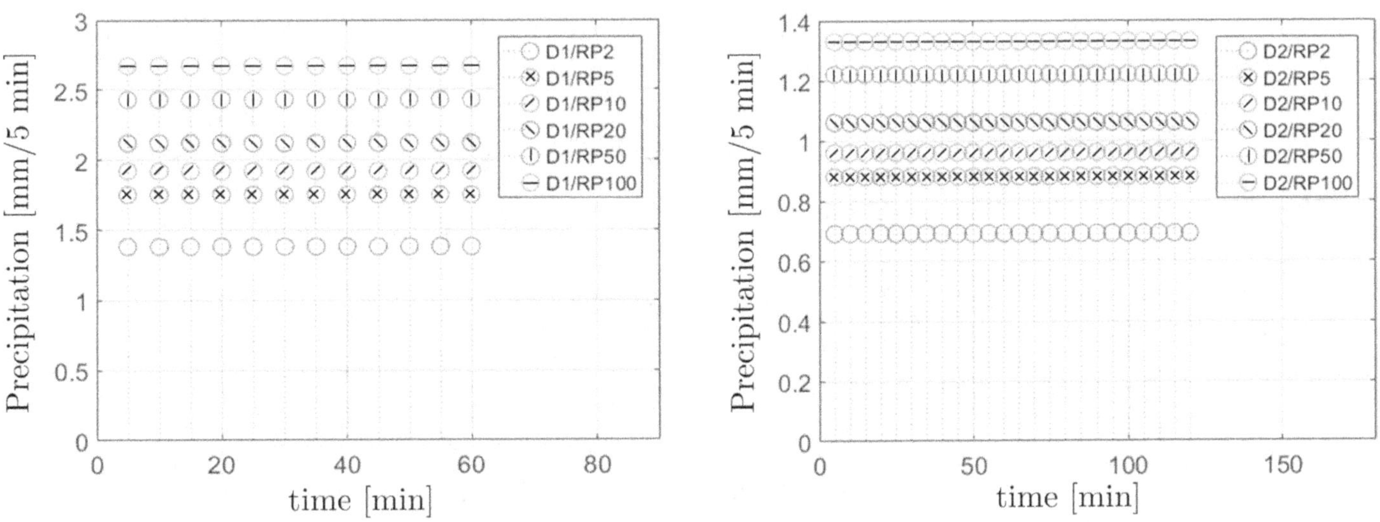

Figure 5.16 Block rain events used for the case study of pipe flow and subsurface flow. Events have durations of 1 hour (D1) and 2 hours (D2) and return periods of 2 years (RP2), 5 years (RP5), 10 years (RP10), 20 years (RP20), 50 years (RP50), and 100 years (RP100)

synthetic block rain events with the durations of 1 and 2 hours and the return periods between 2 and 100 years. All simulations were terminated 30 minutes after the end of each rain event. All block rain events are depicted in Figure 5.16.

Results of this case study in form of cumulative pipe exfiltration volume $V_{\text{leak, EXF}}$ [L^3] and infiltration volume $V_{\text{leak, INF}}$ [L^3] are given in Table 5.1. Despite of the fact that 70% of pipes in the test area are above the groundwater table (and thus prone to pipe water exfiltration), infiltration is significantly higher than exfiltration in all scenarios. This is due to the absence of a colmation layer below the pipe rupture (Karpf, 2012) and due to the fully saturated soil in the pipe vicinity under groundwater infiltration conditions. While leakage in the form of pipe exfiltration increases with the increasing return period and duration, the leakage in the form of groundwater infiltration shows the reverse behaviour. Explanations about these mechanisms are given by Peche et al. (2017).

Figure 5.17 illustrates the dynamic leakage at two exemplary pipes for the rain event D2/RP100. In Figure 5.16, the values for pipe water exfiltration $Q_{\text{leak, EXF}}$ are positive and the values for groundwater infiltration $Q_{\text{leak, INF}}$ are negative, which is due to the respective assignation as a source/sink term in the groundwater model ($Q_{\text{leak, EXF}}$ as a source and $Q_{\text{leak, INF}}$ as a sink). Clearly, leakage in both cases correlates qualitatively with the temporal pattern of the rain event. Groundwater infiltration $Q_{\text{leak, INF}}$ exceeds pipe water exfiltration $Q_{\text{leak, EXF}}$ by two orders of magnitude.

Table 5.1 *Cumulative leakage volume for block rain events of different duration and return period.*

$V_{\text{leak,INF}}$ [m^3]		return period [a]					
$V_{\text{leak,EXF}}$ [m^3]		2	5	10	20	50	100
duration [h]	1	0.0447	0.0441	0.0439	0.0436	0.0434	0.0427
	1	0.0150	0.0194	0.0227	0.0257	0.0277	0.0338
	2	0.0714	0.0706	0.0705	0.0704	0.0702	0.0699
	2	0.0109	0.0222	0.0231	0.0243	0.0263	0.0280

5.5.3 Contaminant Transport

We show the simulation of prediction of contaminant transport in the urban subcatchment. Velocity fields in pipes and on the triangle surface grid are calculated with HE2D and stored in a database with time resolution of 5 minutes. For the particle transport model, velocities are read in from the database and interpolated linearly for every necessary time step. Time integration is carried out with an explicit Euler scheme.

To show uncertainties in dynamic flow field, the spill time of a contaminant is varied from 0 to 30 minutes after rain start in three injection steps. Each injection lasts 1 minute followed by a 14-minute pause. These pulses make contaminated areas easier to distinguish in a transient simulation.

A spilled contaminant does not follow an a priori suspected path estimated based on a gravity-driven drainage system. Due to hydraulic shortcuts in the pipe system and on the surface, the spilled solute does not follow the gravity-based direction (Figure 5.18).

Directly at the location of the spill, the contaminant is directed to the drainage system. At many positions downstream, the pollutant spills out to the surface and moves on (Figure 5.19). When drained back to the pipe system, new pipes are contaminated even though they are not connected to polluted water from upstream pipes.

Spilled contaminants do not affect all surface areas with the same intensity. The spreading area is highly sensitive to the

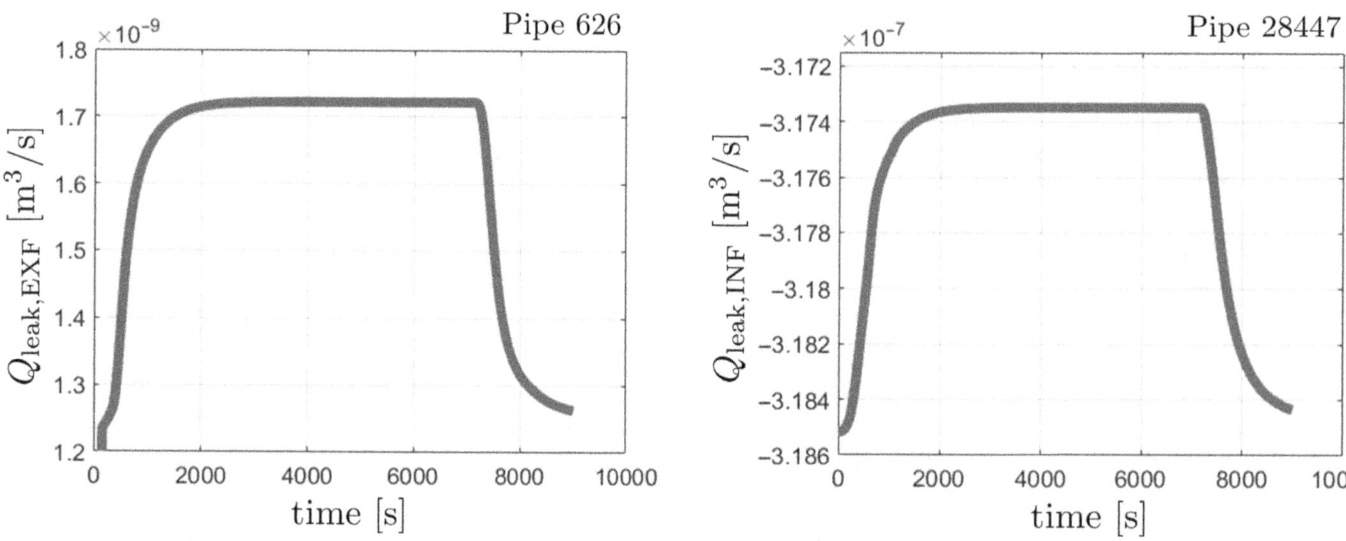

Figure 5.17 Leakage flow over time at two exemplary pipes

Figure 5.18 Estimated and predicted contaminant routing

Figure 5.19 Contamination prediction on the surface

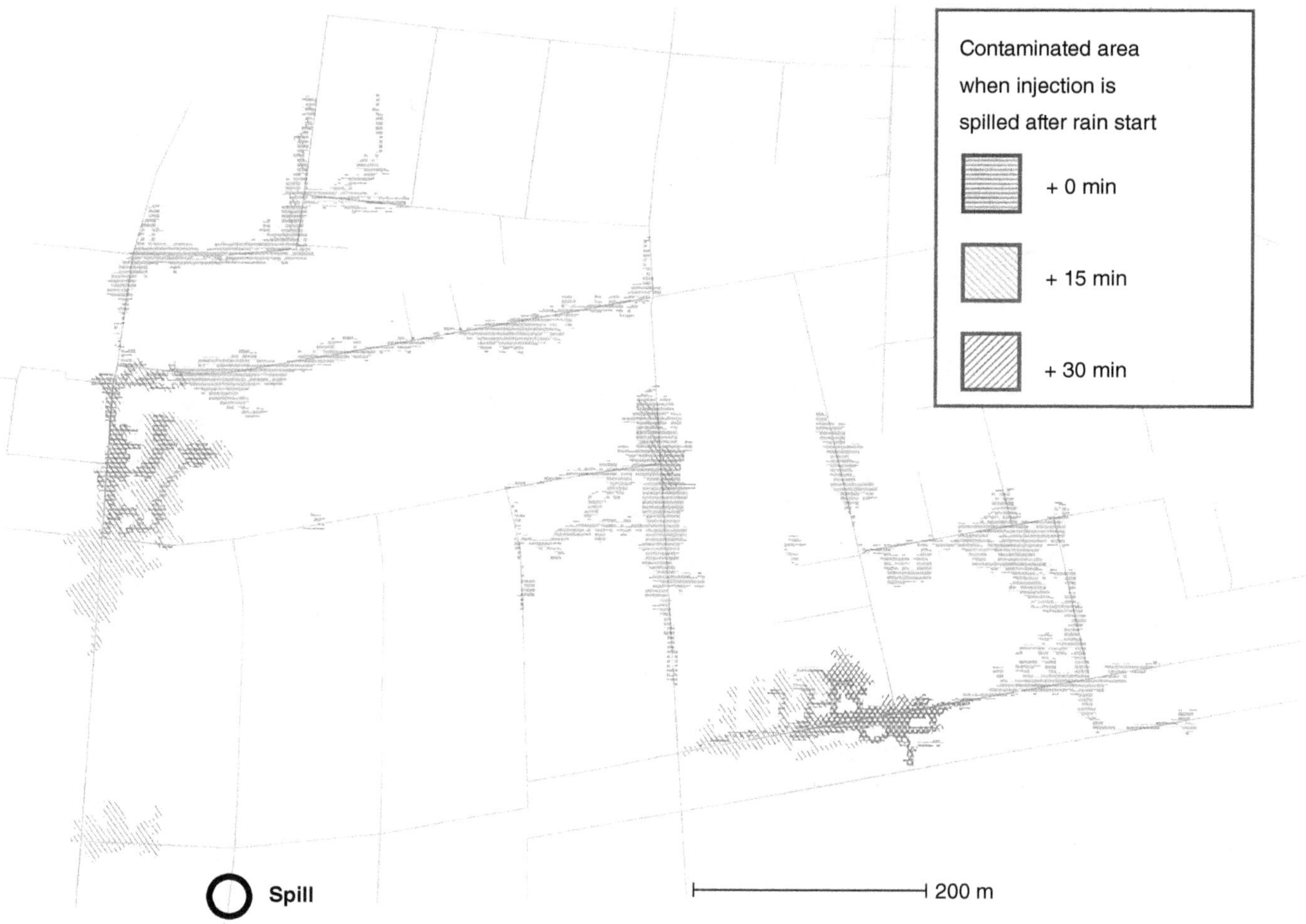

Figure 5.20 Superposition of surface paths with different contamination injection time

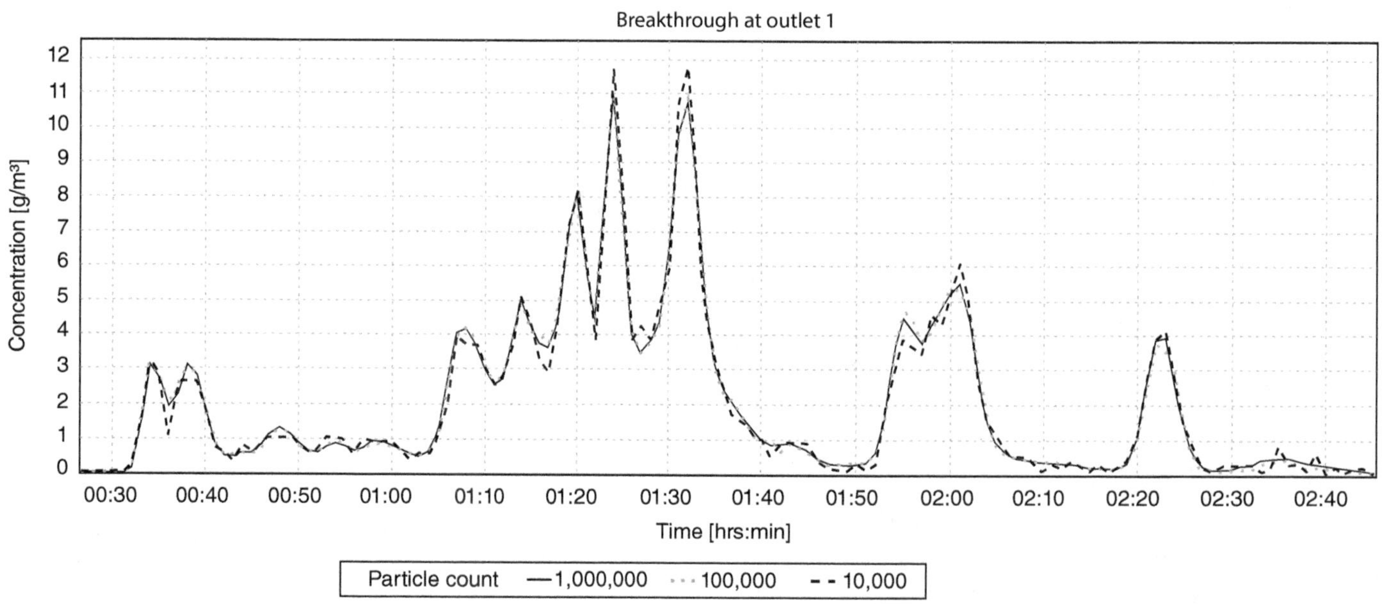

Figure 5.21 Concentration for different stages of quality over time at southern outlet 1

injection time. To show the influence of varying spill time, different patterns of transport paths on the surface of contaminants with different injection times are superimposed on the map in Figure 5.20. First flush pollution is spilled out at positions with larger distance (horizontal pattern). Contaminants, spilled with the peak flow, are flushed to the surface earlier (decreasing diagonals pattern). Areas affected by both are coloured in yellow. Injections after 30 minutes (increasing diagonal pattern) appear only rarely in all time flooded areas and are therefore coloured in white.

The computational time varies with the number of particles and is therefore dependent on the quality requirement. Figure 5.21 shows breakthrough curves of contaminants at the outfall to receiving waters generated from three simulations with identical flow dynamic but different number of particles (10,000/100,000/1,000,000 particles). Using 1,000,000 particles as the reference simulation, decreasing number of particles leads to differences in particle movement due to the random walk scheme. Figure 5.21 shows the breakthrough curve at the southern outlet 2. Difference of factor 10 between 1,000,000 and 100,000 particles results in an error of 0.0017 for total particle flow at the end of simulation (not shown). Concentration peaks at the time of maximum concentration (at time 01 h 24 min) slightly differ with a relative deviation of 0.003. Decreasing number of particles to 10,000 results in a deviation of 0.005 for flow and 0.06 in peak concentration.

Without particle interaction, the computational effort increases linearly with the number of particles. On a multicore desktop PC with 3.6 GHz, this 3-hour scenario requires 10 minutes for 1 million particles, 1 minute for 100,000, and 6 seconds for 10,000 particles when the simulation time step is 1 second.

5.6 CONCLUSIONS AND OUTLOOK

The chapter discusses the need of early warning systems to predict pluvial flood events in urban environments. Relevant hydraulic interactions of the urban hydraulic flow system are described. Urban areas are threatened not only by pluvial floods caused by heavy rainfall but also by the subsequent release and transport of hazardous substances. Existing numerical models are discussed that attempt to address the problem of flow simulation and contaminant transport in an urban area. To continue this field of research, the chapter also presents a fully coupled urban flow model along with a transport model to assess the migration of contaminants during and after a pluvial flood event.

Hydrodynamic urban flood models, consisting of a 2D overland flow model coupled to a 1D sewer network flow model, are considered to be the best practice methods for performing flood hazard analyses with regard to pluvial floods in urban areas (DWA, 2017). Although the underground drainage system has a decreasing influence on the surface flood dynamics with increasing return period of the rainfall event, neglecting the pipe system can result in significant errors of the predicted flood extent and intensity (Fuchs et al., 2015). Given the commonly short lead times of these event on the one hand and the high computational burden associated to these models on the other, application of 2D hydraulic models within forecasting systems for rainfall-induced floods is not yet possible. Instead, using these models for generating the data for fast data-driven models, currently seems to be the best approach to the forecasting of pluvial floods. When properly corrected, radar data can supply an adequate representation of the temporal and spatial variability of high-intensity rainfall events.

In potentially defect pipe networks, the upscaling with the LF enables a significant reduction in computation time with only little drawback in result accuracy. The LF approach was successfully applied to our test area (on the scale of an urban catchment), where simulations are based on block rains with different durations and return periods. Results show that groundwater infiltration significantly exceeds pipe water exfiltration despite the fact that only 30% of pipes are prone to groundwater infiltration in our test area. Large return periods and small durations result in largest pipe water exfiltration. In case of contaminated pipe water, these events might lead to largest subsurface contamination. In future work, we plan to implement a further coupling from HE2D surface flow to groundwater flow and we plan to implement the coupled model for the entire city of Hannover.

Coupling of Eulerian and Lagrangian methods is an efficient opportunity to speed up forecasting processes. Ensembles of possible flow fields can generate a range of areas threatened by contamination. Results can be generated within minutes on city-scale areas based on the spatial and temporal resolutions and the number of used particles. As the selection of flow fields from the ensemble of historical data cannot represent reality, one has to keep in mind that results have to be considered as worst-case scenarios.

The presented model for water level prediction was successfully tested for the test case in the urban subcatchment. As shown in Berkhahn et al. (2017), the ANN-based model is suitable for spatially uniform rain events. The RMSE for a test set of 14 synthetic rain events was 0.56 centimetre. The computation time for a simulation was about 25 seconds on a desktop PC. Further investigations have to be conducted on real rain events and spatially distributed rain events. The model has to be tested for different topographies, for example, steep slopes. For the test area, the gap between accuracy and computation time was closed with the presented ensemble ANN approach.

References

Abderrezzak, K. E. K, Paquier, A., Mignot, E. (2009). Modelling flash flood propagation in urban areas using a two-dimensional numerical model. *Natural Hazards*, 50, 433–460.

Abdullah, A. F. B. (2013). A Methodology for Processing Raw LiDAR Data to Support Urban Flood Modelling Framework: UNESCO-IHE, Aronica, G. T., Franza, F., Bates, P. D., Neal, J. C. (2012). Probabilistic evaluation of flood hazard in urban areas using Monte Carlo simulation. *Hydrological Processes*, 26, 3962–3972.

Achleitner, S., Fuchs, S., Einfalt, T., Rauch, W. (2009). Nowcasting of rainfall and combined sewage flow in urban drainage systems. *Water Science and Technology*, 59 (6), 1145–1151.

Alho, P., Hyyppä, H., Hyyppä, J. (2009). Consequence of DTM precision for flood hazard mapping: A case study in SW Finland. *Nordic Journal of Surveying and Real Estate Research*, 6 (1), 21–39.

Aronica, G. T., Franza, F., Bates, P. D., Neal, J. C. (2012). Probabilistic evaluation of flood hazard in urban areas using Monte Carlo simulation. *Hydrological Processes*, 26, 3962–3972.

Banton, O., Delay, F., Porel, G. (1997). A new time domain random walk method for solute transport in 1-D heterogeneous media. *Ground Water*, 35 (6), 1008–1013.

Basnayaka, A. P., Sarukkalige, P. R. (2013). Impact of urbanization on flood vulnerability in shallow groundwater catchment. In A. Chavoshian and K. Takeuchi (eds.), *Floods – From Risk to Opportunity*, 255–262. IAHS Press, Wallingford, Oxfordshire, OX10 8BB, UK.

Bates, P. D., Anderson, M. G., Price, D. A., Hardy, R. J., Smith, C. N. (1996). Analysis and development of hydraulic models for floodplain flow. In M. G. Anderson, D. E. Walling, and P. D. Bates (eds.), *Floodplain Processes*, 215–254. John Wiley & Sons Inc.

Berkhahn, S., Fuchs, L., Neuweiler, I. (2017). Artificial neural network for two-dimensional urban flood simulation and real-time forecasting. Proc. 14th ICUD, Prague.

Berne, A., Delrieu, G., Creutin, J., Obled, C. (2004). Temporal and spatial resolution of rainfall measurements required for urban hydrology. *Journal of Hydrology*, 299, 166–179.

Beyer, K.-D. (2003). Erhalt der Gebäudesicherheit – Sofortmaßnahmen und Dauerlösung Beispiel St.-Benno Gymnasium Dresden. In: LH DD and DGFZ (2003): Hochwassernachsorge Grundwasser Dresden. Tagungsband zum Statusseminar am 08.10.2003, 63–68.

Blunt, M. J., Bijeljic, B., Dong, H., Gharbi, O., Iglauer, S., Mostaghimi, P., Pentland, C. (2013). Pore-scale imaging and modelling. *Advances in Water Resources*, 51, 197–216.

Boonya-Aroonnet, S., Maksimovic, C., Prodanovic, D., Djordjevic, S. (2007). Urban pluvial flooding: development of GIS based pathway model for surface flooding and interface with surcharged sewer model. NOVATECH 2007.

Buchholz, O., Schwanenberg, D., Sommer, B. (2013). Anforderungen an moderne Hochwasser-Vorhersagesysteme – Beispiele für die Umsetzung mit Delft-FEWS [eng.: Requirements of modern flood prediction systems – Examples for the implementation with Delft-FEWS]. 36. Dresdner Wasserbaukolloquium 2013.

Bustamante, M. L. (2007). *Learning WCF: A Hands-on Guide*. O'Reilly Media, Inc.

Cambez, M. J., Pinho, J., David, L. M. (2008). Using SWMM 5 in the continuous modelling of stormwater hydraulics and quality. In 11th International Conference on Urban Drainage, Edinburgh, Scotland, UK.

Casas, A., Benito, G., Thorndycraft, V. R., Rico, M. (2006). The topographic data source of digital terrain models as a key element in the accuracy of hydraulic flood modelling. *Earth Surface Processes and Landforms*, 31 (4), 444–456.

CRED (2013). www.emdat.be/

Dehghan, M. (2004). Numerical solution of the three-dimensional advection-diffusion equation. *Applied Mathematics and Computation*, 150 (1), 5–19.

Delay, F., Ackerer, P., Danquigny, C. (2005). Simulating solute transport in porous or fractured formations using random walk particle tracking. *Vadose Zone Journal*, 4 (2), 360–379.

Duncan, A., Chen, A. S., Keedwell, E., Djordjevic, S., Savic, D. (2011). Urban flood prediction in real-time from weather radar and rainfall data using artificial neural networks.

DWA (2016). Merkblatt DWA-M 119 – Risikomanagement in der kommunalen Überflutungsvorsorge für Entwässerungssysteme bei Starkregen [eng.: Pamphlet DWA-M 119 – Risk managemnt in the municipal flood prevention for drainage systems during heavy rainfall events]. DWA, Hennef.

Egodawatta, P., Thomas, E., Goonetilleke, A. (2007). Mathematical interpretation of pollutant wash-off from urban road surfaces using simulated rainfall. *Water Research*, 41 (13), 3025–3031.

Fuchs, L., Beeneken, T. (2005). Development and implementation of real-time control strategy for the sewer system of the Vienna City. *Water Science and Technology*, 52 (5), IWA Publishing, London.

Fuchs, L., Schmidt, N. (2015). Comparison of three different modelling approaches for the simulation of flooding in urban areas. 10th International Urbane Drainage Modelling Conference, Québec, Canada

GDV (2011). Gesamtverband der Deutschen Versicherungswirtschaft e. V. Auswirkungen des Klimawandels auf die Schadensituation in der deutschen Versicherungswirtschaft [eng.: Impacts of climate change on the damage situation in the German assurance economy].

Gires, A., Onof, C., Maksimović, Č., Schertzer, D., Tchiguirinskaia, I. and Simões, N. (2012). Quantifying the impact of small scale unmeasured rainfall variability on urban hydrology through multifractal downscaling: A case study. *Journal of Hydrology*, 442, 117–128.

Göbel, P., Stubbe, H., Weinert, M., Zimmermann, J., Fach, S., Dierkes, C., Kories, H., Messer, J., Mertsche, V., Geiger, W. F., Coldeweya, W. G. (2004). Near-natural stormwater management and its effects on the water budget and groundwater surface in urban areas taking account of the hydrogeological conditions. *Journal of Hydrology*, 299, 267–283.

Grischek, T., Nestler, N., Piechniczek, D., Fischer, T. (1996). Urban groundwater in Dresden, Germany. *Hydrogeology Journal*, 4, 48–63.

Grünewald, U. (2009). Gutachten zu Entstehung und Verlauf des extremen Niederschlag-Abfluss-Ereignisses am 26.07.2008 im Stadtgebiet von Dortmund, Web-Adresse: http://tiefbauamt.dortmund.de/upload/binarydata_do4ud4cms/46/32/18/00/00/00/183246/gutachten_neu.pdf.

Han, D., Griffith, R. J., Cluckie, I. D. (1998). Radar based rainfall forecasting. In: Radar hydrology for real time flood forecasting. Proc. of the Advanced Study Course, University of Bristol, June 24th–July 3rd. 1998

Hansen, L. K., Liisberg, C., Salamon, P. (1992). Ensemble methods for handwritten digit recognition. In Neural Networks for Signal Processing [1992], Proceedings of the 1992 IEEE-SP Workshop (pp. 333–342). IEEE.

Harder-Lauridsen, N. M., Kuhn, K. G., Erichsen, A. C., Mølbak, K., Ethelberg, S. (2013). Gastrointestinal illness among triathletes swimming in non-polluted versus polluted seawater affected by heavy rainfall, Denmark, 2010–2011. *PloS one*, 8 (11), e78371.

Henonin, J., Russo, B., Mark, O., Gourbesville, P. (2013). Real-time urban flood forecasting and modelling–a state of the art. *Journal of Hydroinformatics*, 15 (3), 717–736.

Hitschfeld, W., Bordan, J. (1954). Errors inherent in the radar measurement of rainfall at attenuating wavelengths. *Journal of Meteorology*, 11, 58–67.

Hunter, N. M., Bates, P. D., Neelz, S., Pender, G., Villanueva, I., Wright, N. G., Liang, D., Falconer, R. A., Lin, B., Waller, S., Crossley, A. J., Mason, D. C. (2008). Benchmarking 2D hydraulic models for urban flooding. *Water Management*, 161, 13–30.

hvz.baden-württemberg.de

itwh (2014). Überflutungsberechnung; HYSTEM-EXTRAN 2D Modellbeschreibung, Institut für technisch-wissenschaftliche Hydrologie GmbH Hannover.

Jaakkola, A., Hyyppä, J., Hyyppä, H., Kukko, A. (2008). Retrieval algorithms for road surface modelling using laser-based mobile mapping. *Sensors* 8 (9), 5238–5249.

Jacobi, S., Heistermann, M. (2016). Benchmarking attenuation correction procedures for six years of single-polarized C-band weather radar observations in South-West Germany. *Geomatics Natural Hazards & Risk*, 7(6), 1785–1799.

Karpf, C., Krebs, P. (2004). Sewers as drainage systems – quantification of groundwater infiltration. Proceedings 5th international conference NOVATECH, Lyon, 969–976.

Karpf, C., Krebs, P. (2013). Modelling of groundwater infiltration into sewer systems. *Urban Water Journal*, 10 (4), 221–229. doi:10.1080/1573062X.2012.724077.

Kolditz, O., Bauer, S., Bilke, L., Böttcher, N., Delfs, J. O., Fischer, T., … Zehner, B. (2012). OpenGeoSys: An open-source initiative for numerical simulation of thermo-hydro-mechanical/chemical (THM/C) processes in porous media. *Environmental Earth Sciences*, 67 (2), 589–599. https://doi.org/10.1007/s12665-012-1546-x.

Krajewski, W. F., Smith, J. (2002). Radar hydrology: Rainfall estimation. *Advances in Water Resources*, 25, 1387–1394.

Krämer, S. (2008). Quantitative Radardatenaufbereitung für die Niederschlagsvorhersage und die Siedlungsentwässerung [eng.: Quantitative radar data processing for rainfall forecasting and urban drainage] – Mitteilungen des Institutes für Wasserwirtschaft, Hydrologie und landwirtschaftlichen Wasserbau, Universität Hannover, Heft 92.

Krämer, S., Fuchs, L., Verworn, H.-R. (2007). Aspects of radar rainfall forecasts and their effectiveness for real time control – the example of the sewer system of the city of Vienna, *Water Practice and Technology*, 2 (2), 2007 – doi10.2166/wpt.2007.0042.

Krämer, S., Verworn, H.-R. (2009). Improved radar data processing algorithms for quantitative rainfall estimation in real time. *Water Science Technology*, 60, 175–184.

Kurri, M., Huskoonen, A. (2008). Measurements of the transmission loss of a radome at different rain intensities. *Journal of Atmospheric and Oceanic Technology*, 25, 1590–1599.

Laszewski, T., Nauduri, P. (2011). *Migrating to the Cloud: Oracle Client/Server Modernization*. Elsevier, Waltham, MA.

Leandro, J., Chen, A. S., Djordjević, S., Savić, D. A. (2009). Comparison of 1D/1D and 1D/2D coupled (sewer/surface) hydraulic models for urban flood simulation. *Journal of Hydraulic Engineering*, 135 (6), 495–504.

Leitão, J. P., Boonya-Aroonnet, S., Prodanović, D., Maksimović, C. (2009). The Influence of digital elevation model resolution on overland flow networks for modelling urban pluvial flooding. *Water Science and Technology*, 60 (12), 3137–3149.

Lerner, D. N. (1986). Leaking pipes recharge groundwater. *Ground Water*, 24 (5), 654–662.

Lerner, D. N. (2002). Identifying and quantifying urban recharge: A review. *Hydrogeology Journal*, 10, 143–152.

Li, L. Q., Yin, C. Q., He, Q. C., Kong, L. L. (2007). First flush of storm runoff pollution from an urban catchment in China. *Journal of Environmental Sciences*, 19 (3), 295–299.

Liang, Q., Xia, X., Hou, J. (2015). Efficient urban flood simulation using a GPU-accelerated SPH model. *Environmental Earth Sciences*, 74 (11), 7285–7294.

Liguori S., Rico-Ramirez, M. A., Schellart, A. N. A., Saul, A. J. (2012). Using probabilistic radar rainfall nowcasts and NWP forecasts for flow prediction in urban catchments. *Atmospheric Research*, 103, 80–95.

Löwe, R., Thorndahl, S., Mikkelsen, P. S., Rasmussen, M. R., Madsen, H. (2014). Probabilistic online runoff forecasting for urban catchments using inputs from rain gauges as well as statistically adjusted weather radar. *Journal of Hydrology*, 512, 397–407.

Lucas, B. D., Kanade, T. (1981). An iterative image registration technique with an application to stereo vision. Proceedings of Imaging understanding workshop, 121–130.

Mandlburger, G. (2015). Topographische Modelle für Anwendungen in Hydraulik und Hydrologie: Aufbau und Aufbereitung Digitaler Wasserlauf-Geländemodelle [eng.: Topographic models for applications in hydraulics and hydrology: design and preprocessing of digital waterway elevation models]. Südwestdeutscher Verlag Für Hochschulschriften Ag.

Mansell, M., Wang, S. (2010). Water balance modelling in Glasgow and Beijing. *Proceedings of the Institution of Civil Engineers: Water Management*, 163 (5), 219–226.

Manz, A. (2001). Radome influence on weather radar systems principles and calibration issues, Radar Calibration and Validation, Specialty Meeting, Amer. Met. Soc., Albuquerque, New Mexico.

Marre, D., Walther, W., Ullrich, K. (2005). Influence of the flood 2002 on groundwater chemistry in Dresden, Germany (in German). *Grundwasser*, 10, 146–156.

McCulloch, W. S., Pitts, W. (1943). A logical calculus of the ideas immanent in nervous activity. *The Bulletin of Mathematical Biophysics*, 5 (4), 115–133.

Nguyen, H. Q., Radhakrishnan, M., Huynh, T. T. N., Baino-Salingay, M. L., Ho, L. P., Steen, P. V. D., Pathirana, A. (2017). Water quality dynamics of urban water bodies during flooding in can Tho City, Vietnam. *Water*, 9 (4), 260.

Niu, J., Phanikumar, M. S. (2015). Modeling watershed-scale solute transport using an integrated, process-based hydrologic model with applications to bacterial fate and transport. *Journal of Hydrology*, 529, 35–48.

Nix, S. J. (1994). *Urban Stormwater Modeling and Simulation*. CRC Press.

Pathirana, A., Maheng Dikman, M., Brdjanovic, D. (2011, September). A Two-dimensional pollutant transport model for sewer overflow impact simulation. In Proceedings: 12th International Conference on Urban Drainage, Porto Alegre/Brazil (pp. 10–15).

Peche, A., Graf, T., Fuchs, L., Neuweiler, I. (2017). A coupled approach for the three-dimensional simulation of pipe leakage in variably saturated soil. *Journal of Hydrology*, doi.org/10.1016/j.jhydrol.2017.10.050.

Pedersen, L., Jensen, N. E., Christensen, L. E., Madsen, H. (2010). Quantification of the spatial variability of rainfall based on a dense network of rain gauges. *Atmospheric Research*, 95, 441–454.

Preissmann, A. (1961). Propagation des intumescences dans les canaux et rivières [eng.: Propagation of translation waves in canals and rivers]. 1st Congress of the French Association for Computation (AFCALTI), Grenoble, September.

Rahimi, A. R., Holt, A. R., Upton, G. J. G., Krämer, S., Redder, A., Verworn, H.-R. (2006). Attenuation calibration of an X-band weather radar using a microwave link. *Journal of Atmospheric and Oceanic Technology*, 23, 395–405.

Rauch, W., Bertrand-Krajewski, J.-L., Krebs, P., Mark, O., Schilling, W., Schütze, M. Vanrolleghem, P. A. (2002). Deterministic modelling of integrated urban drainage systems. *Water Science and Technology*, 45 (3), 81–94.

Richards, L. A. (1931). Capillary conduction of liquids through porous mediums. *Journal of Applied Physics*, 1 (5), 318–333, doi:10.1063=1.1745010.

Rosenblatt, F. (1958). The perceptron: A probabilistic model for information storage and organization in the brain. *Psychological Review*, 65 (6), 386.

Rossa, A., Liechti, K., Zappa, M., Bruen, M., Germann, U., Haase, G., Keil, C., Krahe, P. (2011). The COST 731 action: A review on uncertainty propagation in advanced hydro-meteorological forecast systems. *Atmospheric Research*, 100, 150–167.

Rözer, V., Müller, M., Bubeck, P., Kienzler, S., Thieken, A., Pech, I., ... Kreibich, H. (2016). Coping with pluvial floods by private households. *Water*, 8 (7), 304.

Saint-Venant, A. (1871). Théorie du mouvement non-permanent des eaux avec application aux crues des rivières et à l'introduction de marées dans leurs lits [eng.: Theory of non-permanent water movement with application to river flooding and the introduction of the tides in their tidal beds]. Comptes rendus hebdomadaires des séances de l'Académie des sciences, ISSN 0001-4036; 73, 1871.

Salamon, P., Fernàndez-Garcia, D., Gómez-Hernández, J. J. (2006). A review and numerical assessment of the random walk particle tracking method. *Journal of Contaminant Hydrology*, 87 (3), 277–305.

Sales-Ortells, H., Medema, G. (2015). Microbial health risks associated with exposure to stormwater in a water plaza. *Water Research*, 74, 34–46.

Sansalone, J. J., Buchberger, S. G. (1997). Partitioning and first flush of metals in urban roadway storm water. *Journal of Environmental Engineering*, 123 (2), 134–143.

Schellart, A., Ligouri, S., Krämer, S., Saul, A., Rico-Ramirez, M. A. (2014). Comparing quantitative precipitation forecast methods for prediction of sewer flows in a small urban area. *Hydrological Sciences Journal*, 59, 1418–1436.

Schellart, A., Wang, L., Onof, C. (2012). High resolution rainfall measurement and analysis in a small urban catchment, 9th International Workshop on Precipitation in Urban Areas, 6–9 December 2012, St. Moritz, Switzerland, 115–120.

Schilling, W. (1991). Rainfall data for urban hydrology: What do we need? *Journal of Atmospheric Research*, 27, 5–21.

Schirmer, M., Leschik, L., Musolff, A. (2013). Current research in urban hydrogeology – A review. *Advances in Water Resources*, 51, 280–291.

Semadeni-Davies, A., Hernebring, C., Svensson, G., Gustafsson, L. G. (2008). The impacts of climate change and urbanisation on drainage in Helsingborg, Sweden: Combined sewer system. *Journal of Hydrology*, 350 (1), 100–113.

Sempere-Torres, D, Berenguer M., Corral C., Sanchez-Diezma R. (2005). Improvements on flow forecasting using precipitation nowcasting based in radar advection techniques: assessment of predictability and uncertainty propagation. 32nd Conference on Radar Meteorology, Albuquerque, New Mexico, US, Oct. 24th–29th, 2005.

Sen, S. (2013). Real-time rainfall monitoring and flood inundation forecasting for the city of Kolkata. *Journal of Hydraulic Engineering*, 19 (2), 137–144.

Sommer, T. (2011). Chapter 10: Groundwater – The subterranean part of flood risk. In A. H. Schumann (ed.), *Flood Risk Assessment and Management*. Springer Science + Business Media B.V.

Spekkers, M., Rözer, V., Thieken, A., ten Veldhuis, M. C., Kreibich, H. (2017). A comparative survey of the impacts of extreme rainfall in two international case studies. *Natural Hazards and Earth System Sciences*, 17 (8), 1337–1355.

Steiner, M., Houze, R., Yuter, S. (1994). Climatological characterization of three dimensional storm structure from operational radar and rain gauge data. *Journal of Applied Meteorology*, 34, 1978–2007.

Thorndahl, S., Nielsen, J. E., Jensen, D. G., Jepsen, M. T. L., Larsen, M. K., Rasmussen, M. R. (2015). Can we predict the next urban flood? A case study of rain nowcasting and real-time flood forecasting. Proc. 10th UDM, Québec, Ca.

van Genuchten, M. T. (1980). A Closed-form equation for predicting the hydraulic conductivity of unsaturated soils. *Soil Science Society of America Journal*. doi:10.2136 =sssaj1980.03615995004400050002x.
van Metre, P. C., Mahler, B. J., Furlong, E. T. (2000). Urban sprawl leaves its PAH signature. *Environmental Science and Technology*, 34 (19), 4064–4070.
Vaze, J., Chiew, F. H. (2002). Experimental study of pollutant accumulation on an urban road surface. *Urban Water*, 4 (4), 379–389.
Versteeg, H. K., Malalasekera, W. (2007). *An Introduction to Computational Fluid Dynamics: The Finite Volume Method*. Pearson Education.
Verworn, H.-R., Krämer, S. (2008). Radar based nowcasting of rainfall events- analysis and assessment of a one-year continuum. In Samuels, P. (ed.), *Flood Risk Management: Research and Practice*. CRC Press, Leiden.
Visser, A. W. (1997). Using random walk models to simulate the vertical distribution of particles in a turbulent water column. *Marine Ecology Progress Series*, 275–281.
Vojinovic, Z., Tutulic, D. (2009). On the use of 1D and coupled 1D-2D modelling approaches for assessment of flood damage in urban areas. *Urban Water Journal*, 6 (3), 183–199.
Wang, S. (2010). Water balance modelling in Glasgow and Beijing. *Proceedings of the Institution of Civil Engineers. Water Management*, 163 (5), 219–226.
Wildi, W., Dominik, J., Loizeau, J. L., Thomas, R. L., Favarger, P. Y., Haller, L., Peytremann, C. (2004). River, reservoir and lake sediment contamination by heavy metals downstream from urban areas of Switzerland. *Lakes and Reservoirs: Research and Management*, 9 (1), 75–87.
Yan, H., Kouyi, G. L., Gonzalez-Merchan, C., Becouze-Lareure, C., Sebastian, C., Barraud, S., Bertrand-Krajewski, J. L. (2014). Computational fluid dynamics modelling of flow and particulate contaminants sedimentation in an urban stormwater detention and settling basin. *Environmental Science and Pollution Research*, 21 (8), 5347–5356.
Yeung, P. K., Pope, S. B. (1989). Lagrangian statistics from direct numerical simulations of isotropic turbulence. *Journal of Fluid Mechanics*, 207, 531–586.
Zawadzki, I. (1975). On radar-rain gauge comparison. *Journal of Applied Meteorology*, 14, 1430–1436.
Zhang, J., Hua, P., Krebs, P. (2017). Influences of land use and antecedent dry-weather period on pollution level and ecological risk of heavy metals in road-deposited sediment. *Environmental Pollution*, 228, 158–168.
Zhang, J., Wang, J., Hua, P., Krebs, P. (2015). The qualitative and quantitative source apportionments of polycyclic aromatic hydrocarbons in size dependent road deposited sediment. *Science of the Total Environment*, 505, 90–101.

6 Identifying the Contaminant Levels in River Floods That Occurred in the Past

Anna Lintern, Lauren MacDonald, Brent B. Wolfe, Roland I. Hall, Ana Deletic, Paul Leahy, Hendrik Heijnis, Atun Zawadzki, Patricia Gadd, and David T. McCarthy

6.1 INTRODUCTION

We currently have limited understanding of the dispersal of contaminants to urban and natural areas due to flooding by creeks and rivers (Wheater, 2002; Zonta et al., 2005; Sauer et al., 2007). This is largely because of the lack of data of water and sediment quality during and immediately after floods. Not only are large floods infrequent (Walling, He and Nicholas, 1996), but it can also be difficult to obtain accurate measurements safely during and immediately after flood events (Zonta et al., 2005; Zak, Rohovec and Navratil, 2009). In addition, due to the difficulties with forecasting flood events, it can be challenging to mobilise a sampling team in time to sample floodwaters and flood sediments. Whilst there are studies that have sampled the water quality of floods (Caetano, Vale and Falcão, 2006), or contaminant levels of flood-deposited sediments (Albering et al., 1999; Cobb et al., 2006; Rotkin-Ellman et al., 2010; Brown and Chanson, 2012; Hall et al., 2012), these studies generally include only a small number of flood events. In addition, existing datasets of water quality of flood waters and sediments cover a short time period (e.g., one flood event (Abraham and Wenderoth, 2005; Cobb et al., 2006; Pulkrabová et al., 2008; Wölz et al., 2011), or one decade (Hilscherova et al., 2007)).

In the absence of sufficient data, models could be used to predict contaminant dispersal in flood-inundated regions both in and downstream of urban and industrialised areas. However, there is a limited number of models that can do this (e.g., Telemac2D (Schulz et al., 2009) and TREX (Velleux, England Jr and Julien, 2008)). Furthermore, extensive field data are needed to implement and calibrate these models. Previous studies (Letcher et al., 2002; Newham and Drewry, 2006) have highlighted that we currently lack data for developing, calibrating and using models of flood contaminant dispersal and deposition.

Even if we implement monitoring projects now, because floods are infrequent (Walling, He and Nicholas, 1996), it may be many years before a representative number of samples of flood-deposited sediments can be measured. Another problem is that long-term monitoring projects can be financially prohibitive and discontinued with changes in economic or political climates (Caughlan and Oakley, 2001). As such, we need an alternative method for developing long-term flood quality datasets that can be utilised: (i) as calibration and validation datasets when modelling the quality of flood-deposited sediments and (ii) to better understand contaminant transport and deposition by floods. This is critical due to the expected increase in flooding of rivers and creeks, in both urban and natural areas in the future due to climate change (Jonkman and Dawson, 2012; Miller and Hutchins, 2017).

Long-term flood water and sediment quality datasets can be developed using sediment cores to trace flood and pollution history (Walling, He and Nicholas, 1996; Brown et al., 2000; Wolfe et al., 2006; Hall et al., 2012; Lintern et al., 2016a). Sediment cores from aquatic systems (e.g., lakes and estuaries) have been used to reconstruct past environmental changes in a number of studies, such as fire history (e.g., Whitlock et al., 2010), and for retrospective studies of ecology of waterways (Leahy et al., 2005). Deposits of floods that occurred in the past can also be preserved in bed sediments of aquatic systems (Benito et al., 2004; Wolfe et al., 2006; Baker, 2008). Furthermore, the contaminants often found in urban waterways such as heavy metals and persistent organic contaminants (polycyclic aromatic compounds, pesticides) can be preserved in these bed sediments (Ensminger et al., 2013; Lintern et al., 2015, 2018; Allinson et al., 2017). Sediment cores from floodplain lakes in urban and downstream floodplain environments can therefore be used to determine the contaminant levels and temporal changes within sediments deposited by river floods. The reconstruction of contaminants in past flood deposits can be used to augment existing data of contemporary flood deposit contaminant levels. The contaminant levels of flood deposits could potentially be a good indicator of urban and downstream flood water quality and the risks associated with poor flood water quality due to: (1) the adsorption of contaminants (e.g., metals and persistent organic pollutants) to sediments; (2) the fact that aquatic organisms are often more affected by sediment contaminant levels than contaminants in the water column and (3) measurements of sediment contaminant levels can be more temporally consistent (Wiklund et al., 2014). As such, using sediment cores from floodplain lakes in urban and downstream regions to identify the contaminant levels in sediments deposited by floods that occurred in the past is a potential solution to the problem of the lack of urban and downstream flood water quality data. These investigations can also provide access to baseline, reference conditions of natural contaminant concentrations, which are often unknown because monitoring programmes are frequently initiated after potential disturbance by urbanisation, industrialisation and mining (Hall et al., 2012; Wiklund et al., 2014; Lintern et al., 2016c).

This chapter discusses: (1) the theoretical background of identifying historical flood contaminant levels from sediment cores; (2) two case studies of the use of sediment cores for identifying past changes in flood water quality; one in an urbanised area of Australia and the other from a remote, natural floodplain landscape located downstream of industrialised areas in Canada and (3) a discussion of the future work required to further develop this technique for use in flood water quality management practice.

6.2 THEORETICAL BACKGROUND

To identify the contaminant concentrations in historical flood deposits, it is necessary to first (1) identify the historical flood deposits in sedimentary records and then to (2) determine the contaminant concentrations contained within these flood deposits.

6.2.1 Identifying Historical Flood Deposits in Sediment Cores

Sediment cores have previously been used to determine long-term records of past flood events in rivers (Bábek et al., 2011), lakes (Arnaud et al., 2005; Wolfe et al., 2006; Moreno et al., 2008; Brock et al., 2010) and estuaries (Malamud-Roam et al., 2006; Zong et al., 2011). Previous studies have identified past flood deposits in the sedimentary record using physical, chemical and biological characteristics of the sediments in the cores. These sediment characteristics are summarised in Figure 6.1 and discussed in Section 6.2.2. These sediment characteristics can be used to identify flood-deposited sediments in the sediment cores because they indicate both where the sediments originated and how they were deposited in the aquatic system (Harrison, Heijnis and Caprarelli, 2003).

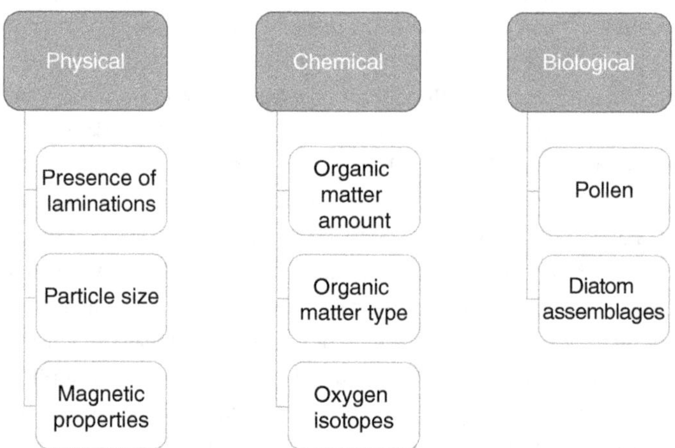

Figure 6.1 Characteristics that can be used to identify flood-deposited sediments in sediment cores

6.2.2 Sediment Characteristics Used to Identify Flood-Deposited Sediments in Sediment Cores

6.2.2.1 PHYSICAL PROPERTIES

Lamination presence. Laminations, layers of sediment that are visually distinct, can be used to identify flood-deposited sediments in sediment cores (Thorndycraft et al., 1998; Wolfe et al., 2006; Moreno et al., 2008; Bábek et al., 2011; D'Haen, Verstraeten and Degryse, 2012; Schillereff et al., 2014). The thickness of these laminations can be as little as one millimetre (Arnaud et al., 2002; Schiefer, Gilbert and Hassan, 2011) or up to several centimetres (Moreno et al., 2008). The colour of these sediment layers can indicate the origin and the type of sediment (Wolfe et al., 2006). As such, a change in the colour of sediment between distinct layers may indicate a change in source of sediment from autochthonous (within-lake) to allochthonous (from outside of the lake) sources and vice versa. Laminations may not necessarily be visible by the naked eye, but may be identified using x-ray analysis (e.g., Lintern et al., 2016b).

Particle size distribution. The particle size distribution of sediments is another proxy that is used to identify flood-deposited sediments in cores (Brown et al., 2000; Boe et al., 2006; Augustinus et al., 2008; Daessle et al., 2009; Bábek et al., 2011). High-energy flows (e.g., river floods) are able to mobilise and deliver large particles over long distances (Charlton, 2007). As such, coarse grains (e.g., sands) in sediment cores have often been transported by high flows (i.e., floods) (Brown et al., 2000; Bábek et al., 2011; Hollins et al., 2011; Vasskog et al., 2011; Schillereff et al., 2014). Sediment cores can also contain a number of sub-layers of differently sized particles (fining upward graded beds) within the sediments deposited by one flood event (Arnaud et al., 2002; Sandercock and Wyrwoll, 2005). This is a due to the coarse particles being deposited out of the water column first (due to their greater settling velocity), and the finer particles being deposited more slowly out of flood waters.

The particle size of sediments in lake sediment cores has been used to identify past flood events and this has been verified using measured flow data (Winter et al., 2001; Wolfe et al., 2006). For example, Winter et al. (2001) found that there were increases in average particle size at two points in 35-cm-long sediment cores from a floodplain lake in the UK. Daily discharge data showed high river flows in 1961–1962, and in 1976, which correlated to the dates assigned to the sediment core depths where there were increases in particle size.

Magnetic properties. Sediment magnetic properties have also been used to identify flood deposits in cores. This is because the source of the sediment can strongly affect the deposit's magnetic properties (Foster et al., 1998; Dearing, 1999; Brown et al., 2000; Sandgren and Snowball, 2001; Arnaud et al., 2005; Boe et al., 2006; Cremer et al., 2010; Bábek et al., 2011). Several measures are used to indicate the magnetic properties of sediments, including magnetic susceptibility, magnetic ratios and magnetic remanence (Foster et al., 1998; Sandgren and Snowball, 2001). Magnetic susceptibility, which is defined as the magnetisation of the material per unit of applied magnetic field, is commonly used to infer sources of sediments in cores because it is a non-destructive and rapid method to identify variations in magnetic properties through a sediment core (Sandgren and Snowball, 2001; Michael and Friedrich, 2003). Sediments with low magnetic susceptibility are usually sands (Sandgren and Snowball, 2001; Cremer et al., 2010) or sediments with high organic matter content (Bradshaw and Thompson, 1985). Clays and silts with a terrestrial origin may have high magnetic susceptibility levels (Moreno et al., 2008; Francu et al., 2010; Bellucci et al., 2012). Wolfe et al. (2006) verified magnetic susceptibility of sediments as an indicator of flood history by comparing magnetic susceptibility levels in sediment cores to historical records of river levels and flows. Further investigations were used to reconstruct a 600-year flood history of the Peace-Athabasca Delta (Alberta, Canada) that aligned with climate-driven changes in quantity and seasonality of river discharge (Wolfe et al., 2008).

6.2.2.2 CHEMICAL PROPERTIES

The amount of organic matter in sediment deposits can indicate the sediments' origins. The organic matter content can be described in terms of total organic nitrogen or carbon content, loss on ignition or the ratio of incoherent to coherent scatter (inc/coh) from micro-X-ray fluorescence (micro-XRF) scan measurements (Meyers and Teranes, 2001; Wolfe et al., 2006; Guyard et al., 2007; Croudace and Rothwell, 2015). Sediments deposited by floods are generally low in organic matter relative to inorganic matter (Brown et al., 2000; Nesje et al., 2001; Leng et al., 2006). This is because the amount of inorganic sediments carried into aquatic systems or eroded from river beds and banks increases during flood events (Rodbell et al., 1999). A high level of inorganic matter in sediment deposits is often an indication that there has been greater hydrologic activity in the aquatic system, as inorganic matter generally has terrestrial (or lithogenic) sources. Sediments are mobilised and transported from the catchment into aquatic systems when there is greater surface runoff (and hence, increased streamflow) (Kylander et al., 2013; Corella et al., 2014).

The ratio of carbon to nitrogen (C/N) and organic carbon isotope ratios ($\delta^{13}C$) can also be used to infer the type or origin of organic matter (Meyers and Teranes, 2001). Typically, lower C/N ratios (4–10) indicate that the organic matter was produced within the lake environment, and higher ratios (>20) indicate that the organic matter originates from terrestrial plants (Meyers and Teranes, 2001; Leng et al., 2006). Brock et al. (2010) used C/N ratios measured on Slave River flood sediment and a lake sediment core to reconstruct 80 years of flood history in the Slave River Delta (SRD) (Northwest Territories, Canada). Carbon isotope compositions ($\delta^{13}C$) may be high in sediments deposited during flood events because terrestrial plants often have higher $\delta^{13}C$ values (−25 to −30‰) than aquatic plants (−29 to −34‰) (Bierman et al., 1997; Brown et al., 2000; Wolfe et al., 2006). Measurements of the oxygen isotope ratios ($\delta^{18}O$) in various fractions (e.g., carbonate shells, diatom silica, aquatic plant cellulose) within lake sediment deposits have also been previously used to identify past changes in lake hydrology (Wolfe et al., 2001; Leng et al., 2006).

6.2.2.3 BIOLOGICAL CHEMICAL PROPERTIES

Biological indicators archived in sedimentary records, such as diatom assemblages and pollen grains, can also be an indication of the depositional environment (i.e., high- vs low-energy state) of the sediments in which the biological indicators are contained (Kenyon and Rutherfurd, 1999; Moreno et al., 2008; Brock et al., 2010). Moreno et al. (2008) used the extent of preservation of pollen in a sediment core to infer whether sediments were flood derived or not. Pollen grains that were deposited in high-energy turbulent conditions generally have degraded more than pollen deposited in less turbulent low-energy environments.

Diatoms are unicellular eukaryotic organisms and their diversity in aquatic systems is significantly affected by environmental conditions, including lake water depth, turbidity, pH and nutrient levels (Battarbee et al., 2001; Sonneman et al., 2001; Grundell et al., 2011). In the absence of flooding, diatom communities in rivers are often distinct from those in adjacent floodplain lakes, and floodwaters introduce river-transported diatom taxa into the lakes (Sokal, Hall and Wolfe, 2010). As such, combining an understanding of the effect of flood events on the pH, lake water depth, turbidity and nutrient levels in an aquatic system and an understanding of how diatom assemblages in bed sediments have changed over time can help to pinpoint historical flood deposits in sediment cores. Analysis of modern assemblages and paleolimnological reconstructions of epiphytic diatom communities have shown to be especially promising for reconstructing flood events in floodplain lakes (Wiklund et al., 2010).

6.2.2.4 THE IMPORTANCE OF ADOPTING A MULTI-PROXY APPROACH

Trends in the sediment characteristics listed in Figure 6.1 may not always be due to fluvial floods. For example, laminations visible in sediment cores can be caused also by landslides (Moernaut et al., 2007) or mass land clearing events in the catchment (Enters, Lücke and Zolitschka, 2006). Similarly, changes in particle size in sediment cores can result from various factors including aridity, land clearing, mass wasting, increase in wind strength, waste disposal or the erosion of river banks (Teller and Last, 1990; Moreno et al., 2008). Additionally, magnetic particles found in the bed sediments of aquatic systems do not always have terrestrial sources. They can be produced within the aquatic system by magnetite-producing bacteria and by the formation of magnetic iron sulphides (Dearing, 1999; Sandgren and Snowball, 2001).

The amount and type of organic and inorganic materials within sediment cores can also be affected by factors other than riverine flood events. The organic matter content in sediments is affected not only by the amount of organic matter that first entered the aquatic environment but also on the degradation and transformation after deposition (Meyers and Teranes, 2001). Similarly, some elements found in inorganic sediments (e.g., Ca, Mn and Fe) can come from both terrestrial sources and within the aquatic system itself (Kylander et al., 2013). It is therefore important to determine whether fluctuations in the level of inorganic materials are due to flood events or changes in chemistry in the aquatic system (e.g., via changes in dissolved oxygen).

Given these uncertainties, multiple sediment characteristics should ideally be used to identify past flood deposits in sediment cores (Wolfe et al., 2012; Schillereff et al., 2014). It is also important to understand the history of the catchment, which can be used to validate the flood record developed using the sediment core (Schillereff et al., 2014). Specifically, observed historical measurements of streamflow and water level, or anecdotal records of flooding within the catchment can be used to validate the flood events detected in the sediment cores using multiple proxies (Wolfe et al., 2006; Brock et al., 2010).

6.2.3 Determining Contaminant Concentrations in Flood Deposits Identified in Sediment Cores

6.2.3.1 APPROACHES USED TO IDENTIFY HISTORICAL CONTAMINANT CONCENTRATIONS

Conservative contaminants such as heavy metals and persistent organic pollutants (POPs) accumulate on the beds of aquatic environments because they have low solubility in water and thus bind to sediments (Förstner and Wittman, 1981; Kolak et al., 1998; Blais and Muir, 2001; Bábek et al., 2011; Chon, Ohandja and Voulvoulis, 2012). As such, previous studies have used sediment cores from lakes, oceans, floodplains and estuaries to understand past changes in the level of these contaminants in aquatic systems (Cundy et al., 1997; Yamashita et al., 2000; Latimer et al., 2003; Moriwaki et al., 2005). Persistent organic pollutants that have been studied in such investigations include polycyclic aromatic hydrocarbons (PAHs), polychlorinated biphenyls (PCBs) and organochlorine pesticides (Gevao et al., 1997; Gevao, Harner and Jones, 1999; Latimer et al., 2003; Man et al., 2004; Taylor, Birch and Links, 2004; Moon et al., 2009; Vane et al., 2011). These studies use sediment cores extracted from the beds of lentic environments and to develop dated vertical contaminant profiles (e.g., Figure 6.2). For example, Bellucci et al. (2012) used dated sediment cores to assess past changes in mercury concentrations in August Bay (Italy). They found that the Hg levels in the sediment core were greatest in the 1970s.

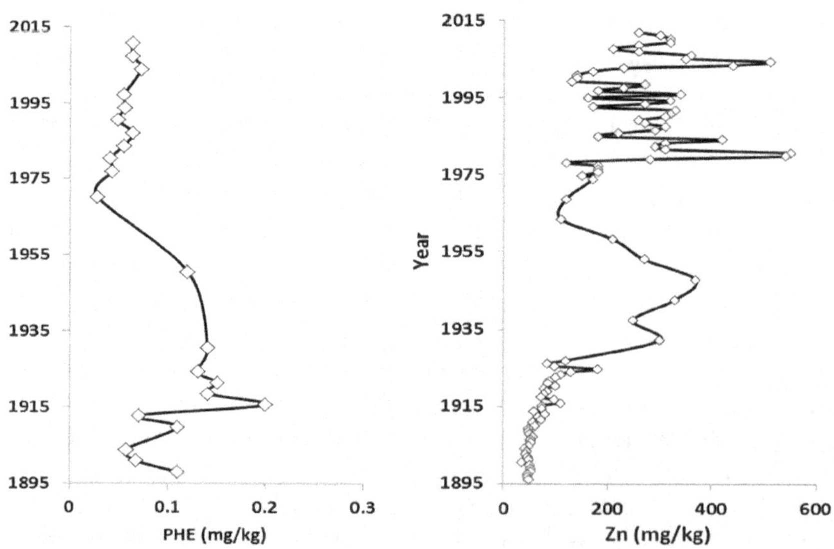

Figure 6.2 Examples of sediment core contaminant profiles for POPs (phenanthrene; PHE) and heavy metals (zinc; Zn) (adapted from Lintern et al., 2015)

Table 6.1 *Previous studies that have used sedimentary records to identify past contaminant concentrations in fluvial flood deposits*

Location	Methods used to identify flood deposits	Pollutants measured	Methods used to analyse pollutant levels	Flood deposits validated using existing flow records	Flood pollutant deposit concentrations validated using existing water quality records	Key findings	Citation
Floodplain lakes of River Morava, Czech Republic	Sediment density, presence of laminations, magnetic susceptibility, grain size	Metals	ICP-MS	Yes	No	Decrease in heavy metal concentrations in flood deposits with high sand contents.	Bábek et al., 2011
Colorado River remnant delta, USA	Grain size	Metals	ICP-AES XRF	Yes	No	As, Pb, Cu and Ba are associated with flood layers.	Daessle et al., 2009
Floodplain of Lower Rhone River, France	Grain size, visual analysis of laminations	Metals	ICP-AES ICP-MS	Yes	Yes	Higher heavy metal concentrations in sediments deposited by low to medium discharge flows compared to strong floods.	Ferrand et al., 2012
Oxbow lake (Kiss-Janosne-Holt Tisza) of River Tisza, Hungary	None	Metals	ICP-AES	No flood deposits identified	No	Assumed that metal concentration increases resulted from riverine floods.	Nguyen et al., 2009
Villerest Dam, Upper Loire River, France	Visual analysis of laminations, grain size	Metals	ICP-AES ICP-MS	Yes	No	High metal concentrations deposited by major flood events.	Dhivert et al., 2015
Lower Athabasca River and Peace-Athabasca Delta, Canada	Organic matter content, oxygen isotope ratios in cellulose, diatoms	PACs, metals, mercury	GC-MS	Yes	No	PAH compositions and normalised metal concentrations in sediments deposited by floods did not differ significantly after oil sands mining began.	Hall et al., 2012; Klemt et al., 2020, 2021; Kay et al., 2020, 2022, 2023

The decrease in Hg after the 1970s was hypothesised to be due to improved management of mercury emission, such as the installation of a demercurisation plant and improvements to waste treatment at factories within the bay's catchment (Bellucci et al., 2012). In a study of PCB contamination, Ruiz-Fernandez et al. (2012) found that there was an increase in PCB levels after 1940 in Las Matas lagoon in Mexico, which was likely due to higher usage rates of synthetic PCBs. Levels decreased after 1990, likely a result of the introduction of environmental regulations in the Las Matas lagoon area.

Having identified the flood-deposited sediments in sedimentary records, contaminant levels in flood deposits can be determined by high-resolution sampling of sedimentary records and comparing this high-resolution contaminant profile with the record of flood deposits. These high-resolution profiles of heavy metals can be developed using techniques such as inductively coupled plasma mass spectroscopy (ICP-MS) (Duan et al., 2014; Lintern et al., 2015), conventional XRF (Guo and Yang, 2016; Hosono, Alvarez and Kuwae, 2016), inductively coupled plasma atomic emission spectroscopy (ICP-AES) (Xu et al., 2014; Sharley et al., 2016) and micro-XRF core scanning (Croudace, Rindby and Rothwell, 2006; Miller et al., 2015; Lintern et al., 2016b); and profiles of POPs can be developed using gas chromatography-mass spectrometry (GC-MS) (Lebeuf and Nunes, 2005; Lorgeoux et al., 2016).

6.2.3.2 EXAMPLES OF STUDIES THAT HAVE IDENTIFIED THE CONTAMINANT LEVELS OF FLOOD EVENTS

Several studies have identified contaminant levels in flood-transported materials that are preserved in sediment cores retrieved from floodplains or floodplain lakes (Table 6.1). Floodplains and floodplain lakes, as opposed to rivers, serve as depositional zones and generally preserve a coherent stratigraphic record of past hydrological conditions and contaminant deposition. These systems receive sediment deposits during times of fluvial flooding and can therefore provide an archive of riverine floods. The studies listed in Table 6.1 have investigated aquatic environments downstream of urban centres or mining sites and have all studied contaminants in flood deposits. A myriad of proxies have been used including laminations in sediment cores (Bábek et al., 2011), particle size (Daessle et al., 2009; Ferrand et al., 2012) and sediment density (Bábek et al., 2011) to identify flood layers in the sediment core. Depending on the final use of the data, it may be sufficient to assess contaminant concentrations in periods of high flood frequency rather than in individual discrete flood deposits. For example, not all of these studies in Table 6.1 identified the contaminant levels in discrete flood deposits in sediment cores, but instead have studied general (or average) trends in hydrology and identified contaminant levels in times of high flood frequency (e.g., Ferrand et al., 2012).

The majority of studies have validated the flood-deposited layers with existing discharge records, and it appears that few of these studies have validated the flood contaminant data from sediment cores using water and sediment quality data. This is likely due to the unavailability of these data, and such comparisons would likely increase our confidence in these data. Finally, these studies have identified differing interactions between flooding and contamination of aquatic systems and soils. Whilst Bábek et al. (2011) and Ferrand et al. (2012) found decreasing heavy metal concentrations in flood deposits, others (Daessle et al., 2009; Dhivert et al., 2015) have identified the opposite. This may be due to the different geomorphologies, contaminant sources and the type of sediment deposited by floods in these aquatic systems. For example, whilst the River Morava flood-deposited sands (Bábek et al., 2011), which have generally lower metal concentrations, the Colorado River floods deposits had high clay contents (Daessle et al., 2009). To account for changing energy conditions, which influences grain size and consequently contaminant concentrations including metals due to their preferential adsorption to clay-sized sediment, a number of studies have incorporated geochemical normalisation approaches (e.g., Loring, 1991; Kersten and Smedes, 2002; Wiklund et al., 2014; MacDonald et al., 2016). This is strongly recommended for studies on floodplains and other depositional environments that are subject to highly dynamic hydrological conditions.

6.3 CASE STUDIES THAT HAVE USED SEDIMENT CORES TO IDENTIFY CONTAMINANT LEVELS IN FLOODPLAIN LAKE DEPOSITS

Two case studies that have used floodplain lake sediment records for identifying evidence of contamination are discussed in Sections 6.3.1 and 6.3.2. The first example of the urban floodplain lakes in the Yarra River catchment (Australia) describes how identifying the contaminant concentrations in individual flood deposits can assist in identifying the quality of sediments deposited by flood events in inundated areas. Understanding the distribution of heavy metal concentrations in past river flood deposits could potentially be used for statistical prediction of heavy metals conveyed by future riverine floods. The second case study of the SRD (Canada), a remote northern landscape downstream of a mining and industrialised area, outlines how sediment cores can be used and analysed to identify intervals of high and low flood influence, and correspondingly, contaminants delivered by fluvial, atmospheric and local runoff pathways. The sediment record provides critically missing knowledge of baseline, reference conditions, which are required to distinguish contaminants derived from natural processes versus those supplied by industrial developments.

6.3.1 Urban Floodplain Lakes of the Yarra River Catchment, Victoria, Australia

Willsmere (37°27′23″S, 145°2′3″E) and Bolin billabongs (37°46′14″S, 145°4′43″E) are floodplain lakes of the Yarra River. The river travels approximately 240 km from the Great Dividing Range through the city of Melbourne, and discharges into Port Phillip Bay (Figure 6.3). Both floodplain lakes are within metropolitan Melbourne. They are relatively small, with bank-full areas of 1.9 and 3.4 ha for Willsmere and Bolin billabongs, respectively (Leahy, 2007). Both systems receive overland runoff from their local catchments, in addition to overbank river flows from the Yarra River. The Yarra River flows enter the lakes when water surface elevations are at 6 m (Australian Height Datum; AHD) at Bolin Billabong and at 3 m at Willsmere Billabong (Vic SES, 2012, 2013). Overbank flows of the river at these locations occur approximately every 3 to 4 years. However, it has been estimated that overbank flows occurred more frequently (approximately once a year) before the mid-20th century (Sinclair Knight Merz, 2005).

Sediment cores obtained from these floodplain lakes contain sediment deposits dating back to approximately 1890 (for Willsmere Billabong) and 1908 (for Bolin Billabong). The cores were dated using radioisotopic methods (^{210}Pb and ^{137}Cs) (as detailed in Lintern et al., 2016b). Flood-deposited sediments in these cores were identified using proxies commonly used in literature including the presence of high-density sediments in the core (Gilbert et al., 2006), high levels (i.e., maxima) in magnetic susceptibility (Arnaud et al., 2005), low levels (i.e., minima) in organic matter (Nesje et al., 2001) and fining upward sequences in particle size (Ambers, 2001). The identification of flood deposits in the Willsmere and Bolin cores using these methods is detailed by Lintern et al. (2016b).

The heavy metal concentrations within these identified sediment deposits were then determined from micro-XRF core scanning of the Willsmere and Bolin Billabong cores (Lintern et al., 2016a). The micro-XRF units at 1 cm intervals were converted to absolute heavy metal concentrations. This was done by correlating the micro-XRF core scan heavy metal concentrations to ICP-MS derived heavy metal profiles (Lintern et al., 2018). It should be noted that in these examples, the heavy metal concentrations are not normalised using particle size or a conservative element, which are often used to account for variations in heavy metal association to particulates due to particle size differences

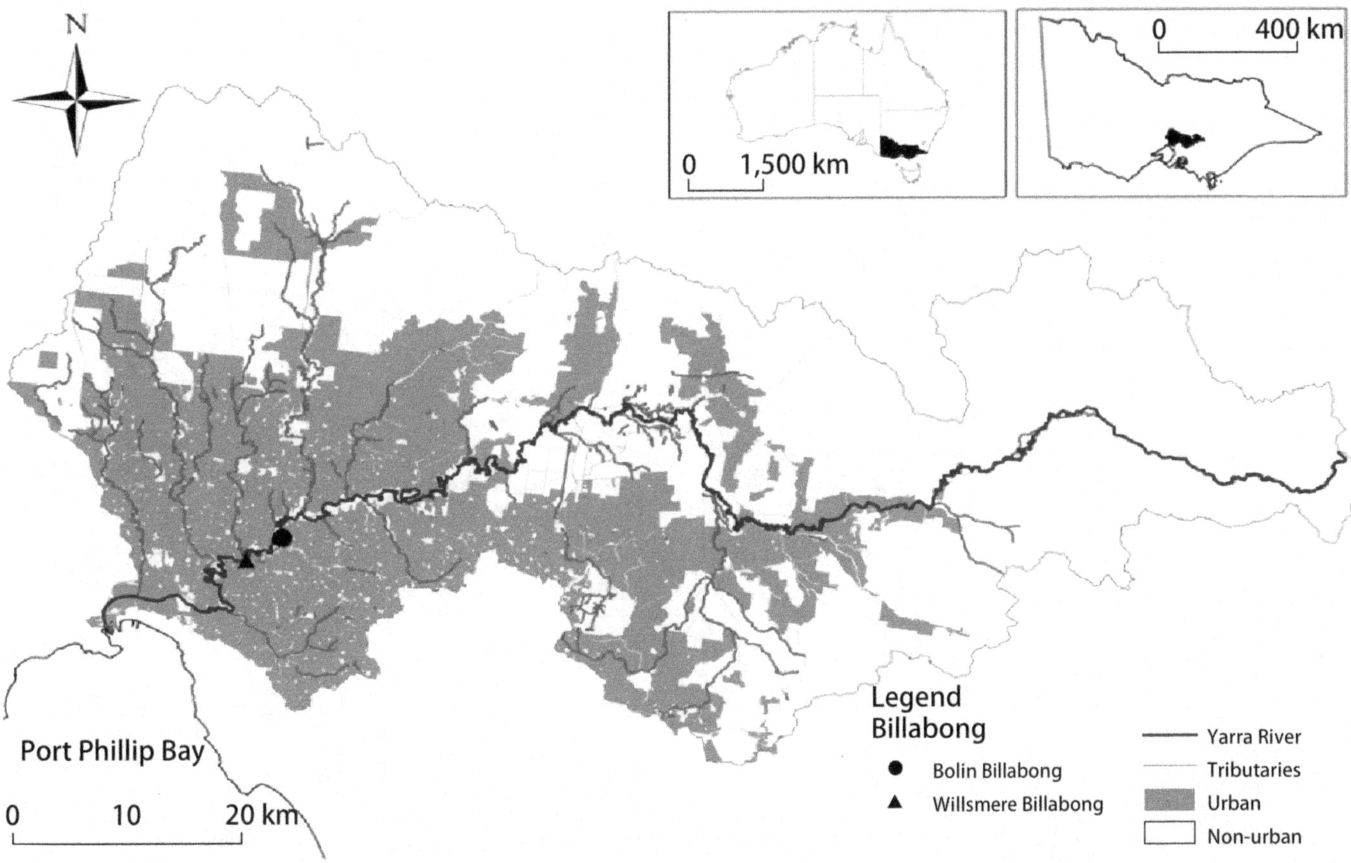

Figure 6.3 The Yarra River and two urban floodplain lakes, Willsmere and Bolin Billabong. Urban regions shown represent urbanised areas of metropolitan Melbourne. The inserts show the location of Victoria in Australia, and the location of the Yarra River catchment in the state of Victoria (adapted from Lintern et al., 2016a)

(Loring and Rantala, 1992). Here, the heavy metal concentrations remain un-normalised so that the concentrations can be compared to existing sediment quality guidelines.

Pb concentrations in sediments deposited by floods were not significantly different to those in sediments not deposited by floods for both Billabongs (Mann-Whitney U-test, $p > 0.05$ for both). This lack of statistical significance is likely influenced by the small number of flood events detected in the Willsmere Billabong core. Visually, Figure 6.4 suggests that in Willsmere Billabong absolute Pb concentrations are greater in sediments not deposited by floods compared to flood-deposited sediments. This could be suggesting that the key source of Pb into this billabong is either stormwater runoff from the local catchment or atmospheric deposition. This could also be a result of the fact that flood sediments are generally coarse, thereby having lower concentration of heavy metals. The 75th percentile Pb concentrations in flood deposits in both

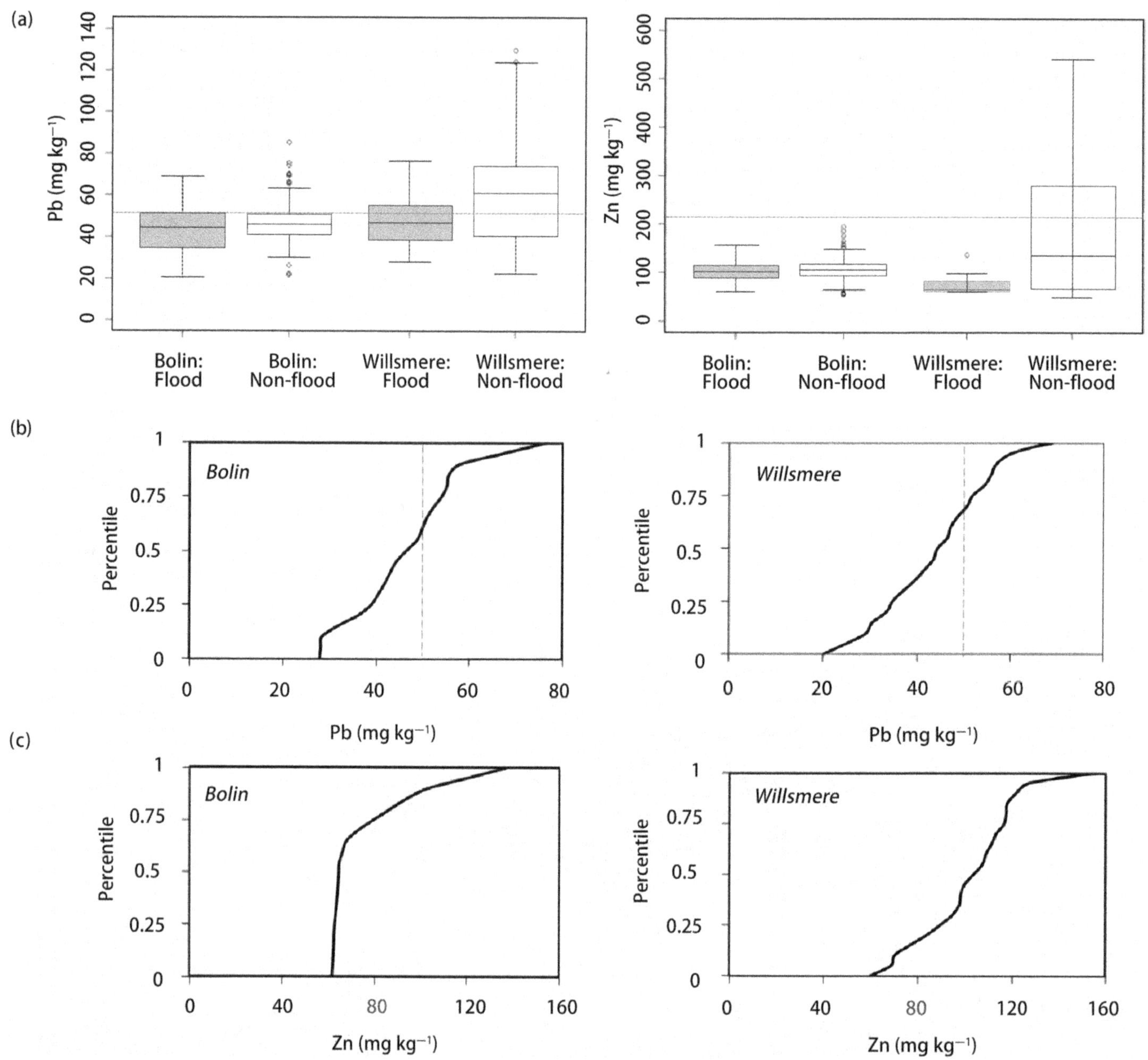

Figure 6.4 Pb and Zn in flood and non-flood sediment deposits of the Yarra River billabongs. (a) The Pb and Zn concentration distributions in sediments in the middle Yarra River catchment at Bolin Billabong and Willsmere Billabong. One outlier in the Willsmere Billabong core, non-flood deposits for Zn at 800 mg/kg is not shown. Horizontal dashed lines represent the Australian and New Zealand guideline trigger values representing low probability of toxic effects. (b) Cumulative density functions of Pb and (c) Zn concentrations in sediments deposits by river floods in Bolin Billabong and Willsmere Billabong (adapted from Lintern et al., 2016a)

Willsmere Billabong and Bolin Billabong are above 50 mg/kg, which is the Australian sediment quality trigger value (ANZECC/ARMCANZ, 2000) that represents low probability of toxic effects. As such, the sediments deposited by floods in the two floodplain lakes have concentrations exceeding recommended current Australian sediment quality guideline trigger values. This is also the case for sediments deposited in periods without river flooding.

Absolute Zn concentrations in flood deposits are significantly less than in non-flood deposits for Willsmere Billabong (Mann-Whitney U-test, $p < 0.05$). However, there is no significant difference between absolute Zn concentrations in flood and non-flood deposits in Bolin Billabong ($p = 0.29$). Zn concentrations in the flood deposits in both billabongs are below 210 mg/kg; the Australian and New Zealand guideline trigger values that represents low probability of toxic effects provided by ANZECC/ARMCANZ (2000) (Figure 6.4).

The cumulative density functions (Figure 6.4) indicate that the 90th percentile Pb concentration in flood-deposited sediments at both Willsmere and Bolin Billabongs is 58 mg/kg (above the Australian and New Zealand trigger value of 50 mg/kg for low probability of toxic effects). The 90th percentile Zn concentrations in flood deposits are 120 and 100 mg/kg for Bolin Billabong and Willsmere Billabong, respectively (Figure 6.4). These are both less than the sediment quality trigger values for low probability toxic effects used in Australia (ANZECC/ARMCANZ, 2000). These cumulative density functions could be used to identify the probability at which certain heavy metal concentrations will occur in flood deposits. It is important to note that with all statistical analyses based on sediment core data, the accuracy of the frequency analyses relies on the assumption of statistical stationarity in the data. For Willsmere and Bolin Billabongs, it has been previously identified that the Pb and Zn trends exhibit non-stationary behaviour (Lintern et al., 2016a). However, these principles and methods could be applied to sediment cores that have been taken from locations with stationary heavy metal trends, or to truncated time-periods with stationary trends.

The concentration of heavy metals (As, Cd, Cr, Cu, Hg, Ni, Pb and Zn) has been measured in the Yarra River at Bolin Billabong approximately on a weekly basis between 1984–1986 and 1990–1997 (Department of Sustainability and Environment, 2013). However, only two of these grab samples coincided with river floods. As such, the relationships shown above could not be validated using traditionally sampled flood sediment and water quality measurements. Although the reconstructed heavy metal concentrations in flood deposits in Willsmere and Bolin Billabongs are not yet validated, the demonstrated methodology can be applied to other sites. This would help enhance our understanding of the quality of flood-borne sediments deposited in inundated areas and the risk that these flood deposits pose to humans and the natural environment, as demonstrated by the creation of 20th-century cumulative distribution functions of flood deposit quality.

6.3.2 Slave River Delta, Northwest Territories, Canada

The SRD is a large (400 km^2) hydrologically dynamic floodplain landscape located in the Northwest Territories, Canada, where the Slave River enters Great Slave Lake (Brock et al., 2007). Slave River discharge is derived mainly from the Peace River (~2/3) and the Athabasca River and its tributaries (~1/3). The SRD provides important habitat for wildlife and key resources that help support the traditional lifestyle of the mainly First Nation community of Fort Resolution (English et al., 1997; Brock et al., 2010). Previously, local land users expressed concerns over reduced river and lake water levels in the SRD. Research conducted on floodplain lake sediment cores to assess these concerns identified previous intervals of both high and low flood influence during the past century and varying degrees of importance of different hydrological processes (i.e., flooding and evaporation) on lakes throughout the SRD (Brock et al., 2007; Wolfe et al., 2007; Sokal, Hall and Wolfe, 2008; Brock et al., 2010). Recently, studies documenting pollution from the Alberta oil sands region (Kelly et al., 2009, 2010; Timoney and Lee, 2009; Kurek et al., 2013) have sparked new concerns in communities located downstream, including Fort Resolution, regarding the potential for long-range transport of contaminants and the consequential effects on water quality and human health (Campbell and Spitzer, 2007; Wesche, 2007, 2009). Unfortunately, long-term environmental monitoring programmes do not exist, which confounds ability to distinguish industrially derived contaminants and their pathways from natural sources and processes. As part of the Slave River and Delta Partnership, a paleolimnological study was conducted to address community concerns regarding the potential effects of upstream oil sands mining activities.

MacDonald et al. (2016) utilised a detailed pulse-flood record from a small (~1.2 km^2, maximum depth ~1.5 m) floodplain lake ('SD2'; 61°16′56.21″N, 113°34′55.51″W; Figure 6.5a) in the SRD to assess if there was recent transport and deposition of metal contaminants from the Alberta oil sands above expected natural background levels. Previous paleolimnological studies indicated that this lake contains an exceptionally detailed record of 20th-century Slave River flood history (Brock et al., 2010). Metal deposition via fluvial processes was assessed by analysing sediment intervals deposited during periods of high flood influence (defined by high C/N ratios) and metal deposition by atmospheric transport was assessed by analysing sediment intervals deposited during periods of low flood influence (defined by low C/N ratios). To account for variations in grain size and mineralogy that occurs in varying-energy floodplain systems and other depositional environments, MacDonald et al. (2016) normalised metal concentrations to lithium (Li), as recommended in prior studies (Loring, 1991; Kersten and Smedes, 2002; Wiklund et al., 2014). Linear relations between metal concentrations relative to Li concentrations for pre-1967 (year commercial oil

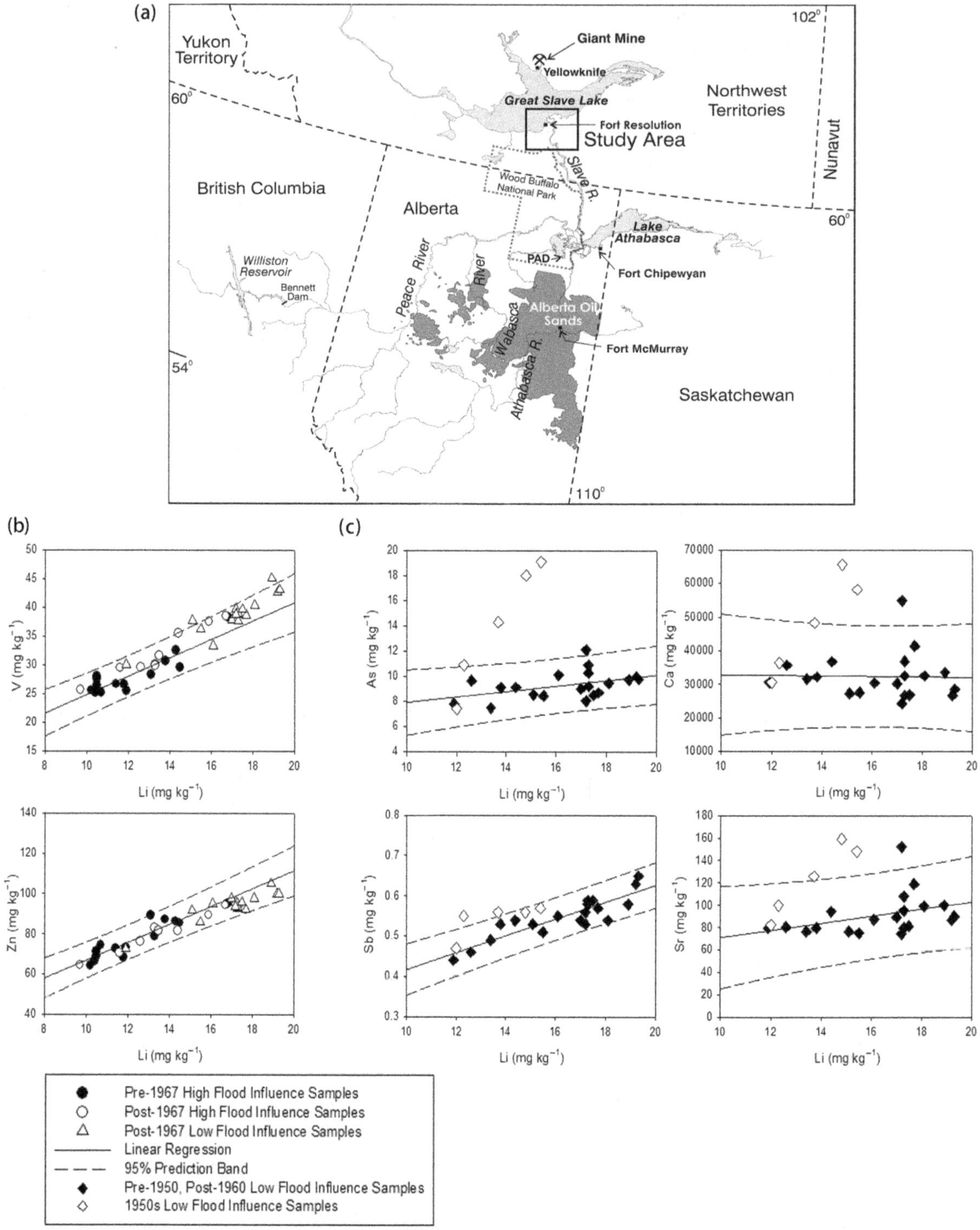

Figure 6.5 Assessment of metal contaminants at floodplain lake SD2 within the SRD, Northwest Territories, Canada (modified from MacDonald et al., 2016). (a) Location of the study area. (b) Examples of scatterplots for the assessment of oil sands pollution, showing the linear relations between concentrations of metals of interest (V, Zn) and Li and the range of natural variability. (c) Scatterplots of the linear relations of Li and key metals for the assessment of pollution from Giant Mine and the range of natural variability

sands development started; Gosselin et al., 2010) high flood influence intervals were determined and 95% prediction intervals (PI) were calculated based on these relations to define the natural range of variation of normalised metal concentrations in sediments supplied by the Slave River. MacDonald et al. (2016) then evaluated post-1967 high flood influence sediment metal concentrations and post-1967 low flood influence sediment metal concentrations (due to possibility of flood deposits in the catchment being remobilised during snowmelt and rainfall runoff) against this established natural range of variation to test for industrial-derived sources of contamination (indicated when >2.5% of samples plot above the upper 95% PI; Figure 6.5b). Results showed that <2.5% of post-1967 high flood influence sediment samples and low flood influence sediment samples plotted above the upper 95% PI, which indicated little to no evidence of industrially derived contamination from the Alberta oil sands post-1967. Rather, presence and variability in metal concentrations post-1967 in floodplain lake SD2 likely continue to be driven by natural processes.

Unexpectedly, MacDonald et al. (2016) found unusually high arsenic (As) concentrations in sediments of lake SD2 deposited during the 1950s. This was a low flood influence period suggesting the lake sediment profile archived a different source of industrially derived atmospheric pollution during this time, as this predates the development of oil sands. To further probe these results, linear relations between As relative to Li concentrations for pre-1950 and post-1960 low flood influence deposited sediment were determined and PIs were calculated based on these relations to define the natural range of variation for atmospherically deposited As (Figure 6.5c). MacDonald et al. (2016) found that 80% of samples deposited during the 1950s plotted above the 95% PI providing evidence that As concentrations became enriched during this time. MacDonald et al. (2016) identified that a possible source of As was emissions from a gold mine in Yellowknife, NWT (Giant Mine, ~140 km N of SD2). Giant Mine, which converted arsenopyrite ore to gold while releasing a highly toxic As vapour as a by-product, was developing rapidly in the 1950s without significant environment protection procedures and equipment (Hocking et al., 1978). Furthermore, sediments containing elevated As concentrations were also high in antimony, calcium and strontium concentrations, which are major volatile elements within the mineralogy of the Giant Mine ore deposit (Figure 5c; Silke, 2013). These findings suggested a legacy of previously unrecognised far-field atmospheric pollution from Giant Mine. Notably, use of paleohydrological knowledge was critical to inform accurate interpretations of pathways and sources of contaminant deposition at floodplain lake SD2. Further studies of As in lake sediment cores from the SRD, and the region in the vicinity of Giant Mine, are reviewed by Kay et al. (2023).

6.4 SUMMARY AND FUTURE OUTLOOK

We currently have a lack of data and understanding of the contaminant concentrations in flood deposits of rivers and creeks in urban areas and environments downstream of urbanisation, industrialisation and mining. Whilst ongoing data collection and sampling of urban and downstream flood deposits is important, it may take many years to establish a dataset of pollution in flood deposits. As such, we must develop alternative methods to determine the concentrations of contaminants in flood-deposited sediments in urban and downstream areas. The above discussion indicated how these data can and have been collected using sedimentary records from aquatic systems. Data from sediment cores can also provide an insight into flood sediment quality prior to urbanisation and industrialisation. The analysis of sediment core data can therefore provide waterway managers and engineers with a unique understanding of urban and downstream flood water quality and the factors potentially affecting it.

However, despite the potential that sediment cores have to broaden our understanding of urban and downstream flood sediment quality, further work is required before these data can be effectively implemented and applied to flood water quality management. Additionally, further exploration could be conducted to increase the usefulness of these data obtained from these sedimentary records for water quality and flood managers.

6.4.1 Uncertainties That Must Be Addressed

First, exploration of the uncertainties associated with reconstructing past changes in sediment quality using sediment cores is critical. There are important assumptions and uncertainties inherent in the study of sedimentary records (Boyle, 2001; Smol et al., 2012) such as uncertainties in the analysis of contaminant concentrations, uncertainties in the development of the age-depth models, spatial variability in contamination deposition on the bed of the aquatic deposition and post-depositional transformations in contaminants within the bed sediment. It is these assumptions and uncertainties that will most likely challenge the practical application of data collected from sediment cores for developing a better understanding of and modelling water quality of urban river floods. Preliminary work has been conducted to quantify age-depth model uncertainties, analytical uncertainties in sediment core reconstructions and the extent of post-depositional transformations of heavy metals (Binford, 1990; Telford et al., 2004; Blaauw and Heegaard, 2012; Maher et al., 2012; Walker, 2012; Lintern et al., 2018). However, further work is necessary to better understand and model the impact of chemical, biological and physical characteristics of the aquatic system on the deposition of contaminants found in urban and downstream waters (e.g., heavy metals, POPs, pharmaceuticals) on the sediment bed and preservation in the bed sediment. This would greatly assist accurate interpretation of the sediment core data, and aid in effective decision-making.

Second, we require current urban and downstream flood water quality models to be able to deal with uncertain data, such as those obtained from sedimentary records. Previous studies have identified errors in heavy metal pollution records of up to 105% due to uncertainties in the development of the sediment chronology (Lintern et al., 2018). Current predictive models of flood water quality such as Telemac2D (Schulz et al., 2009) and TREX (Velleux, England Jr and Julien, 2008) are not designed to be able to accommodate for data with potentially large uncertainties such as those collected from sedimentary records. Adapting these models to deal with uncertain data would facilitate the practical implementation of these data into existing model structures, and hence the use of these data for predicting and managing the water quality of floods in the future.

6.4.2 Avenues for Further Exploration

There is still a need to further develop our understanding of how flood characteristics affect contaminant deposition in flood-inundated urban and downstream areas. Ferrand et al. (2012) previously explored this question using sediment cores from floodplains, and they found higher heavy metal concentrations in sediment deposits within the core that were deposited by low to medium discharge flows compared to layers deposited by strong floods. However, a quantitative link between metrics of urban river and downstream floods (e.g., streamflow, velocity and inundation area) and contaminant concentrations in flood-deposited sediments is required to further our understanding of the processes affecting contaminant deposition after and during river floods, and to predict the quality of sediments deposited by future floods. Additionally, accounting for variation in the particle size distribution of urban and downstream flood deposits by normalisation could elucidate whether particle size is the key driver of contaminant concentrations in flood-borne deposits or whether other factors (e.g., land use, flood origin, flow rates) are important.

6.4.3 Final Thoughts

Sediment quality and hydrologic data have been collected from sedimentary records by palaeoenvironmental scientists for decades, although water managers and engineers have yet to implement these data into practical mitigation strategies and into the models. Further research is still necessary before these palaeoenvironmental data can be incorporated into models and into management frameworks. However, these data have the potential to increase our understanding about water quality processes in urban rivers and downstream floodplain environments and about the risks that extreme events such as flooding of these waterways pose to humans, which cannot be obtained from other sources.

6.5 ACKNOWLEDGEMENTS

Portions of this chapter have been adapted from a component of a thesis submitted in fulfilment of the degree of Doctor of Philosophy (Lintern, 2016).

References

Abraham, W.-R., Wenderoth, D. (2005). 'Fate of Facultative Pathogenic Microorganisms during and after the Flood of the Elbe and Mulde Rivers in August 2002', *Acta hydrochimica et hydrobiologica*, 33, pp. 449–454. 10.1002/aheh.200400587.

Albering, H. J., van Leusen, S. M., Moonen, E. J., Hoogewerff, J. A., Kleinjans, J. C. (1999). 'Human health risk assessment: A case study involving heavy metal soil contamination after the flooding of the river Meuse during the winter of 1993–1994', *Environmental Health Perspectives*, 107(1), pp. 37–43. https://doi.org/10.1289/ehp.9910737

Allinson, M., Zhang, P., Bui, A. T., Myers, J. H., Pettigrove, V., Rose, G., Salzman, S., Walters, R. D., Allinson, G. (2017). 'Herbicides and trace metals in urban waters in Melbourne, Australia (2011–12): Concentrations and potential impact', *Environmental Science and Pollution Research*, 24, pp. 7274–7284.

Ambers, R. K. (2001). 'Using the sediment record in a western Oregon Flood – Control reservoir to assess the influence of storm history and logging on sediment yield', *Journal of Hydrology*, 244, pp. 181–200.

ANZECC/ARMCANZ. (2000). *Australian and New Zealand Guidelines for Fresh and Marine Water Quality. Volume 1. The Guidelines*. Canberra, Australia: ANZECC and ARMCANZ.

Arnaud, F., Lignier, V., Revel, M., Desmet, M., Beck, C., Pourchet, M., Charlet, F., Trentesaux, A., Tribovillard, N. (2002). 'Flood and earthquake disturbance of 210Pb geochronology (Lake Anterne, NW Alps)', *Terra Nova*, 14.

Arnaud, F., Revel, M., Chapron, E., Desmet, M., Tribovillard, N. (2005). '7200 of Rhone river flooding activity in Lake Le Bourget, France: A high-resolution sediment record of NW Alps hydrology', *The Holocene*, 15, pp. 420–430.

Augustinus, P., Bleakley, N., Deng, Y., Shane, P., Cochran, U. (2008). 'Rapid change in early Holocene environments inferred from Lake Pupuke, Auckland City, New Zealand', *Journal of Quaternary Science*. John Wiley & Sons, Ltd., 23(5), pp. 435–447. doi: 10.1002/jqs.1153.

Bábek, O., Faměra, M., Hilscherová, K., Kalvoda, J., Dobrovolný, P., Sedláček, J., Machát, J., Holoubek, I. (2011). 'Geochemical traces of flood layers in the fluvial sedimentary archive; implications for contamination history analyses', *CATENA*, 87, pp. 281–290.

Baker, V. R. (2008). 'Paleoflood hydrology: Origin, progress, prospects', *Geomorphology*, 101, pp. 1–13.

Battarbee, R. W., Jones, V. J., Flower, R., J., Cameron, N. G., Bennion, H., Carvalho, L., Juggins, S. (2001). 'Diatoms', in Last, W. M. and Smol, J. P. (eds) *Tracking Environmental Change Using Lake Sediments. Volume 3: Terrestrial, Algal and Siliceous Indicators*. Dordrecht: Kluwer Academic Publishers, pp. 155–190.

Bellucci, L. G., Giuliani, S., Romano, S., Albertazzi, S., Mugnai, C., Frignani, M. (2012). 'An integrated approach to the assessment of pollutant delivery chronologies to impacted areas: Hg in the Augusta Bay (Italy)', *Environmental Science & Technology*, 46(4), pp. 2040–2046. doi: 10.1021/es203054c.

Benito, G., Lang, M., Barriendos, M., Llasat, M. C., Frances, F., Ouarda, T., Thorndycraft, V., Enzel, Y., Bardossy, A., Couer, D., Bobee, B. (2004). 'Use of systematic, palaeoflood and historical data for the improvement of flood risk estimation. Review of scientific methods', *Natural Hazards*, 31, pp. 623–643.

Bierman, P., Lini, A., Zehfuss, P. H., Church, A., Davis, P., Southon, J. (1997). 'Postglacial ponds and alluvial fans: Recorders of Holocene landscape history', *Geological Society of America*, 7(10), pp. 1–8.

Binford, M. (1990). 'Calculation and uncertainty analysis of 210Pb dates for PIRLA project lake sediment cores', *Journal of Paleolimnology*. Kluwer Academic Publishers, 3(3), pp. 253–267. doi: 10.1007/BF00219461.

Blaauw, M., Heegaard, E. (2012). 'Estimation of age-depth relationships', in Birks, H. J. B. et al. (eds) *Tracking Environmental Change Using Lake Sediments: Volume 5: Data Handling and Numerical Techniques*. Dordrecht: Springer Science & Business Media, pp. 379–414.

Blais, J. M., Muir, D. C. G. (2001). 'Paleolimnological methods and applications for persistent organic pollutants', in Last, W. M. and Smol, J. P. (eds) *Tracking Environmental Change Using Lake Sediments Volume 2: Physical and Geochemical Methods*. Dordrecht: Kluwer Academic Publishers, pp. 271–298.

Boe, A.-G., Dahl, S. O., Lie, O., Nesje, A. (2006). 'Holocene river floods in the upper Glomma catchment, southern Norway: A high-resolution multiproxy record from lacustrine sediments', *The Holocene*, 16(3), pp. 445–455.

Boyle, J. (2001). 'Inorganic geochemcial methods in palaeolimnology', in Last, W. M. and Smol, J. P. (eds) *Tracking Environmental Change Using Lake Sediments Volume 2: Physical and Geochemical Methods*. Dordrecht: Kluwer Academic Publishers, pp. 83–141.

Bradshaw, R., Thompson, R. O. Y. (1985). 'The use of magnetic measurements to investigate the mineralogy of Icelandic lake sediments and to study catchment processes', *Boreas*. Blackwell Publishing Ltd, 14(3), pp. 203–215. doi: 10.1111/j.1502-3885.1985.tb00720.x.

Brock, B. E., Martin, M. E., Mongeon, C. L., Sokal, M. A., Wesche, S., Armitage, D., Wolfe, B., Hall, R., Edwards, T. (2010). 'Flood frequency variability during the past 80 years in the Slave River Delta, NWT, as determined from multi-proxy paleolimnological analysis', *Canadian Water Resources Journal*, 35(3), pp. 281–300.

Brock, B. E., Wolfe, B. B., Edwards, T. W. D. (2007). 'Characterizing the hydrology of Shallow Floodplain Lakes in the Slave River Delta, NWT, Canada, using water isotope tracers', *Arctic, Antarctic, and Alpine Research*. Institute of Arctic and Alpine Research (INSTAAR), University of Colorado, 39(3), pp. 388–401. doi: 10.1657/1523-0430(06-026)[BROCK]2.0.CO;2.

Brown, R., Chanson, H. (2012). 'Suspended sediment properties and suspended sediment flux estimates in an inundated urban environment during a major flood event', *Water Resources Research*, 48(11), p. W11523. doi: 10.1029/2012WR012381.

Brown, S. L., Bierman, P. R., Lini, A., Southon, J. (2000). '10 000 yr record of extreme hydrologic events', *Geology*, 28(4), pp. 335–338.

Caetano, M., Vale, C., Falcão, M. (2006). 'Particulate metal distribution in Guadiana estuary punctuated by flood episodes', *Estuarine, Coastal and Shelf Science*, 70(1–2), pp. 109–116. http://dx.doi.org/10.1016/j.ecss.2006.05.030.

Campbell, D., Spitzer, A. (2007). 'High and dry', *Up here: explore Canada's far north*.

Caughlan, L., Oakley, K. L. (2001). 'Cost considerations for long-term ecological monitoring', *Ecological Indicators*, 1(2), pp. 123–134. http://dx.doi.org/10.1016/S1470-160X(01)00015-2.

Charlton, R. (2007). *Fundamentals of Fluvial Geomorphology*. Hoboken: Routledge.

Chon, H.-S., Ohandja, D.-G., Voulvoulis, N. (2012). 'The role of sediments as a source of metals in river catchments', *Chemosphere*, 88(10), pp. 1250–1256. Available at: www.sciencedirect.com/science/article/pii/S0045653512004870.

Cobb, G. P., Abel, M. T., Rainwater, M. R., Austin, G. P., Cox, S. B., Kendall, R. J., Marsland, E. J., Anderson, T. A., Leftwich, B. D., Zak, J. C., Presley, S. M. (2006). 'Metal distributions in New Orleans following Hurricanes Katrina and Rita: A continuation study', *Environmental Science & Technology*. American Chemical Society, 40(15), pp. 4571–4577. doi: 10.1021/es060041g.

Corella, J. P., Benito, G., Rodriguez-Lloveras, X., Brauer, A., Valero-Garces, B. L. (2014). 'Annually-resolved lake record of extreme hydro-meteorological events since AD 1347 in NE Iberian Peninsula', *Quaternary Science Reviews*, 93(0), pp. 77–90. doi: http://dx.doi.org/10.1016/j.quascirev.2014.03.020.

Cremer, H., Bunnik, F. P. M., Donders, T. H., Hoek, W. A., Koolen-Eekhout, M., Koolmees, H. H., Lavooi, E. (2010). 'River flooding and landscape changes impact ecological conditions of a scour hole lake in the Rhine-Meuse delta, The Netherlands', *Journal of Paleolimnology*, 44(3), pp. 789–801. doi: 10.1007/s10933-010-9452-2.

Croudace, I. W., Rindby, A., Rothwell, R. G. (2006). 'ITRAX: Description and evaluation of a new multi-function X-ray core scanner', *Geological Society, London, Special Publications*, 267(1), pp. 51–63. doi: 10.1144/gsl.sp.2006.267.01.04.

Croudace, I. W., Rothwell, R. G. (2015). *Micro-XRF Studies of Sediment Cores*. Dordrecht: Springer Netherlands.

Cundy, A. B., Croudace, I., Thomson, J., Lewis, J. L. (1997). 'Reliability of Salt Marshes as "Geochemical Recorders" of pollution input: A case study from contrasting Estuaries in Southern England', *Environmental Science & Technology*. American Chemical Society, 31(4), pp. 1093–1101. doi: 10.1021/es960622d.

D'Haen, K., Verstraeten, G., Degryse, P. (2012). 'Fingerprinting historical fluvial sediment fluxes', *Progress in Physical Geography*, 36(2), pp. 154–186.

Daessle, L. W., Lugo-Ibarra, K. C., Tobschall, H. J., Melo, M., Gutierrez-Galindo, E. A., Garcia-Hernandez, J., Alvarez, L. G. (2009). 'Accumulation of As, Pb, and Cu associated with the recent sedimentary processes in the Colorado Delta, South of the United States-Mexico Boundary', *Archives of Environmental Contamination and Toxicology*, 56(4), pp. 680–692. Available at: www.springeronline.com.

Dearing, J. A. (1999). *Environmental Magnetic Susceptibility: Using the Bartington MS2 System*. 2nd edn. Kenilworth: Chi Publishing.

Department of Sustainability and Environment. (2013). Victorian Water Resources Data Warehouse. Available at: www.vicwaterdata.net/ (Accessed: 10 June 2013).

Dhivert, E., Grosbois, C., Coynel, A., Lefevre, I., Desmet, M. (2015). 'Influences of major flood sediment inputs on sedimentary and geochemical signals archived in a reservoir core (Upper Loire Basin, France)', *CATENA*, 126, pp. 75–85. https://doi.org/10.1016/j.catena.2014.10.030.

Duan, D., Ran, Y., Cheng, H., Chen, J., Wan, G. (2014). 'Contamination trends of trace metals and coupling with algal productivity in sediment cores in Pearl River Delta, South China', *Chemosphere*, 103, pp. 35–43. https://doi.org/10.1016/j.chemosphere.2013.11.011.

English, M. C., Hill, R. B., Stone, M. A., Ormson, R. (1997). 'Geomorphological and botanical change on the Outer Slave River Delta, NWT, before and after impoundment of the Peace River', *Hydrological Processes*. John Wiley & Sons, Ltd, 11(13), pp. 1707–1724. doi: 10.1002/(SICI)1099-1085(19971030)11:13<1707::AID-HYP600>3.0.CO;2-O.

Ensminger, M. P., Budd, R., Kelley, K., C., Goh, K. S. (2013). 'Pesticide occurrence and aquatic benchmark exceedances in urban surface waters and sediments in three urban areas of California, USA, 2008–2011', *Environmental Monitoring and Assessment*, 185(5), pp. 3697–3710. doi: 10.1007/s10661-012-2821-8.

Enters, D., Lücke, A., Zolitschka, B. (2006). 'Effects of land-use change on deposition and composition of organic matter in Frickenhauser See, northern Bavaria, Germany', *Science of the Total Environment*, 369(1–3), pp. 178–187. Available at: www.sciencedirect.com/science/article/pii/S0048969706004141.

Ferrand, E., Eyrolle, F., Radakovitch, O., Provansal, M., Dufour, S., Vella, C., Raccasi, G., Gurriaran, R. (2012). 'Historical levels of heavy metals and artificial radionuclides reconstructed from overbank sediment records in lower Rhone River (South-East France)', *Geochimica et Cosmochimica Acta*, 82, pp. 163–182. Available at: www.elsevier.com.

Förstner, U., Wittman, G. T. W. (1981). *Metal Pollution in the Aquatic Environment*. Berlin: Springer-Verlag.

Foster, I. D. L., Lees, J. A., Owens, P. N., Walling, D. E. (1998). 'Mineral magnetic characterization of sediment sources from an analysis of lake and floodplain sediments in the catchments of the Old Mill reservoir and Slapton Ley, South Devon, UK', *Earth Surface Processes and Landforms*. John Wiley & Sons, Ltd, 23(8), pp. 685–703.

Francu, E., Schwarzbauer, J., Lana, R., Nyvlt, D., Nehyba, S. (2010). 'Historical changes in levels of organic pollutants in sediment cores from Brno Reservoir, Czech Republic', *Water Air and Soil Pollution*, 209(1–4), pp. 81–91. Available at: www.springeronline.com.

Gevao, B., Hamilton-Taylor, J., Murdoch, C., Jones, K. C., Kelly, M., Tabner, B. J. (1997). 'Depositional time trends and remobilization of PCBs in Lake sediments', *Environmental Science & Technology*. American Chemical Society, 31(11), pp. 3274–3280. doi: 10.1021/es970276f.

Gevao, B., Harner, T., Jones, K. C. (1999). 'Sedimentary record of Polychlorinated Naphthalene concentrations and deposition fluxes in a Dated Lake Core', *Environmental Science & Technology*. American Chemical Society, 34(1), pp. 33–38. doi: 10.1021/es990663k.

Gilbert, R., Crookshanks, S., Hodder, K. R., Spagnol, J., Stull, R. B. (2006). 'The record of an extreme flood in the sediments of Montane Lillooet Lake, British Columbia: Implications for paleoenvironmental assessment', *Journal of Paleolimnology*. Kluwer Academic Publishers, 35(4), pp. 737–745. doi: 10.1007/s10933-005-5152-8.

Gosselin, P., Hrudey, S. E., Naeth, M. A., Plourde, A., Therrien, R., Van Der Kraak, Gl., Xu, Z. (2010). 'Environmental and health impacts of Canada's oil sands industry', *Royal Society of Canada Expert panel report, Ottawa, ON*.

Grundell, R., Gell, P., Mills, K., Zawadzki, A. (2011). 'Interaction between a river and its wetland: Evidence from the Murray River for spatial variability in diatom and radioisotope records', *Journal of Paleolimnology*, 47(2), pp. 205–219. doi: 10.1007/s10933-011-9572-3.

Guo, Y., Yang, S. (2016). 'Heavy metal enrichments in the Changjiang (Yangtze River) catchment and on the inner shelf of the East China Sea over the last 150 years', *Science of the Total Environment*, 543, Part, pp. 105–115. doi: http://doi.org/10.1016/j.scitotenv.2015.11.012.

Guyard, H., Chapron, E., St-Onge, G., Anselmetti, F. S., Arnaud, F., Magand, O., Francus, P., Melieres, M-A. (2007). 'High-altitude varve records of abrupt environmental changes and mining activity over the last 4000 years in the Western French Alps (Lake Bramant, Grandes Rousses Massif)', *Quaternary Science Reviews*, 26(19–21), pp. 2644–2660. doi: http://dx.doi.org/10.1016/j.quascirev.2007.07.007.

Hall, R. I., Wolfe, B. B., Wiklund, J. A., Edwards, T. W. D., Farwell, A. J., Dixon, D. G. (2012). 'Has Alberta oil sands development altered delivery of polycyclic aromatic compounds to the Peace-Athabasca Delta?', *PLoS ONE*, 7(9), p. e46089. doi: 10.1371/journal.pone.0046089.

Harrison, J., Heijnis, H., Caprarelli, G. (2003). 'Historical pollution variability from abandoned mine sites, Greater Blue Mountains World Heritage Area, New South Wales, Australia', *Environmental Geology*, 43, pp. 680–687.

Hilscherova, K., Dusek, L., Kubik, V., Cupr, P., Hofman, J., Klanova, J., Holoubek, I. (2007). 'Redistribution of organic pollutants in river sediments and alluvial soils related to major floods', *Journal of Soils and Sediments*. Springer Berlin / Heidelberg, 7(3), pp. 167–177. doi: 10.1065/jss2007.04.222.

Hocking, D., Kuchar, P., Plambeck, J. A., Smith, R. A. (1978). 'The impact of gold smelter emissions on vegetation and soils of a sub-Arctic Forest-Tundra transition ecosystem', *Journal of the Air Pollution Control Association*. Taylor & Francis, 28(2), pp. 133–137. doi: 10.1080/00022470.1978.10470580.

Hollins, S. E., Harrison, J. J., Jones, B. G., Zawadzki, A., Heijnis, H., Hankin, S. (2011). 'Reconstructing recent sedimentation in two urbanised coastal lagoons (NSW, Australia) using radioisotopes and geochemistry', *Journal of Paleolimnology*, 46(4), pp. 579–596. https://doi.org/10.1007/s10933-011-9555-4.

Hosono, T., Alvarez, K., Kuwae, M. (2016). 'Lead isotope ratios in six lake sediment cores from Japan Archipelago: Historical record of trans-boundary pollution sources', *Science of the Total Environment*, 559, pp. 24–37. https://doi.org/10.1016/j.scitotenv.2016.03.138.

Jonkman, N. S., Dawson, J. R. (2012). 'Issues and challenges in flood risk management—Editorial for the special issue on flood risk management', *Water*. doi: 10.3390/w4040785.

Kay, M. L., Wiklund, J. A., Remmer, C. R., Owca, T. J., Klemt, W. H., Neary, L. K., Brown, K. C., MacDonald, E., Thomson, K., Vucic, J. M., Wesenberg, K., Hall, R. I., Wolfe. B. B. (2020). 'Evaluating temporal patterns of metals concentrations in floodplain lakes of the Athabasca Delta (Canada) relative to pre-industrial baselines', *Science of the Total Environment*, 704, pp. 135309. https://doi.org/10.1016/j.scitotenv.2019.135309.

Kay, M. L., Sun, X., Wiklund, J. A., Savage, C. A. M., Adams, J. K., MacDonald, L. A., Klemt, W. H., Brown, K. C., Hall, R. I., Wolfe, B. B. (2022). 'Assessment of mercury enrichment in lake sediment records from Alberta Oil Sands development via fluvial and atmospheric pathways', *Frontiers in Environmental Science*, 10, pp. 949339. https://doi.org/10.3389/fenvs.2022.949339.

Kay, M. L., Jasiak, I., Klemt, W. H., Wiklund, J. A., Faber, J. A., MacDonald, L. A., Telford, J. V. K., Savage, C. A. M., Cooke, C. A., Wolfe, B. B., Hall, R. I. (2023). Paleolimnological evaluation of metal(loid) enrichment from oil sands and gold mining operations in northwestern Canada', *Environmental Research*, 216, pp. 114439. doi: 10.1016/j.envres.2022.114439.

Kelly, E. N., Short, J. W., Schindler, D. W., Hodson, P. V., Ma, M., Kwan, A. K., Fortin, B. L. (2009). 'Oil sands development contributes polycyclic aromatic compounds to the Athabasca River and its tributaries', *Proceedings of the National Academy of Sciences*, 106(52), pp. 22346–22351. doi: 10.1073/pnas.0912050106.

Kelly, E. N., Schindler, D. W., Hodson, P. V., Short, J. W., Radmanovich, R., Nielsen, C. C. (2010). 'Oil sands development contributes elements toxic at low concentrations to the Athabasca River and its tributaries', *Proceedings of the National Academy of Sciences*, 107(37), pp. 16178–16183. doi: 10.1073/pnas.1008754107.

Kenyon, C., Rutherfurd, I. D. (1999). 'Preliminary evidence for pollen as an indicator of recent floodplain accumulation rates and vegetation changes: The Barmah-Millewa Forest, SE Australia', *Environmental Management*. Springer-Verlag, 24(3), pp. 359–367. doi: 10.1007/s002679900239.

Kersten, M., Smedes, F. (2002). 'Normalization procedures for sediment contaminants in spatial and temporal trend monitoring', *Journal of Environmental Monitoring*. The Royal Society of Chemistry, 4(1), pp. 109–115. doi: 10.1039/B108102K.

Klemt, W. H., Kay, M. L., Wiklund, J. A., Wolfe, B. B., Hall, R. I. (2020). Assessment of vanadium and nickel enrichment in Lower Athabasca River floodplain lake sediment within the Athabasca Oil Sands Region (Canada)', *Environmental Pollution*, 265, Part A, pp. 114920. https://doi.org/10.1016/j.envpol.2020.114920.

Klemt, W. H, Brua, R. B., Culp, J. M., Hicks, K., Wolfe, B. B., Hall, R. I. (2021). 'Evaluating lower Athabasca River sediment metal concentrations from Alberta Oil Sands monitoring programs using pre-development baselines'. *Environmental Science & Technology*, 55, pp. 8817–8828. https://doi.org/10.1021/acs.est.1c01761.

Kolak, J. J., Long, D. T., Beals, T. M., Eisenrich, S. J., Swackhamer, D. L. (1998). 'Anthropogenic inventories and historical and present accumulation rates of copper in Great Lakes sediments', *Applied Geochemistry*, 13(1), pp. 59–75. http://dx.doi.org/10.1016/S0883-2927(97)00054-1.

Kurek, J., Kirk, J. L., Muir, D. C. G., Wang, X., Evans, M. S., Smol, J. P. (2013). 'Legacy of a half century of Athabasca oil sands development recorded by lake ecosystems', *Proceedings of the National Academy of Sciences*, 110(5), pp. 1761–1766. doi: 10.1073/pnas.1217675110.

Kylander, M., Klaminder, J., Wohlfarth, B., Lowemark, L. (2013). 'Geochemical responses to paleoclimatic changes in southern Sweden since the late glacial: the Hässeldala Port lake sediment record', *Journal of Paleolimnology*. Springer Netherlands, 50(1), pp. 57–70. doi: 10.1007/s10933-013-9704-z.

Latimer, J. S., Boothman, W. S., Pesch, C. E., Chmura, G. L., Pospelova, V., Jayaraman, S. (2003). 'Environmental stress and recovery: The geochemical record of human disturbance in New Bedford Harbor and Apponagansett Bay, Massachusetts (USA)', *Science of the Total Environment*, 313(1–3), pp. 153–176.

Leahy, P. J., Tibby, J., Kershaw, P., Heijis, H., Kershaw, J. (2005). 'The impact of European settlement on Bolin Billabong, a Yarra river floodplain lake, Melbourne, Australia', *River Research and Applications*, 21, pp. 131–149.

Leahy, P. J. (2007). *The Palaeolimnology and Current Status of Yarra River Billabongs (Melbourne, Victoria, Australia)*. Clayton, Australia: PhD thesis, School of Geography and Environmental Science, Monash University.

Lebeuf, M., Nunes, T. (2005). 'PCBs and OCPs in Sediment Cores from the Lower St. Lawrence Estuary, Canada: Evidence of Fluvial Inputs and Time Lag in Delivery to Coring Sites', *Environmental Science & Technology*. American Chemical Society, 39(6), pp. 1470–1478. doi: 10.1021/es049051c.

Leng, M. J., Lamb, A. L., Heaton, T. H. E., Marshall, J. D., Wolfe, B. B., Jones, M. D., Holmes, J. A., Arrowsmith, C. (2006). 'Isotopes in lake sediments', in Leng, M. J. (ed.), *Isotopes in Palaeoenvironmental Research*. Dordrecht: Springer, pp. 147–184.

Letcher, R. A., Jakeman, A. J., Calfas, M., Linforth, S., Baginska, B., Lawrence, I. (2002). 'A comparison of catchment water quality models and direct estimation techniques', *Environmental Modelling & Software*, 17(1), pp. 77–85. doi: http://dx.doi.org/10.1016/S1364-8152(01)00054-8.

Lintern, A., Deletic, A., Leahy, P., McCarthy, D. (2015). 'Digging up the dirty past: Evidence for stormwater's contribution to pollution of an urban floodplain lake', *Marine and Freshwater Research*, 66(7), pp. 596–608.

Lintern, A. (2016). *Using Sediment Cores to Reconstruct Historical Pollution Records: Digging Up the Yarra's Dirty Past*. Clayton, Australia: PhD thesis, Department of Civil Engineering, Monash University.

Lintern, A., Leahy, P. J., Heijnis, H., Zawadzki, A., Gadd, P., Jacobsen, G., Deletic, A., McCarthy, D. (2016a). 'Identifying heavy metal levels in historical flood water deposits using sediment cores', *Water Research*, 105, pp. 34–46. doi: http://dx.doi.org/10.1016/j.watres.2016.08.041.

Lintern, A., Leahy, P. J., Zawadzki, A., Gadd, P., Heijnis, H., Jacobsen, G., Connor, S., Deletic, A., McCarthy, D. (2016b). 'Sediment cores as archives of historical changes in floodplain lake hydrology', *Science of the Total Environment*, 544, pp. 1008–1019.

Lintern, A., Anderson, M., Leahy, P., Deletic, A., McCarthy, D. (2016c). 'Using sediment cores to establish targets for the remediation of aquatic environments', *Water Science and Technology*, 73(3), pp. 628–635.

Lintern, A., Leahy, P., Deletic, A., Heijnis, H., Zawadzki, A., Gadd, P., McCarthy, D. (2018). 'Uncertainties in historical pollution data from sedimentary records from an Australian urban floodplain lake', *Journal of Hydrology*, 560, pp. 560–571. https://doi.org/10.1016/j.jhydrol.2018.03.055.

Lorgeoux, C., Moilleron, R., Gasperi, J., Ayrault, S., Bonte, P., Lefevre, I., Tassin, B. (2016). 'Temporal trends of persistent organic pollutants in dated sediment cores: Chemical fingerprinting of the anthropogenic impacts in the Seine River basin, Paris', *Science of the Total Environment*, 541, pp. 1355–1363. https://doi.org/10.1016/j.scitotenv.2015.09.147.

Loring, D. H. (1991). 'Normalization of heavy-metal data from estuarine and coastal sediments', *ICES Journal of Marine Science*, 48(1), pp. 101–115. Available at: http://dx.doi.org/10.1093/icesjms/48.1.101.

Loring, D. H., Rantala, R. T. T. (1992). 'Manual for the geochemical analyses of marine sediments and suspended particulate matter', *Earth-Science Reviews*, 32(4), pp. 235–283. Available at: www.sciencedirect.com/science/article/pii/001282529290001A.

MacDonald, L. A., Wiklund, J. A., Elmes, M. C., Wolfe, B. B., Hall, R. I. (2016). 'Paleolimnological assessment of riverine and atmospheric pathways and sources of metal deposition at a floodplain lake (Slave River Delta, Northwest Territories, Canada)', *Science of the Total Environment*, 544, pp. 811–823.

Maher, L. J., Heiri, O., Lotter, A. F. (2012). 'Assessment of uncertainties associated with paleolimnological laboratory methods and microfosssil analysis', in Birks, H. J. B. et al. (eds) *Tracking Environmental Change Using Lake Sediments: Volume 5: Data Handling and Numerical Techniques*. Dordrecht: Springer Science & Business Media, pp. 143–166.

Malamud-Roam, F. P., Ingram, B. L., Hughes, M., Florsheim, J. L. (2006). 'Holocene paleoclimate records from a large California estuarine system and its watershed region: linking watershed climate and bay conditions', *Quaternary Science Reviews*, 25(13–14), pp. 1570–1598. doi: http://dx.doi.org/10.1016/j.quascirev.2005.11.012.

Man, K.-W., Zheng, J., Leung, A. P. K., Lam, P. K. S., Lam, M. H-W., Yen, Y-F. (2004). 'Distribution and behavior of trace metals in the sediment and porewater of a tropical coastal wetland', *Science of the Total Environment*, 327(1–3), pp. 295–314. doi: http://dx.doi.org/10.1016/j.scitotenv.2004.01.023.

Meyers, P. A., Teranes, J. L. (2001). 'Sediment organic matter', in Last, W. M. and Smol, J. P. (eds) *Tracking Environmental Change Using Lake Sediments Volume 2: Physical and Geochemical Methods*. Dordrecht: Kluwer Academic Publishers, pp. 239–269.

Michael, E. E., Friedrich, H. (2003). *Environmental Magnetism: Principles and Applications of Enviromagnetics, International Geophysics*. New York: Academic Press. Available at: www.sciencedirect.com/science/article/pii/S0074614203803157.

Miller, H., Croudace, I. W., Bull, J. M., Cotterill, C. J., Dix, J. K., Taylor, R. N. (2015). 'Modern Pollution Signals in Sediments from Windermere, NW England, Determined by Micro-XRF and Lead Isotope Analysis BT – Micro-XRF Studies of Sediment Cores: Applications of a non-destructive tool for the environmental sciences', in Croudace, I. W. and Rothwell, R. G. (eds) *Micro-XRF Studies of Sediment Cores*. Dordrecht: Springer Netherlands, pp. 423–442. doi: 10.1007/978-94-017-9849-5_16.

Miller, J. D., Hutchins, M. (2017). 'The impacts of urbanisation and climate change on urban flooding and urban water quality: A review of the evidence concerning the United Kingdom', *Journal of Hydrology: Regional Studies*, 12, pp. 345–362. doi: https://doi.org/10.1016/j.ejrh.2017.06.006.

Moernaut, J., Batist, M. D., Charlet, F., Heirman, K., Chapron, E., Pino, M., Brummer, R., Urrutia, R. (2007). 'Giant earthquakes in South-Central Chile revealed by Holocene mass-wasting events in Lake Puyehue', *Sedimentary Geology*, 195(3–4), pp. 239–256. Available at: www.sciencedirect.com/science/article/pii/S0037073806002284.

Moon, H.-B., Choi, M., Choi, H.-G., Ok, G., Kannan, K. (2009). 'Historical trends of PCDDs, PCDFs, dioxin-like PCBs and nonylphenols in dated sediment cores from a semi-enclosed bay in Korea: Tracking the sources', *Chemosphere*, 75(5), pp. 565–571. Available at: www.sciencedirect.com/science/article/pii/S0045653509001155.

Moreno, A., Valero-Garces, B. L., Gonzalez-Samperiz, P., Rico, M. (2008). 'Flood response to rainfall variability during the last 2000 years inferred from the Taravilla Lake record (Central Iberian Range, Spain)', *Journal of Paleolimnology*. Springer Netherlands, 40(3), pp. 943–961. doi: 10.1007/s10933-008-9209-3.

Moriwaki, H., Katahira, K., Yamamoto, O., Fukuyama, J., Kamiura, T., Yamazaki, H., Yoshikawa, S. (2005). 'Historical trends of polycyclic aromatic hydrocarbons in the reservoir sediment core at Osaka', *Atmospheric Environment*, 39(6), pp. 1019–1025. Available at: www.sciencedirect.com/science/article/pii/S1352231004009938.

Nesje, A., Dahl, S. O., Matthews, J. A., Berrisford, M. S. (2001). 'A ~4500 yr record of river floods obtained from a sediment core in Lake Atnsjøen, eastern Norway', *Journal of Paleolimnology*. Springer Netherlands, 25(3), pp. 329–342. doi: 10.1023/a:1011197507174.

Newham, L. T. H., Drewry, J. J. (2006). *Modelling Catchment-scale Nutrient Generation, Technical Report 28/05, National River Contaminants Program of Land and Water Australia*. Canberra: CSIRO Land and Water.

Nguyen, H. L., Braun, M., Szaloki, I. et al. (2009). 'Tracing the metal pollution history of the Tisza River through the analysis of a sediment depth profile', *Water, Air, & Soil Pollution*, 200, pp. 119–132. https://doi.org/10.1007/s11270-008-9898-2.

Pulkrabová, J., Suchanova, M., Tomaniova, M., Kocourek, V., Hajslova, J. (2008). 'Organic pollutants in areas impacted by flooding in 2002: A 4-Year Survey', *Bulletin of Environmental Contamination and Toxicology*, 81(3), pp. 299–304. doi: 10.1007/s00128-008-9486-6.

Rodbell, D. T., Seltzer, G. O., Anderson, D. M., Abbott, M. B., Enfield, D. B., Newman, J. H. (1999). 'An ~15,000-year record of El Nino-driven alluviation in Southwestern Ecuador', *Science*, 283(5401), pp. 516–520. https://doi.org/10.1126/science.283.5401.516.

Rotkin-Ellman, M., Solomon, G., Gonzales, C. R., Agwaramgbo, L., Mielke, H. W. (2010). 'Arsenic contamination in New Orleans soil: Temporal changes associated with flooding', *Environmental Research*, 110(1), pp. 19–25. doi: https://doi.org/10.1016/j.envres.2009.09.004.

Ruiz-Fernández, A. C., Sanchez-Cabeza, J.-A., Alonso-Hernandez, C., Martinez-Herrera, V., Perez-Bernal, L. H., Preda, M., Hillaire-Marcel, C., Gastaud, J., Quejido-Cabezas, A. J. (2012). 'Effects of land use change and sediment mobilization on coastal contamination (Coatzacoalcos River, Mexico)', *Continental Shelf Research*, 37(0), pp. 57–65. https://doi.org/10.1016/j.csr.2012.02.005.

Sandercock, P., Wyrwoll, K.-H. (2005). 'The historical and palaeoflood record of Katherine river, northern Australia: Evaluating the likelihood of extreme discharge events in the context of the 1998 flood', *Hydrological Processes*. John Wiley & Sons, Ltd., 19(20), pp. 4107–4120. doi: 10.1002/hyp.5875.

Sandgren, P., Snowball, I. (2001). 'Application of mineral magnetic techniques to paleolimnology', in Last, W. M. and Smol, J. P. (eds) *Tracking Environmental Change Using Lake Sediments*. Dordrecht: Kluwer Academic Publishers, pp. 217–238.

Sauer, A., Schanze, J., Walz, U. (2007). 'Development of a GIS-based risk assessment methodology for flood pollutants. Information technologies in environmental Engineering', in Gómez, J. M. et al. (eds) *Information Technologies in Environmental Engineering*. Springer, Berlin Heidelberg, pp. 357–366. doi: 10.1007/978-3-540-71335-7_37.

Schiefer, E., Gilbert, R., Hassan, M. (2011). 'A lake sediment-based proxy of floods in the Rocky Mountain Front Ranges, Canada', *Journal of Paleolimnology*. Springer Netherlands, 45(2), pp. 137–149. doi: 10.1007/s10933-010-9485-6.

Schillereff, D. N., Chiverrell, R. C., Macdonald, N., Hooke, J. M. (2014). 'Flood stratigraphies in lake sediments: A review', *Earth-Science Reviews*, 135(0), pp. 17–37. http://dx.doi.org/10.1016/j.earscirev.2014.03.011.

Schulz, M., Buttner, O., Baborowski, M., Bohme, M., Matthies, M., von Tumpling, W. (2009). 'A dynamic model to simulate Arsenic, Lead, and Mercury contamination in the terrestrial environment during extreme floods of rivers', *Clean-Soil Air Water*, 37(3), pp. 209–217. Available at: www.wileyinterscience.com.

Sharley, D. J., Sharp, S., Bourgues, S., Pettigrove, V. J. (2016). 'Detecting long-term temporal trends in sediment-bound trace metals from urbanised catchments', *Environmental Pollution*, 219, pp. 705–713. doi: https://doi.org/10.1016/j.envpol.2016.06.072.

Silke, R. (2013). *Giant Mine Milling and Roasting Process, Yellowknife, NWT: A Historical Summary*. Yellowknife, NT: Giant Mine Remediation Team Directorate, Aboriginal Affairs and Northern Development Canada.

Sinclair Knight Merz. (2005). *Determination of the Minimum Environmental Water Requirement for the Yarra River. Minimum Environmental Water Requirements and Complentary Works Recommendations*. Melbourne, Australia: Melbourne Water.

Smol, J. P., Birks, H. J. B., Lotter, A. F., Juggins, S. (2012). 'The march towards the quantitative analysis of palaeolimnological data', in Birks, H. J. B. et al. (eds) *Tracking Environmental Change Using Lake Sediments: Volume 5: Data Handling and Numerical Techniques*. Dordrecht: Springer Science & Business Media, pp. 3–18.

Sokal, M. A., Hall, R. I., Wolfe, B. B. (2008). 'Relationships between hydrological and limnological conditions in lakes of the Slave River Delta (NWT, Canada) and quantification of their roles on sedimentary diatom assemblages', *Journal of Paleolimnology*, 39(4), pp. 533–550. doi: 10.1007/s10933-007-9128-8.

Sokal, M. A., Hall, R. I., Wolfe, B. B. (2010). 'The role of flooding on inter-annual and seasonal variability of lake water chemistry, phytoplankton diatom communities and macrophyte biomass in the Slave River Delta (Northwest Territories, Canada)', *Ecohydrology*. John Wiley & Sons, Ltd., 3(1), pp. 41–54. doi: 10.1002/eco.93.

Sonneman, J. A., Walsh, C. J., Breen, P. F., Sharpe, A. K. (2001). 'Effects of urbanization on streams of the Melbourne region, Victoria, Australia. II. Benthic diatom communities', *Freshwater Biology*, 46(4), pp. 553–565. doi: 10.1046/j.1365-2427.2001.00689.x.

Taylor, S. E., Birch, G. F., Links, F. (2004). 'Historical catchment changes and temporal impact on sediment of the receiving basin, Port Jackson, New South Wales', *Australian Journal of Earth Sciences*. Blackwell Science Pty, 51(2), pp. 233–246. doi: 10.1111/j.1400-0952.2004.01054.x.

Telford, R. J., Heegaard, E., Birks, H. J. B. (2004). 'All age–depth models are wrong: but how badly?', *Quaternary Science Reviews*, 23(1–2), pp. 1–5. http://dx.doi.org/10.1016/j.quascirev.2003.11.003.

Teller, J. T., Last, W. M. (1990). 'Paleohydrological indicators in playas and salt lakes, with examples from Canada, Australia, and Africa', *Palaeogeography, Palaeoclimatology, Palaeoecology*, 76, pp. 215–240.

Thorndycraft, V. R., Hu, Y., Oldfield, F., Crooks, P. R. J., Appleby, P. G. (1998). 'Individual flood events detected in the recent sediments of the Petit Lac d'Annecy, eastern France', *The Holocene*, 8(6), pp. 741–746. doi: https://doi.org/10.1191/095968398668590504.

Timoney, K. P., Lee, P. (2009). 'Does the Alberta tar sands industry pollute? The scientific evidence', *The Open Conservation Biology Journal*, 3(2009), pp. 65–81.

Vane, C. H., Chenery, S. R., Harrison, I., Kim, A. W., Moss-Hayes, V., Jones, D. G. (2011). 'Chemical signatures of the Anthropocene in the Clyde estuary, UK: sediment-hosted Pb, 207/206Pb, total petroleum hydrocarbon, polyaromatic hydrocarbon and polychlorinated biphenyl pollution records', *Philosophical Transactions of the Royal Society A: Mathematical, Physical and Engineering Sciences*, 369(1938), pp. 1085–1111. doi: 10.1098/rsta.2010.0298.

Vasskog, K., Nesje, A., Storen, E. N., Waldmann, N., Chapron, E., Ariztegui, D. (2011). 'A Holocene record of snow-avalanche and flood activity reconstructed from a lacustrine sedimentary sequence in Oldevatnet, western Norway', *The Holocene*. London, United States, London, 21(4), pp. 597–614. doi: 10.1177/0959683610391316.

Velleux, M. L., England Jr, J. F., Julien, P. Y. (2008). 'TREX: Spatially distributed model to assess watershed contaminant transport and fate', *Science of the Total Environment*, 404(1), pp. 113–128. doi: http://dx.doi.org/10.1016/j.scitotenv.2008.05.053.

Vic SES. (2012). *Manningham City Countil Flood Emergency Plan: A Sub-plan of the Municipal Emergency Management Plan*. Melbourne, Australia.

Vic SES. (2013). *Boroondara City Council Flood Emergency Plan: A Sub-plan of the Municipal Emergency Management Plan*. Melbourne, Australia. Available at: www.ses.vic.gov.au.

Walker, M. (2012). 'Dating environmental change and constructing chronologies', in Matthews, J. A. (eds) *The SAGE Handbook of Environmental Change. Volume 1: Approaches, Evidence and Causes*. London: SAGE Publications.

Walling, D. E., He, Q., Nicholas, A. P. (1996). 'Floodplains as suspended sediment sinks', in Anderson, M. G., Walling, D. E., and Bates, P. D. (eds) *Floodplain Processes*. Chichester: John Wiley & Sons, pp. 399–440.

Wesche, S. (2007). 'Adapting to change in Canada's north: Voices from Fort Resolution, NWT', *Meridian Spring/Summer*, 2007, pp. 19–25.

Wesche, S. (2009). *Responding to Change in a Northern Aboriginal Community: Linking Social and Ecological Perspectives*. Waterloo, Ontario, Canada: Wilfrid Laurier University.

Wheater, H. S. (2002). 'Progress in and prospects for fluvial flood modelling', *Philosophical Transactions of the Royal Society of London. Series A: Mathematical, Physical and Engineering Sciences*, 360(1796), pp. 1409–1431. doi: 10.1098/rsta.2002.1007.

Whitlock, C., Higuera, P. E., McWethy, D. B., Briles, C. (2010). 'Paleoecological perspectives on fire ecology: revisiting the fire-regime concept', *The Open Ecology Journal*, 3, pp. 6–23.

Wiklund, J. A., Bozinovski, N., Hall, R. I., Wolfe, B. B. (2010). 'Epiphytic diatoms as flood indicators', *Journal of Paleolimnology*, 44(1), pp. 25–42. doi: 10.1007/s10933-009-9383-y.

Wiklund, J. A., Hall, R. I., Wolfe, B. B., Edwards, T. W. D., Farwell, A. J., Dixon, D. G. (2014). 'Use of pre-industrial floodplain lake sediments to establish baseline river metal concentrations downstream of Alberta oil sands: a new approach for detecting pollution of rivers', *Environmental Research Letters*, 9(12), p. 124019. Available at: http://stacks.iop.org/1748-9326/9/i=12/a=124019.

Winter, L. T., Foster, I. D. L., Charlesworth, S. M., Lees, J. A. (2001). 'Floodplain lakes as sinks for sediment-associated contaminants – a new source of proxy hydrological data?', *Science of the Total Environment*, 266(1–3), pp. 187–194. Available at: www.sciencedirect.com/science/article/pii/S0048969700007452.

Wolfe, B. B., Edwards, T. W. D., Elgood, R. J., Beuning, K. R. M. (2001). 'Carbon and oxygen isotope analysis of lake sediment cellulose: Methods and applications BT – Tracking environmental change using lake sediments: Physical and geochemical methods', in Last, W. M. and Smol, J. P. (eds) Dordrecht: Springer Netherlands, pp. 373–400. doi: 10.1007/0-306-47670-3_14.

Wolfe, B. B., Hall, R. I., Last, W. M., Edwards, T. W. D., English, M. C., Karst-Riddoch, T. L., Paterson, A., Palmini, R. (2006). 'Reconstruction of multi-century flood histories from oxbow lake sediments, Peace-Athabasca Delta, Canada', *Hydrological Processes*, 20, pp. 4131–4153.

Wolfe, B. B. et al. (2007). 'Progress in isotope paleohydrology using lake sediment cellulose', *Journal of Paleolimnology*, 37(2), pp. 221–231. doi: 10.1007/s10933-006-9015-8.

Wolfe, B. B., Falcone, M. D., Clogg-Wright, K. P., Mongeon, C. L., Yi, Y., Brock, B. E., St. Amour, N. A., Mark, W. A., Edwards, T. W. D. (2008). 'Climate-driven shifts in quantity and seasonality of river discharge over the past 1000 years from the hydrographic apex of North America', *Geophysical Research Letters*, 35(24), p. L24402. doi: 10.1029/2008GL036125.

Wolfe, B. B., Hall, R. I., Edwards, T. W. D., Johnston, J. W. (2012). 'Developing temporal hydroecological perspectives to inform stewardship of a northern floodplain landscape subject to multiple stressors: Paleolimnological investigations of the Peace-Athabasca Delta', *Environmental Reviews*. Canadian Science Publishing, 20(3), pp. 191–210. doi: http://dx.doi.org/10.1139/a2012-008.

Wölz, J., Schulze, T., Varel, U. R., Fleig, M., Reifferscheid, G., Brack, W., Kuhlers, D., Braunbeck, T., Hollert, H. (2011). 'Investigation on soil contamination at recently inundated and non-inundated sites', *Journal of Soils and Sediments*, 11, pp. 82–92.

Xu, Y., Sun, Q., Yi, L., Yin, X., Wang, A., Li, Y., Chen, J. (2014). 'The source of natural and anthropogenic heavy metals in the sediments of the Minjiang River Estuary (SE China): Implications for historical pollution', *Science of the Total Environment*, 493, pp. 729–736. doi: https://doi.org/10.1016/j.scitotenv.2014.06.046.

Yamashita, N., Kannan, K., Imagawa, T., Villeneuve, D., Hashimoto, S., Miyazaki, A., Giesy, J. P. (2000). 'Vertical Profile of Polychlorinated Dibenzo-p-dioxins, Dibenzofurans, Naphthalenes, Biphenyls, Polycyclic Aromatic Hydrocarbons, and Alkylphenols in a Sediment Core from Tokyo Bay, Japan', *Environmental Science & Technology*. American Chemical Society, 34(17), pp. 3560–3567. doi: 10.1021/es001043i.

Zak, K., Rohovec, J., Navratil, T. (2009). 'Fluxes of heavy metals from a highly polluted watershed during flood events: A case study of the Litavka River, Czech Republic', *Water Air and Soil Pollution*, 203(1–4), pp. 343–358.

Zong, Y., Innes, J. B., Wang, Z., Chen, Z. (2011). 'Mid-Holocene coastal hydrology and salinity changes in the east Taihu area of the lower Yangtze wetlands, China', *Quaternary Research*, 76(1), pp. 69–82. http://dx.doi.org/10.1016/j.yqres.2011.03.005.

Zonta, R., Collavini, F., Zaggia, L., Zuliani, A. (2005). 'The effect of floods on the transport of suspended sediments and contaminants: A case study from the estuary of the Dese River (Venice Lagoon, Italy)', *Environment International*, 31(7), pp. 948–958. https://doi.org/10.1016/j.envint.2005.05.005.

7 Coping with Water Supply and Water Quality Problems during Floods and Their Management

C. K. Jain and Rakesh Kumar

7.1 INTRODUCTION

Water supply during high floods is tagged with flood relief/emergency relief operations and is mainly the responsibility of the State Government and local administration. This chapter discusses the basic principles such as planning, identification of source of supply, its development and transmission of water and quality control from the view of appropriate technology in emergency flood situation.

7.2 PROBLEM OF POTABLE WATER AVAILABILITY

Paradoxically the excess of water causes shortage of potable water. The fury of floods also destroys all infrastructures of civilised society, for example, roads, electricity, telephone lines, public water and sanitation system as well as the source of potable water, which is the basic need of any living being. The spot sources go under water and are sometimes washed away, the ponds become indistinguishable from the vast sheet of muddy contaminated flood water. The age-old traditional potable water sources suddenly disappear and/or become dangerously contaminated.

After shelter, the most sought-after commodity during floods is the potable water. Unlike dry food, supply of water cannot be distributed easily by expensive transportation such as helicopters. People are seen wading through flood water or sailing in boats, rafts and so on to distant potable water sources.

7.3 DRINKING WATER SUPPLY

7.3.1 Quantity

It is difficult to ascertain the minimum per capita water requirement for domestic use including drinking water. Requirements will vary depending upon the prevailing climatic condition. Normally daily requirement for drinking water for an adult in a rainy climate may be about 2.2–3 l/day; apart from this, he or she will require water for other personal use also. A part of it will be obtainable from food. However, for emergency supply during evacuation, we may take per adult requirement for drinking, cooking and dishwashing as 5 l/day temporarily until the time water engineers are able to increase the needs to 10–15 litre per capita per day (lpcd) and more as early as possible. For the camp inmates in temporary shelter, tube wells/tap connection and so on may be provided taking water demand of 15–20 lpcd. For mass feeding centres, the requirement will be about 20–30 lpcd.

This is the barest minimum quantity to be provided under extreme adverse working conditions such as flood for a shorter period, even then, every effort should be made to make more potable water available to the flood relief camp inmates for maintaining sanitary conditions and personal hygiene for a longer period of day. Therefore:

- To start with, provide at least 5 lpcd
- For camps, provide 15–20 lpcd
- For mass feeding centre, provide 20–30 lpcd

7.3.2 Quality

There are three main aspects of water quality, which are physical, chemical and microbiological. Physical quality, such as colour, taste, smell, turbidity, hardness and temperature, is not generally an indication of potential health hazard, but it is of great importance for acceptability to the people. When no means exists to improve physical quality, under scarcity condition, people will tolerate deficiencies in this respect for short period. Chlorine tablets and water purifier packets for in situ purification may be supplied to the people as per demand.

The chemical quality may be described in terms of the potentially toxic substances. Chemical insecticides, herbicides and so on may create a potential health hazard. This should be taken care of by careful selection of sources.

In field condition, the microbiological quality of water is difficult to assess. It requires good laboratory facility, which is almost impossible to obtain during emergency flood conditions. However, microbiological acceptable/non-acceptable quick test is available and should be used. Proper selection of source such as tube well and ring well, and in case of surface source, avoidance of proximity of pit latrine, stoppage of animal sharing of water, bathing at surface source, washing and so on safeguard the microbiological quality and further disinfection will help in improving the quality of the water distributed.

7.3.3 Sources

Generally, in flood situation, conventional sources of drinking water are either out of reach of the people or come to cease because of floods. Therefore, often new sources are to be created by installing new hand pump tube wells in schemes where piped water supply network exists and common water tap connections are to be given to camp inmates. Similarly, provision of chemicals such as halogen tablets and water purifier packets are to be made available, so that even flood/pond water can be used as source. In flood relief camps, a minimum of one hand tube well may be installed per 500 camp inmates, and one tap connection per 250 camp inmates should do in case of intermittent piped supply scheme.

7.3.4 Accessibility

The accessibility of source to the user is an important factor. Average distance of users from source may be taken as an indicator and should be within 100 metres. It is important, however, to take account also of the terrain, since the time taken and effort required in mountainous country can be considerably more than that required in the plains. Water sources should be easily accessible. Normal accessibility is likely to be disrupted during floods. Paths are submerged, bridges are broken/washed away, very often country boats, banana shaft rafts, rubber boats allow accessibility – provision for which has to be kept.

7.3.5 Reliability

Whatever the quantity of water is supplied must be assuredly available as such to the consumer. In case of pipe water supply scheme, proper timing, punctuality and necessary quantity, in case of hand pump, ring well these must be maintained reliably and functioning and in case of supply of halogen tablets, water purifier packets, these must be made available to the people at a pre-appointed time and place. In short it is necessary to maintain:

- Punctuality and regularity of pipe water supply system
- Minimum quantity and quality
- All spot sources like hand tube wells and ring wells in working condition
- Availability of halogen tablets, disinfectant, water purifier packets in prefixed places in adequate quantity
- Adequate stock of spare parts.

7.4 CARING FOR THE WATER SUPPLY SYSTEM

7.4.1 River Sources

Officer-in-charge of river source water supply schemes must take special precautions to protect intake wells and floating barges against flood damage. Where multi-level intake pipes are in use, the lower level pipes should be closed to allow only less turbid water to be collected in intake well. Protection to intake well, infiltration gallery must be checked and a report regarding its collection must be assessed in time and works be completed before the starting of rainy season.

Where floating barges are used, the anchor posts be checked and all iron chains, steel wire ropes etc. be inspected. Special precaution is to be taken so that the barge is not damaged due to high velocity of water in the river. Soundness of the barge structures is to be seen and necessary repairing works to be carried out before the onset of flood. The flexible delivery pipe along with power cable and transformer are to be properly looked after to prevent submergence. Arrangement for round the clock manning of the floating barge during the high flood must be made. Proper tools and plants and accessories should be kept ready.

7.4.2 Reservoir/Lake Source

Special supervision should be exercised over shallow coves or bays. These are areas which provide breeding ground for microorganisms, which will then be carried in to the deeper portion unless precautions are not taken in time. These are associated with taste and odour problems of water.

Reservoir with steep banks and flooded areas usually do not support troublesome growth of weeds. The marginal weeds and other plants can be removed by cutting them during dry period to avoid problems during and after floods. The growth of planktons and taste and odour caused by them after the flood in reservoir can be taken care of by treatment with copper sulphate dose (0.3 mg/l). The dose of copper sulphate is not to be computed from the total volume of water in the reservoir, because the plankton usually gathers in the upper circulating zone (4–9 m).

Most difficult problems are due to the presence of bugs. The problem is of course more psychological than anything else, as no one likes to drink water contaminated with bugs like chironomide, a midget fly. Although it does not cause harm to people not spreads diseases, the larva of the fly is the blood worm, a thin, long hair like creature possess two peculiarities, (i) it can pass through the water treatment plant filter with ease and (ii) it is not affected by chlorination in concentration permissible in potable water. Covered reservoirs do not have this problem. Spraying of strong chlorine solution (500 mg/l) at the bottom of the reservoir when empty will ease the problem.

7.4.3 Distribution System

Special precautions for the distribution system should be taken during high flood. The PVC/HDPE pipelines being lighter have the tendency to float whenever the soil cover on them is eroded and the pipe is empty. The pipeline itself is especially vulnerable to erosion during the first two wet seasons. During the period the back fill and the ridge over the line is less compacted than the surrounding earth, and the surface rainwater quickly expose the week point of the line. Care has to be taken to keep earth cover intact or sand bags may be put to cover the vulnerable section of the pipe line – otherwise the speeding flood water will scour out the pipe line and tear it apart. Pipelines should be planted with grass which should eventually bind the soil. The hanging portion over the small culvert must pass through galvanised iron (GI) and cast iron (CI) pipes or replaced by them to avoid damage by swift flowing swollen flood water through the culverts.

In a distribution system, special care is to be taken to prevent exposure of pipelines due to soil erosion. It occurs along the bank of water courses, along the stretch of metal road where flood water topples the road, near the culverts etc.

Precaution is to be taken also about the pipe crossing over water courses along culverts. Often, culverts flow full and exposed pipeline is subjected to lateral pressure of swift flowing water causing cracks, breakage, splitting and washing away of pipeline. The exposed PVC/HDPE pipes should have casing pipes.

7.4.4 Underground Reservoirs

It is noted that during floods, the ground water level is high and soil remains saturated. The partially/fully underground structures are often subjected to the upward thrust equal to the weight of ground water displaced by the structure. This upward thrust tends to unstable the structure. If not properly designed, the structure may become unstable and lifted. Therefore, it is important to keep the structures always filled up with water.

To determine the dosage of disinfecting powder (bleaching powder) required to disinfect water, a simple test may be carried out in the field and total quantity of bleaching powder required for the reservoir may be calculated. Bleaching powder at 4 mg/l straight way may be used and residual chlorine may be determined at every half an hour. The dose may be increased till the residual chlorine content of water comes to 0.2–0.4 mg/l.

7.4.5 Ponds

In flood-prone area where people mainly depend on surface sources of water like ponds for drinking and no other protected water supply arrangement is available, at least one pond be selected in each such village for maintaining as a source of drinking water during floods. The selected pond is to be such that it does not submerge in flood water and has well protected raised bank. In absence, construction of temporary dike, expedient heightening of dikes, earthen dams, embankments, filling up the beach in flood protection structure with the help of sand bags, earth rock fills or by other available means may be necessary.

The ponds will have to be properly cleaned, disinfected and protected from submergence. Proper chlorinating of an uncovered earthen pond is very difficult. First the pond will have to be cleaned, de weeded, turbidity remove as far as possible by chemical treatment. Care is to be taken for the fish in water. People raise objections (often to save fish) about application of chemicals. Generally toxic level of chlorine dose to fish is below that required for disinfection.

7.4.6 Hand Tube Wells

In flood prone areas the hand pumps may be constructed over raised pedestals, above flood level and provided with steps and railing so that during flood people can collect water wading through flood water by boat or raft. All cracked/broken platforms should be repaired in time to prevent contamination of ground water.

In flood-prone/affected area, sometimes the filter of the tube well is filled with sand and the tube well become inoperational. In this case de-sanding is necessary. The pump head is to be removed first to facilitate sludging operation. As the tube well is 1.5″ in diameter, a smaller diameter sludging pipe, preferably a number of 3-m long ¾″ diameter threaded PVC pipe with 12″ long toothed GI pipe at the tip as cutter is usually used. A ⅓ cut oil barrel is properly mounted on the top of the tube well to hold and provide continuous flow of water as sludging proceeds. With gradual removal of sand, additional sludging pipes are needed till the entire tube well is cleared of sand. After desanding, the tube well must be fully developed by continuously pumping for several hours till sand-free water is ensured. In case of choked-up tube well cannot be restored by desanding, it may be resunk properly after withdraw if economically viable.

7.5 PREPARATION FOR EMERGENCIES

Every public water supply system, whether large or small, should be well prepared for emergency operating procedures, which may be necessitated by catastrophes such as earthquake, flood, war damage or evacuation of populations due to war or sudden catastrophe. It must be remembered that, in such circumstances, water is probably the most urgent need of individuals. If provision is not made quickly for a supply of safe water, people will use any available source, polluted or not, with the consequent hazards of epidemics. In fact, the amount of water available may govern the number of evacuees who may be received and handled in a particular village or town. For this reason, it is recommended that, soon after the start of emergency, or even before, a realistic inventory is made covering all available sources of water supplies, public or private, together with their capacities, location, accessibility and potability. This inventory should also include personnel resources and availability, emergency type, water supply equipment, hand and motor pumps, water-tank trucks, pipe, accessories (especially coupling and jointing materials), mobile or portable filter units, tools, spare parts and chemicals (especially those are used for water disinfection purposes).

It goes without saying that every step should be taken to conserve water, and that every possible sanitary safe guard should be employed to protect existing supplies – especially surface supplies such as lakes, ponds, open reservoirs and springs – from contamination. While ground water supplies from properly constructed wells, infiltration galleries and spring structures may be considered safe, all surface water should be considered of doubtful quality and may be disinfected by boiling, chlorinating or disinfecting by iodine compounds or potassium permanganate. At times of emergencies, the free residual chlorine in reasonably clear water should be not less than 5 mg/l. Free residual chlorine is to be measured after 30 minutes of contact, when the water has not had the benefit of prior filtration. The raw water should be taken from an approved source, and if it has strong taste, odour or colour, it should be avoided as far as possible.

7.5.1 Operation of Treatment Plant

When rivers are in spate/flooded, special problems arise due to back flow of river water, through the sludge disposal pipeline to the clarifier. This happen especially when the treatment plant is on the riverbank or very near to it with sludge disposal in the river or low laying water body. In that case, it is necessary to close the sludge drain with sand bags, and sludge is disposed of by laying separate temporary discharge pipe from the sludge well to push water over the sand bag barrier/embankment into the river/suitable drain. Another method is to create as less sludge as possible. It is prudent to use high-performance chemicals such as polyaluminium chloride or polyelectrolyte that reduces sludge and less volume and to that extent more advantageous.

It is a common scene during floods in rural areas that people row shaft/boat up to the treatment plant and collect water from clarifier/clean water reservoir when plant fails to distribute water due to many reasons such as non-availability of power, breakdown of machinery and submergence of distribution system. Though it is not desirable, often it has to be allowed. The staff should guide proper withdrawal of water and check water quality for free residual chlorine. Care is to be taken to see that a single clean bucket does withdrawal of water. Individual buckets must not be allowed.

7.5.2 Conservation of Chemicals

During floods, the solids load in raw water increases tremendously which translates to a corresponding increase in the demand of chemicals for treatment, and due to sudden blockade of the transportation system/flooding of godown, scarcity of water purification chemicals often occurs and their replenishment becomes difficult.

Every possible step will have to be taken to save chemicals while still adding the necessary amount for effective treatment results. Jar tests help optimise the amount needed to be used when treatment occurs. For a short period of time, filters may have to be overloaded; dual media filters are better in this respect due to their inherent capacities to treat higher turbidity loads with consistent performance.

Use of polyelectrolytes, mostly cationic, further reduces the consumption of chemicals. It reduces the filter cycle thereby the frequency of backwash, saving water and solving sludge disposal problem to considerable extent. This is an important consideration under flood situation. Some of the brand names, just as examples, are Catfloc-T, Setlyte N-10 and Setlyte AP-50.

7.6 TESTING OF WATER IN THE FIELD

Complete physical, chemical and bacteriological testing of water in the field, particularly in flood condition, is not feasible. However, potable water testing kits with prepared reagents and minimum necessary glassware in a single compact box with instructions for use can be used at sectional level to determine pH, turbidity, alkalinity, iron, hardness, residual chlorine and so on.

However, in the absence of testing kits, some inexpensive and easy-to-use water testing facilities should be kept ready at sectional officer level. Since in the flood-affected areas, often the source of water is turbid water, the following simple equipment will be useful:

i) Raw water turbidimeter.
ii) Clear water turbidimeter.
(Comparator box with standard ampoules)
iii) Orthotolidine chlorine testing kit.
iv) Plastic beaker, 250 ml.
v) Graduated plastic cylinder (200 ml).
vi) Sampling bottle (500 ml) with graduated stopper.
vii) Test tubes.

7.6.1 Bacteriological Test

As per the Bureau of Indian Standards, no faecal coliform should be present in 100 ml of a water sample used for drinking purpose. Faecal coliform is the positive indication of the presence of harmful bacteria. It has become the custom to report the results of the coliform test as a most probable number index which denotes the number of coliform bacteria in water sample.

To test bacteriological quality of drinking water, a well-equipped laboratory and services of expert microbiologist is required. Standard coliform or faecal coliform test takes long time and also are quite expensive. For bacteriological safety of water, the disinfected water be tested for its safety. When regular laboratory tests are not possible, simple field tests should be used to ensure the safety of water.

H_2S Strip Test: The presence of coliform in drinking water is consistently associated with hydrogen sulphide. For this reason, in this method, water-borne hydrogen sulphide bacteria act as the indicator of faecal pollution in iron-rich medium which supports the growth of sulphide bacteria present in water. Hydrogen sulphide produced during the growth of bacteria reacts with iron and turns the water samples black, indicating the presence of harmful pathogens in the water sample.

Presence–absence test: Raw water for test is to be poured into the small vial containing appropriate bacterial culture medium up to the mark in the body. The vial is to be kept closed at 30–37°C for 18–24 hours. If pathogenic bacteria are present, the colour of water will turn into black. Such water is to be considered bacteriologically contaminated. The cost of each vial is around 0.2 USD. The field test kits containing both physico-chemical and bacteriological tests supplied by UNICEFE may be suitably used.

7.7 DISINFECTION OF WATER

The water can be disinfected by using bleaching powder. The quantity of bleaching powder can be fixed to achieve 0.4 mg/l of residual chlorine after half-an-hour contact. Bleaching powder consumption (g/m^3) for different sources of water with 25% chlorine content will be as follows:

Source of Water	Chlorine Requirement (mg/l)	Bleaching Powder Consumption (g/m^3)
Rain water	0.5–1.0	2–4
Deep well water	0.5–1.0	2–4
Shallow well water	1.0–1.5	4–6
Spring water	1.5–2.0	6–8
Turbid river water	2.0–2.5	8–10
Pond water (in good environment)	2.0–2.5	8–10
Pond water (in bad environment)	2.5–3.0	10–12
Lightly polluted stored water	2.0–2.5	8–10
Heavily polluted stored water	2.5–3.0	10–12

Bleaching powder consumption (g) for different water depth and diameter of wells will be as follows:

Water Depth (m)	Well Diameter 0.5 m	0.8 m	1.0 m	1.5 m	2.0 m
1	1.60	4.10	6.40	14.40	25.60
2	3.20	8.20	12.80	28.40	51.20
3	4.80	12.30	19.20	43.20	76.20
4	6.40	16.40	25.60	57.60	102.40
5	8.00	20.50	32.00	72.00	128.00
6	9.60	24.60	38.40	86.40	153.60
7	11.20	28.70	44.80	100.80	179.20
8	12.80	32.80	51.20	115.20	204.80
9	14.40	36.90	57.60	129.60	230.40
10	16.00	41.00	64.00	144.00	265.00

The required chlorine is 2 mg of bleaching powder containing 25% free available chlorine of each litre of treated water.

Residual Chlorine Test: Residual chlorine test is important to determine chlorine in water which ensures safety. Orthotolidine test is commonly used for its simplicity in the field. Equipment needed include chloroscope/chlorine test kit. This kit is a small box containing two test tubes with colour slides provided with orthotolidine solution in a small box and a glass rod for stirring.

7.7.1 Well Disinfection

The well to be disinfected should be *pucca* (permanent) well or constructed using RCC ring, which is commonly known as ring well. Water in *kutcha* (temporary) wells cannot be disinfected reliably and therefore should not be selected for disinfection. The well should be covered, and the water to be drawn by hand pump fixed on it. The ring well should be having platform around it, all cracks in the platform must be repaired and joints between the rings be checked and the seepage of water through them be stopped. All ring wells must be checked and repaired before the onset of flood. Ring well water should be continuously and regularly disinfected by bleaching powder using double-pot chlorinator. This method of chlorination also helps to reduce iron content of well water by oxidising and precipitating the dissolved iron in the water and thus improving the quality of water. However, where double-pot chlorinators are not available, dosing by bleaching powder can disinfect the well water.

Process of Adding Bleaching Powder: The disinfecting bleaching powder should be added only in the form of a solution. The powder is first dissolved in a small quantity of water to form a paste and then diluted suitably. This should be poured and dispersed evenly into the well; the water is to be stirred for intimate mixing.

The residual chlorine of the water should be tested half an hour after adding, the residual chlorine should be at least 0.2–0.4 mg/l for successful disinfection (the chlorine demand of the water be first determined for the calculation of the quantity of bleaching powder to be added).

During the time of emergency when it is not possible to work out required chlorine, the dosage of 4 mg/l of bleaching powder can be used straightaway and the result arrived at the end of half an hour, the operation repeated until sufficient residual chlorine is available in the water. Care should be taken to see that the bottom of well is not disturbed as this will lead to wastage of chlorine.

7.7.2 Disinfection of Reservoir and Tanks

Sometimes during floods, contaminated water enters into reservoirs and tanks. It is therefore required to be disinfected before being used. Before disinfecting a reservoir or a storage tank, first, it is essential to remove all dirt, scales and other loose materials from the walls and bottom. The interior should than be flushed out and the water be disinfected by one of the methods detailed bellow.

It is essential for small tanks to compute capacity first and add 1 kg of 25% bleaching powder completely dissolved, per 1,000 gallons (4,500 litres). Fill up the tank with water, allow it to stand 12–24 hrs and finally drain the water through the scour valve to waste.

Make sure that the tank is adequately ventilated before entering into it. Fill the tank after an hour the work is completed. When disinfecting the pressure tanks, it is important to note that it is essential to open the air relief or other valves at the highest point to facilitate the air release. Air should be readmitted before pumping is started. In all cases, a residual chlorine test should be carried out to detect residual chlorine in the water drained out of the tank. In case if there is no residual chlorine detected, the disinfection should be repeated.

7.7.3 Disinfection of Distribution System

Contamination of distribution system in a piped water supply scheme is likely to occur whenever an area is submerged under flood water; especially the intermittent water supply makes the distribution pipe lines more vulnerable. Therefore, it is imperative to disinfect the pipelines before brought into use. Every such pipeline should be swabbed and flushed in order to remove all foreign matters immediately before use. The packing and jointing material should be cleaned and disinfected by immersing them in a 50 mg/l chlorine solution for at least 30 min.

The disinfecting solution is fed through a funnel or a hose to the hydrant or opening made for this purpose at the highest part of the empty line. Since all air valves are placed at this point removing an air valve is a convenient way to provide a point of entry.

7.7.4 Disinfection of Ponds

In flood-affected areas where water from neither spot sources nor piped water scheme is available and reservoirs/ponds are contaminated by flood water, immediate steps are to be taken to decontaminate, disinfect at least one pond/reservoir preferably in central position of each affected village.

It is preferable to select public reservoirs with high banks and protective fencing. If this is not available, a private pond also may be selected through formal governemental agencies/local bodies; in this case, the owner must give clear consent for unconditional utilisation of water by the villagers during and after flood for the period deemed necessary by authority. Ponds/reservoirs utilised as fishery should be avoided for their use as part of the drinking supply systems as far as practically possible.

To start with the pond disinfection, the sources of contamination first be identified and closed. Any breaches on the bank of the pond be closed. Also adding manure, fish food and so on be suspended. The inward slope of bank especially the water line where the pond water and sloped bank meet must be kept clean. Long grass, weeds and so on to be cleaned, and bleaching powder sprinkled on the bank, at least over an area of up to 1-ft. above the water level.

First the quantity of water of the pond is calculated and adequate quantity of lime and alum (30 and 60 mg/l doses) are added before disinfection. For easy solubility, blocks of alum are broken into small pieces and put in a gunny bag and tied with rope; to apply lime a similar procedure is used. Both bags are suspended into water from a small boat, either a banana raft or a bamboo raft.

The raft is dragged to and from throughout the length and breadth of the pond, so that all the chemicals are mixed into the water; the raft or boat is usually propelled in parallel courses about 2.5–3 m apart. Stirring of the water by bamboo pole will enhance the coagulation, but the bottom of the tank should not be disturbed, as this would lead to wastage of the added chlorine (some of the chlorine will be used up in reacting with the disturbed soil).

After ⅘ hours of settling, the water will be cleaner, and the bleaching powder be used carefully, only in the form of solution. The required quantity of bleaching powder is to be calculated. It is mixed with the water of the pond, so that the pond water will contain at least 2 mg/l of chlorine. At a time ½ kg of bleaching powder is to be mixed well with water in a 10-litre plastic bucket. This solution is to be sprinkled on the pond water. With a long bamboo pole water will be stirred to disperse the bleaching powder solution. It is important to avoid high local concentration of chlorine in water during this operation as it will kill the pond's aquatic life; a chlorine concentration exceeding 10 mg/l will kill even the largest fish. This operation is to be repeated in quick succession till the entire required quantity of bleaching powder is mixed with water. After half an hour, the residual chlorine is to be measured; it should be around 0.3–0.4 mg/l. It is not possible to maintain the residual chlorine to this desired level for longer period in every point of pond water especially when the pond is big.

It is desirable to keep the residual chlorine at this level at the water draw-off point; for this purpose, every day a few grams of bleaching powder dissolved in a bucket should be sprinkled and mixed in the pond water at draw-off point. A Panchayat worker or villager, from amongst the users, may be shown the process of utilisation of bleaching powder and this may be checked from time to time.

7.8 PURIFICATION OF WATER

Purification of water is very important in the prevention of diseases. For large-scale water purification in emergency situation, storage is a practical method. In this method, water is stored in a pollution-free large reservoir and kept open in sunlight. Within 24 hours, about 90% of suspended impurities are settled by gravity. Water gets clearer and sunlight with UV ray which is a strong germicide can penetrate better. Certain beneficial chemical changes also take place during storage. The aerobic bacteria oxidise organic matter with the dissolved oxygen in the water.

There are three general methods for the purification of water on an individual or domestic scale, namely, boiling, sunlight and chemical disinfection.

7.8.1 Boiling

Boiling is a satisfactory method for destroying the disease-causing organisms in the water; and it is equally effective whether the water is clear or cloudy, relatively pure or heavily contaminated with organic matter. Boiling destroys all forms of disease-causing organisms usually encountered in the water: bacteria, spores, cercariae, cysts and ova.

The amount of fuel required to boil water varies with the type of fire, stove and vessels used. Under the condition usually associated with the boiling of drinking water, it requires about 1 kg of wood to boil 1 litre of water.

To be safe, water must be brought to a 'rolling' boil. The appearance of bubbles (simmering) is sometimes confused with boiling, as is the appearance of mist or stream over the water. None of these signs is sufficient indication that water has reached the boiling temperature. It is good practice to boil water in the same container in which it is to be cooled and stored, and to use this container for no other purpose.

Boiling alters the taste of water because it drives out the dissolved gases, particularly carbon dioxide. The frequent attempt to aerate water by stirring or by pouring it from one container to another should be avoided, as it poses a serious risk of recontamination during its handling. Water left for a period of several hours, up to a depth, in a partly filled container, where there is a good air surface exposed, even though the mouth of the container is covered, will lose most of the boiled taste. In any case, the taste is harmless and much to be preferred to polluted water.

7.8.2 Sunlight

Sunlight is a good source of UV radiation having strong bactericidal effect. About 99.9% of pathogenic bacteria are killed within an hour of exposure to sunrays. By exposure to sunrays for a day, it is possible to disinfect filtered water. Of course, the exposure has to be done in thin layer of water to allow full length penetration of sunrays. Overcast sky or short winter days however require supplementary treatment with chlorine.

7.8.3 Chemical Disinfection

Bleaching powder: Addition of chlorine through bleaching powder is a practical means of disinfection in emergency due to its easy availability, low cost, transportability, simple way of use, safety in handling and so on. The effectiveness of chlorine as disinfectant increases at relatively low pH.

Bleaching powder or chlorinated lime ($CaOCl_2$) is a white amorphous powder with a pungent smell of chlorine. When freshly made, it contains about 33% of 'available chlorine'. It is however an unstable compound. On exposure to air, light and moisture, it rapidly loses chlorine content. But when mixed with excess of lime, it retains its strength and this is called 'stabilised bleach'. Bleaching powder should be stored in a dark, cool and dry place in closed container that is resistant to corrosion. The chlorine content of bleaching powder stocks should be frequently checked.

Chlorine solution: Chlorine solution may be prepared from bleaching powder. If 4 kg of bleaching powder with 25% available chlorine is mixed with 20 litres of water, it will give 5% solution of chlorine. Like bleaching powder, chlorine solution is subjected to loss on exposure to light or on prolonged storage. The solution should be kept in a dark, cool and dry place in closed containers.

Calcium hypochlorite [$Ca(OCl)_2 4H_2O$]: Calcium hypochlorite is produced from chlorinated slurry of lime and sodium hydroxide. It is solid granules having 70% available chlorine and less than 3% lime.

Sodium hypochlorite ($NaOCl$): Sodium hypochlorite is a compound which contains 5% to 10% available chlorine. It is more stable than bleaching powder and deteriorates much less on storage, and it is also used for disinfecting the water. Being a liquid, it is easy to be mixed with water. It can be prepared by adding bleaching powder to sodium carbonate, allowing the calcium carbonate formed to settle and decanting the clear supernatant liquor.

Chlorine tablet: It is also known as halogen tablet. They are quite good for disinfecting small quantities of water, but costly. Another lacuna is that it cannot remove the turbidity. It is known that in turbid water, disinfection of water is never efficient. Therefore, it should be used in turbidity-free bacteriological contaminated water. Simplicity in use and transportability is its major advantage.

Weight of Tablet (g)	Available Chlorine	Volume of Water Each Tablet Can Purify (Litres)	Details of Container
2.5	300	240	1,000 Tablets in HDPE container
0.5	25	20	2,000 Tablets in HDPE container
0.5	25	20	100 Tablets in Amber Glass Bottle
0.25	5	4	100 Tablets in Amber Glass Bottle
0.125	1.25	1	100 Tablets in Amber Glass Bottle

Iodine: Iodine is an excellent disinfecting agent particularly for small-scale domestic applications. It is too expensive for large-scale use. Two drops of 2% ethanol solution of iodine is sufficient for disinfecting one litre of clear water. A contact time of 20–30 minutes is needed for effective disinfection.

Potassium permanganate: Although a powerful oxidising agent, it is not a satisfactory agent for disinfecting water. It may kill cholera-causing bacterium but is of little use against other disease-causing organisms. It has other drawbacks too, such as altering the colour, smell and taste of the water. People often refuse to drink potassium permanganate-treated water because of these changes. Therefore, potassium permanganate is not recommended for water disinfection.

7.9 MANAGEMENT OF THE PROBLEM

Experience has shown that flood is admittedly an uncertainty. Timing of flood occurrence is outside the management's control. During flood, protection to human life and safety of health are two overriding priorities. We require more adaptability, more resourcefulness and more hedges and retreats to combat non-repetitious situations and a foresight for anticipating obstacles before they arise. Prudent engineers identify these contingencies whose risks are so large that special programmes are justified along with continuously adjustable plans. Coping with disaster is not a one-time job, but it is a continuous process of planning and re-planning, of adjustment and re-adjustments, of learning and unlearning and it can be best done through:

- Preparedness
- Response
- Recovery

7.9.1 Preparedness

- Preparedness means having done the major works and thinking before it is needed.
- Taking measures which will enable government, communities and individuals to respond rapidly and effectively to disaster situations.
- Formulations of viable counter disaster plans, the maintenance of inventories of response and the training of personnel.

7.9.2 Response

- Measures taken immediately prior to and following disaster impact.
- Directed towards saving life-protecting property and to dealing with the immediately damage and other affect caused by the disaster.

7.9.3 Recovery

- This process is for restoration and reconstruction to return to proper level of functioning.
- Funds and human resources are important resources for which planning needs to be done carefully.

Different from usual development situation, during the first phase of a catastrophe or complex crisis, the approach to help people with water and sanitation has to be based on established routine and preparedness structure. Obviously, for preparedness, it would be useful to have some standard manual for public health engineering and a code of practice, which can serve as a guide during floods.

References

Adams, J. (1999), *Managing Water Supply & Sanitation in Emergencies*, Oxfam Skills & Practice Series, London.

APHA (1992), *Standard Methods for the Examination of Water and Waste Waters*, American Public Health Association, 18th Edition, Washington, DC.

AWWA (2007), *Water Supply in Emergencies*, AWWA Publications, USA.

BIS (2012), Specifications for Drinking Water, IS:10500:2012, Bureau of Indian Standards, New Delhi, India.

Davis, J. & Lambert, R. (2002), *Engineering in Emergencies*, ITDG Publishing, London.

Foster, J. (2007), *Water Supply & Sanitation after Flooding*, British Geological Survey, NERC.

8 Impact of Urban Flooding on Water Quality in India with Special Reference to Mumbai

Kapil Gupta

8.1 INTRODUCTION

Indian cities are presently undergoing large-scale expansion due to migration of a large number of people to the cities. India's urban population is expected to increase from 377 million. in 2011 to 600 million by 2031. The number of cities inhabiting million plus people is expected to increase from 50 in 2011 to 87 by 2031, and it is expected that 255 million people will live in these cities alone. The increased population has placed a large burden on the urban water infrastructure. The urban water table has dropped significantly; water has to be pumped from several kilometres away and the sewerage and drainage infrastructure lags significantly. As a consequence, raw sewage is discharged into the open stormwater drains in many of the cities. Added to this is the climatic setting – about 95% of the monsoon rainfall occurs during the months of June to September with very little rain occurring in the remaining months. These three factors, raw sewage in stormwater drains, inadequate drainage capacity and heavy monsoon rainfall result in overflowing of contaminated waters into the surroundings thereby resulting in outbreaks of major diseases such as gastroenteritis, malaria and dengue. This has now become a recurring feature in Indian cities. Taking the example of Mumbai, this chapter reviews causes and mitigation measures to reduce the impact of urban flooding on water quality in India.

The chapter describes the aftermath of the severe urban flooding on water quality in Mumbai due to the 26 July extreme rainfall event which witnessed 944 mm in a 24-hour period. The extent of flooding impacts on water quality and the resulting epidemics are discussed. Subsequently the measures initiated in Mumbai to mitigate the epidemics and flooding to improve the water quality are described. Mumbai can serve as a learning example for other cities. Finally, the major initiatives of the Government of India for the cities in India are mentioned.

8.2 MUMBAI

Mumbai (formerly Bombay) is the capital of Maharashtra State in India. It is a coastal city island located on the west coast of India and (Figure 8.1) has been formed due to the continuous reclamation of the mudflats between seven separate islands and merger with the larger Salsette Island since 1672. Historically, Mumbai developed first as a naval base for the East India Company, then as a textile centre and has now evolved into a major commercial and financial centre of India housing offices of many multinational companies. This has attracted large migrations from all over and has resulted in the rapid and uncontrolled growth of the city. The population increased from 8.19 million in 1981 to 12.44 million in 2011 (Government of India, 2011) in the present municipal area of 437.71 km^2 giving an average population of 28,426 persons per km^2. It is projected to exceed 20 million by 2021. Due to non-availability of land at higher elevations, a large number of informal settlements (approximately 65% of Mumbai's population) came up in the low-lying areas and on the flood plains of the Mithi River. In addition, over 3 million people commute daily to work from nearby towns by Mumbai's suburban railway system.

Mumbai is located on the windward side of the Western Ghats and receives high rainfall due to orographic effect from the Arabian Sea. Due to the reclamations, many parts of the city are just 1 metre above sea level and are vulnerable to flooding, especially when heavy rainfall coincides with high tide (max tidal level is around 5.0 m). Flooding in Mumbai is the most frequently occurring natural disaster due to heavy rainfall during the monsoon months. The average annual rainfall of Mumbai is about 2,354 mm (MCGM, 2017). Over 95% of this rainfall occurs primarily during the monsoon from June to October. About 70% of the average annual rainfall occurs in July and August and about 50% of this occurs in just two or three events (Gupta, 2007). During these – two to three events, it usually rains uniformly over the city and severe flooding occurs in many parts of the city.

The old city has a well-developed underground drainage system which has evolved over the past 200 years. However, the city infrastructure in the suburbs has not kept pace with the rate of population growth. With increasing urbanisation, water supply systems were provided as they were a basic necessity, the development of the corresponding drainage system has

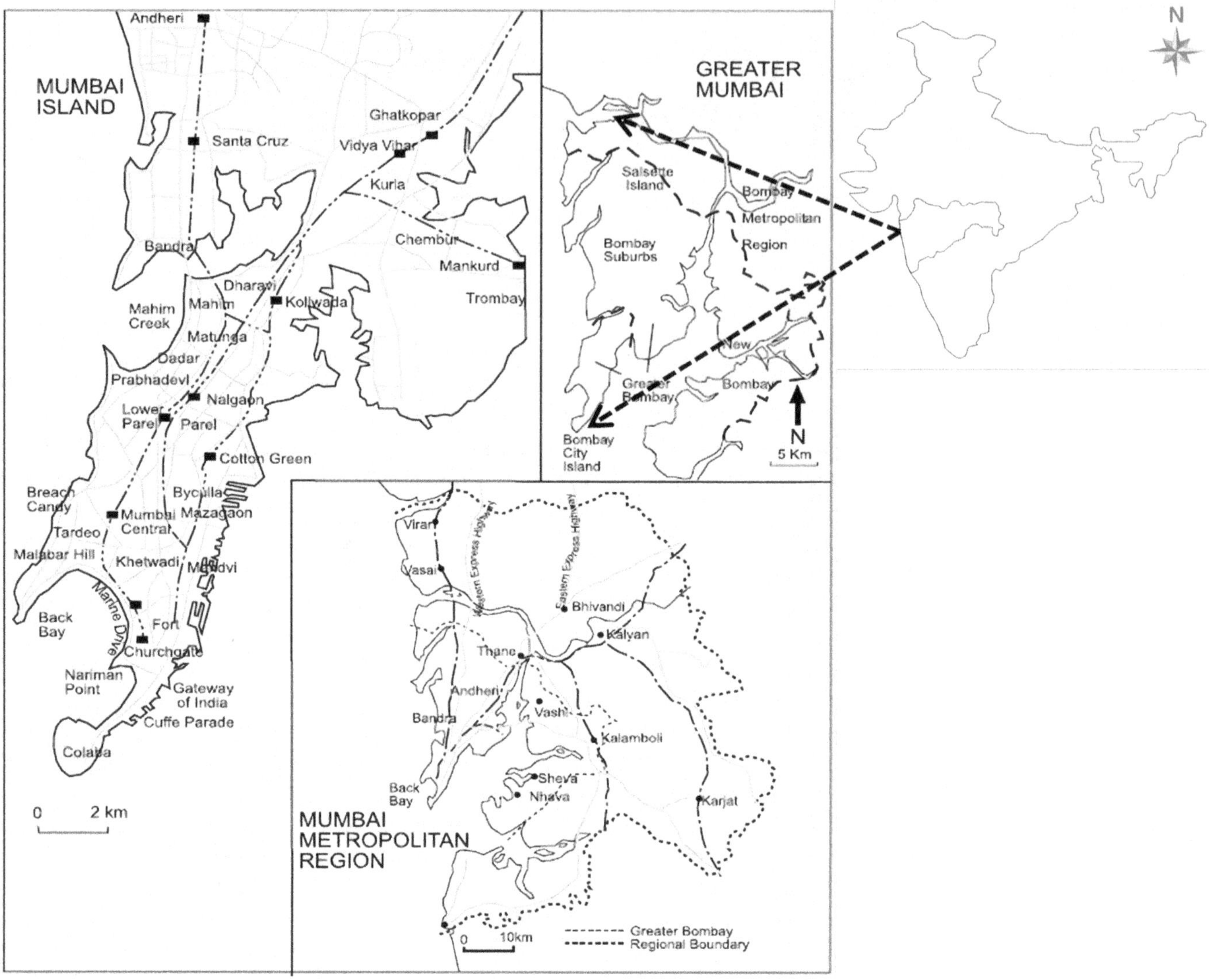

Figure 8.1 Map showing Mumbai Metropolitan region, Greater Mumbai and Mumbai Island (source: MMRDA)

been very slow, thereby resulting in the open stormwater drains receiving increasing quantity of sewage and sullage. Figure 8.2 shows such a drain in Mumbai.

This has resulted in not only non-aesthetic conditions but also health and sanitation problems, which are particularly severe during times of high-intensity rainfall (>50 mm/hr) when overflows result in flooding and outbreak of epidemics due to contaminated waters and loss of lives. This has resulted in local system overloads causing severe flooding at least twice a year, disrupting travel and the daily activities in the city. This usually happens when high-intensity rainfall (>50 mm/hr) occurs continuously for 2–3 hours coinciding with the occurrence of high tides.

To address the problem of flooding, a major study was taken up in 1993, called the BRIMSTOWAD project (Brihanmumbai Stormwater Drainage Project, MCGM, 1993), and it recommended the augmentation of the drainage network. However, on grounds of economy, it recommended the augmentation of the

Figure 8.2 Open stormwater drain receiving sewage and solid waste in Mumbai (photo: Kapil Gupta, 2017)

drainage network for a rainfall intensity of 50 mm/hr corresponding to once in six months return period. This scheme was implemented only partially due to limited funds.

8.3 THE 26 JULY 2005 MEGA FLOOD

Mumbai experienced its worst flooding on 26 July 2005 which caused almost a total shutdown of the city (Gupta, 2007) as over 60% of Mumbai was inundated to various degrees. The Santa Cruz observatory at Mumbai airport recorded 944 mm of rain during the 24 hours ending 0830 am on 27 July 2006, while the Colaba observatory recorded only 74 mm of rain (Jenamani et al., 2006). The rainfall hyetograph for the 26 July 2005 event is shown in Figure 8.3. This event has been classified as "very heavy" (>200 mm/day as per the rainfall classification of the Indian Meteorological Department (IMD)). It can be seen from the figure that at Santa Cruz, heavy rainfall started at 1430 with 481.2 mm falling in just 4 hours between 1430 and 1830 and hourly rainfall exceeded 190 mm/h during 1430 to 1530. The already inadequate drainage system was unable to drain out because of the highest high tide level of the month was 4.48 m at 1550. The event was attributed to a highly localised "offshore vortex." The immediate impact of the heavy rainfall was that there was a total collapse of the rail, road and air transport and communication systems. Disruption of power supply resulted in the failure of sewage pumps and the mixing of sewage with floodwaters resulted in outbreak of epidemics. At least 419 people (and 16,000 cattle) were killed due to the ensuing flash floods and landslides in Mumbai municipal area, and another 216 due to flood-related illnesses. Over 100,000 residential and commercial establishments and 30,000 vehicles were damaged (Govt. of Maharashtra, 2006).

8.4 2005 FLOOD-RELATED DISEASES

After the monsoon deluge of 2005, overflows from the storm drains and Mithi River resulted in contamination of drinking water supplies and outbreak of major diseases such as malaria, leptospirosis, dengue, hepatitis, gastroenteritis and typhoid which resulted in immediate fatalities of 68 (MCGM, 2006). A total of 3,326 people were affected due to these diseases as shown in Table 8.1. During floods, communicable diseases can be classified as (i) water-borne diseases, such as typhoid fever, cholera, leptospirosis and hepatitis A or (ii) vector-borne diseases, such as malaria, dengue and dengue haemorrhagic fever (WHO, 2017).

Immediate action taken by the municipal corporation included cleaning of the Mithi River and the stormwater drains, banning thin plastic bags which choke the drains, organising awareness programmes, training doctors, stocking adequate medicines like doxycycline for leptospirosis. Furthermore, to prevent an epidemic of monsoon-related diseases, 24 metric tonnes of

Table 8.1 *Water-borne and vector-borne diseases after the 2005 Mumbai mega-flood (source: MCGM, 2006)*

S. No.	Disease	No. of patients	No. of fatalities	Remarks
1.	Malaria[2]	1,281	32*	* Probable
2.	Leptospirosis[1]	135*	18	* Probable
3.	Dengue[2]	56*	6	* Probable
4.	Hepatitis	442	7	
5.	Gastroenteritis	1,334	5	
6.	Typhoid	78	0	
	TOTAL	3,326	68	

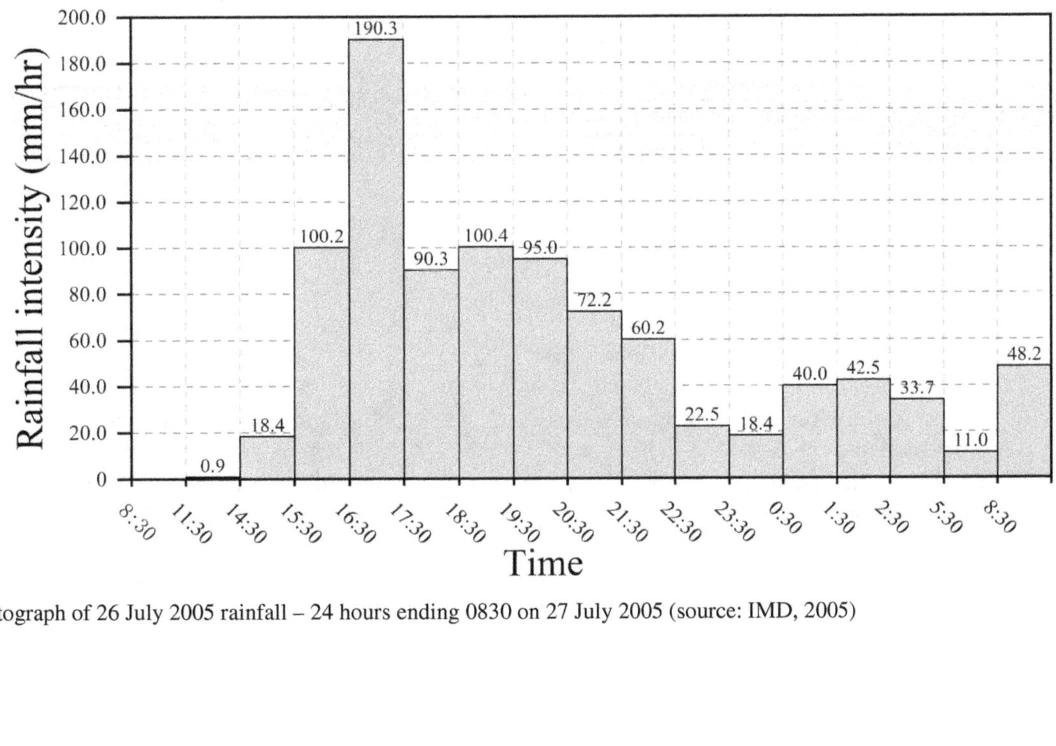

Figure 8.3 Hyetograph of 26 July 2005 rainfall – 24 hours ending 0830 on 27 July 2005 (source: IMD, 2005)

bleaching powder and 2 metric tonnes of disinfectant were used in the city. Fogging was also carried out in the affected areas.

8.5 RECURRENT INCIDENCES

Leptospirosis is a zoonotic bacterial disease and is the only epidemic-prone infection that can be transmitted directly from contaminated water (WHO, 2017). Transmission occurs through the contact of the skin and mucous membranes with water, damp soil or vegetation contaminated with rodent urine. The occurrence of flooding after heavy rainfall facilitates the spread of the organisms due to the proliferation of rodents which shed large amount of leptospires in their urine. One of the main causes of leptospirosis in Mumbai is walking barefoot in water, especially in the areas that are water logged. Hence, open wounds in the legs infect people faster. Leptospirosis is easily curable, and in general there is a rise in cases between July to October. In Mumbai, India, an eight-fold increase in disease incidence was noted after severe flooding in 2005 (Maskey et al., 2006). Table 8.2 shows the cases of leptospirosis during 2010 to 2015.

The outbreaks of common diseases during the monsoon in urban areas are summarised in the Table 8.3. Dengue and leptospirosis are the most recurrent diseases in Mumbai during the monsoon resulting in significant fatalities.

Table 8.2 *Cases of leptospirosis in Mumbai during 2010–15 (source: MCGM)*

S. No.	Year	Cases recorded	Deaths
1	2010	135	18
2	2011	141	6
3	2012	327	3
4	2013	233	3
5	2014	79	4
6	2015	26	12

Table 8.3 *Major monsoon diseases caused due to flooding and their details (adapted from media posters by MCGM)*

S. No.	Disease	Description	Mechanism of infection	Incubation period	Symptoms	Preventive Measures
1	Leptospirosis	An infection that spreads through *Leptospira* bacteria through flood water contaminated with the urine of rats and dogs	Enters body through skin abrasions or the nose	5–10 days	Fever, muscle pain, vomiting, loose motions	Avoid walking through waterlogged areas
2	Gastroenteritis	Infection or inflammation of the digestive tract	Through drinking contaminated water	12–24 hours	Vomiting, loose motions, viral fever	Drink boiled water
3	Viral fever	Viral infection that raises body temperature and causes fever	Through prolonged exposure to rain and wet clothes and shoes	-	Fever, chills, aching body, cough and cold	Change into dry clothes ASAP, avoid air-conditioned rooms
4	Malaria	Infection caused by a parasite and transmitted by the female *Anopheles* mosquito from one person to another	Through bites from mosquitoes breeding in stagnant waters	7–21 days	Fever with chills, joint pain, vomiting, anaemia, convulsions	Prevent stagnation of waters, for example, empty flowerpots, discarded tyres, pots and pans and use mosquito nets
5	Dengue	Infectious disease transmitted by the *Aedes* (tiger) mosquito from person to person – can sometimes be fatal	Through bites from mosquitoes breeding in stagnant waters	3–10 days	High fever (105 degrees), severe headache, falling blood platelet count and joint and muscle pain	– as above –
6	Chikungunya	Transmitted similarly to dengue fever			Illness with an acute febrile phase lasting 2–5 days, followed by a longer period of joint pains in the extremities	

Table 8.3 *(cont.)*

S. No.	Disease	Description	Mechanism of infection	Incubation period	Symptoms	Preventive Measures
7	Jaundice	Disorder of liver caused by hepatitis A or E virus	Through consuming contaminated water and food	15–20 days	Yellow tinge on skin and eyes, nausea, vomiting and fever	Drink boiled water, vaccinations for children below 16 available
8	Typhoid	Caused by the bacterium *Salmonella typhi* that causes inflammation of the intestines	Through consuming contaminated water and food	10 days	Unrelenting fever, rashes and overall weakness	Drink boiled water, wash hands and maintain general personal hygiene
9	Cholera/ diarrhoea	Water-borne bacterial disease	Through consuming contaminated water	5 h to 5 days	Dry skin, dehydration, nausea, abdominal cramps,	Drink only boiled water, oral rehydration therapy

8.6 MITIGATION MEASURES TAKEN BY THE MCGM TO MINIMISE HEALTH HAZARDS DURING THE MONSOON

The preventive measures taken during the monsoon are rodent control drive to get rid of the rodents in the city, fogging areas to exterminate insects and pests and finding and removing stagnated water spots where mosquitoes breed. All required medicines are made available in all BMC (Brihanmumbai Municipal Corporation) dispensaries, and extra beds are added to all BMC hospitals. The OPDs (outpatient department treatment) as well as casualty services are active for 24 hours a day. The civic body also carries out other preventive activities such as fogging drains and gutters with anti-malaria spray. Health officials also visit housing colonies and tell them that they should not allow any kind stagnant water to collect in the area, since these are breeding grounds for malarial mosquitoes. As part of the monsoon preparation, resident doctors of JJ Hospital also embarked on a major vaccination programme to protect their doctors against the H1N1 infection.

Medicines and chemicals are only short-term measures, and proper sanitation and prevention of water stagnation is a must for the prevention of water-borne diseases. Both dengue and chikungunya are transmitted by the mosquito species *Aedes aegypti* though the viruses causing the diseases are different. One infected mosquito can infect several buildings in its lifespan of three weeks. It is imperative that people prevent indoor breeding which is rampant in the city. A simple example is that of a building in Mumbai where residents on the sixth and eighth floors were down with dengue. It was found that an unused bucket filled with water on the seventh floor of the apartment was the breeding ground. Although the civic body regularly fines premises for failing to prevent mosquito breeding, but it continues to be rampant. It is well known that mosquitoes can breed only in stagnant waters and their eggs sink in flowing waters. Hence, a well-drained urban area with no stagnant waters would greatly reduce the breeding of mosquitoes and hence the mosquito-borne diseases.

8.7 ENHANCEMENT OF FLOOD RESPONSE

Following the deluge of 26 July 2005, it was realised that the implementation of structural measures as envisaged in BRIMSTOWAD (1993) was slow due to non-availability of land and limitations of funding. A comprehensive revamping of the drainage system is currently under progress and the system has been upgraded to cope with a design rainfall intensity of 50 mm/h and provision of eight pumping stations in the city.

8.8 INSTITUTIONAL MECHANISMS

The institutional mechanisms have been strengthened – a Disaster Risk Management Master Plan has been finalised by the Municipal Corporation of Greater Mumbai (MCGM). To enable better flood protection and management of the Mithi River, the Mithi River Development Authority has been set up. Municipal Corporation of Greater Mumbai has now been entrusted with providing clearance for all construction works in the Mithi River catchment. To reduce runoff to the drains, rainwater harvesting has been made compulsory for areas greater than 300 square metres.

8.9 DISASTER CONTROL CENTRE

The disaster control centre of the MCGM has been upgraded at an estimated cost of US$2 million and serves as a control room to handle disasters in Mumbai. It is the command and control centre for all rescue operations and coordinates with the various agencies. It has state-of-art systems like communications systems with land lines, hot lines, cellular phones, wireless, VHFs, video conferencing, press rooms, emergency water supplies and rations, and uninterruptible power supply with standby generators.

8.10 RAINFALL AND FLOOD WARNING SYSTEM

The IMD was unable to monitor the rainfall and issued warnings in real time on 26 July 2005. This has been attributed to the lack of state-of-the-art equipment like tipping bucket rain gauges. The Indian Meteorological Department had only two rain gauges in Mumbai of the syphonic type which record data on graph paper attached to clockwork driven drums and were read manually twice a day.

The MCGM (with IIT Bombay) installed 60 automatic weather stations which display rainfall data in real time every 15 minutes during June to October (www.mumbaiflood.in/). A mobile phone application has also been developed to enable this information to be available on smartphones (https://play.google.com/store/apps/details?id=com.mfws_react&hl=en_NZ&pli=1).

8.11 MOBILE APP-BASED MEASURES FOR MONSOON-RELATED DISEASES

In September 2017, the MCGM experimented with the launch of a mobile phone application called "Monsoon Related Diseases" to provide users with the information about the diseases such as malaria, dengue, swine flu and chikungunya. The application was available on the Google Play Store. The mobile application provided information about the symptoms of monsoon-related diseases, enlists precautions to prevent catching diseases and also lists out remedies for the diseases. It had informational about cleanliness measures that could be taken to prevent such diseases and types of foods to eat for recovery. There was a do's and don'ts list on the application, which is extensive and exhaustive. In addition, the app had names, numbers and addresses of all municipal hospitals and clinics across the city.

8.12 CONCLUSIONS

Monsoon-related flooding and the consequent outbreak of epidemics such as dengue, leptospirosis and malaria have now becoming a recurring feature in many cities of India. The Mumbai experience would be helpful for planning response strategies for other large cities to cope with similar events in the future. At the national level, to tackle the issues of urban flooding and water quality issues in India, the Government of India has launched the AMRUT (Atal Mission for Rejuvenation and Urban Transformation) in June 2015 for an initial duration of 5 years. The Government of India had identified 500 cities for the AMRUT. The mission focused on the following thrust areas: (i) water supply, (ii) sewerage facilities and septage management, (iii) storm water drains to reduce flooding, (iv) pedestrian, non-motorised and public transport facilities, parking spaces and (v) enhancing amenity value of cities by creating and upgrading green spaces, parks and recreation centres, especially for children. As it can be observed, urban water management is deemed very crucial for the health of the cities and four of the five objectives are related to water. The Government of India's plans on urban development under Smart Cities Mission and the AMRUT targeted approximately of 500 cities in a 16 billion USD investment. By October 2021, 134 lakh water tap connections and 102 lakh sewer connections (including households covered through Faecal Sludge and Septage Management - FSSM) had been provided through AMRUT & in convergence with other schemes against targeted 139 lakh water connections and 145 lakh sewer connections respectively. AMRUT 2.0, was launched on 1st October, 2021to continue with the needed infrastructure (https://pib.gov.in/PressReleasePage.aspx?PRID=1885837).

8.13 ACKNOWLEDGEMENTS

The authors are grateful to the officers of the MCGM for their help and support in providing the data.

References

Gazetteer of India. (1987). Maharashtra State, History of Mumbai, Modern Period.

Government of India. (2011). Census of India, https://censusindia.gov.in/census.website/ (accessed on 18 September 2017).

Government of Maharashtra. (March 2006). Fact Finding Committee on Mumbai floods.

Gupta, K. (2007). Urban flood resilience planning and management and lessons for the future: A case of study in Mumbai, India, *Urban Water Journal*, 4(3), 183–194.

Jenamani, R. K., Bhan, S. C., and Kalsi, S. R. (2006). Observational/forecasting aspects of the meteorological event that caused a record highest rainfall in Mumbai, *Current Science*, 90, 10.

Maskey, M., Shastri, J. S., Saraswathi, K., Surpam, R., and Vaidya, N. (2006). Leptospirosis in Mumbai: Post-deluge outbreak 2005, *Indian Journal of Medical Microbiology*, 24, 337–8.

MCGM (Municipal Corporation of Greater Mumbai). (1993). BRIMSTOWAD Project, final report (3 volumes).

MCGM (Municipal Corporation of Greater Mumbai). (2017). Flood Preparedness Guidelines, www.mcgm.gov.in/ (accessed on 22 October 2017).

MCGM (Municipal Corporation of Greater Mumbai). (2017). Flood Management Guidelines, www.mcgm.gov.in/ (accessed on 22 October 2017).

WHO. (2017). Flooding and communicable diseases fact sheet https://cdn.who.int/media/docs/default-source/disaster-mngmt/risk-management-communicable-diseases-december2017.pdf?sfvrsn=83724e4a_1&download=true (accessed on 24 October 2017).

9 The 2016 Seine River Flood and Its Impacts to the Paris Sanitation System and on the River's Water Quality

Jean-Pierre Tabuchi, Béatrice Blanchet, Vincent Rocher, and Sheila Aboulouard

9.1 INTRODUCTION

In June 2016 the Seine and the Marne had substantial flooding that was unusual at the time of the year as flooding usually occurs in the winter. This flooding had significant consequences on the operation of the sanitation system at the heart of Île-de-France, whose primary infrastructure is managed by the SIAAP, and the four *départements* of the inner suburbs of Paris (see Figure 9.1). The objective of this contribution is to evaluate, on the one hand, the effects of the flood on the operation of the sanitation system and, on the other hand, the subsequent impact on the quality of the receiving environment: the Marne and the Seine.

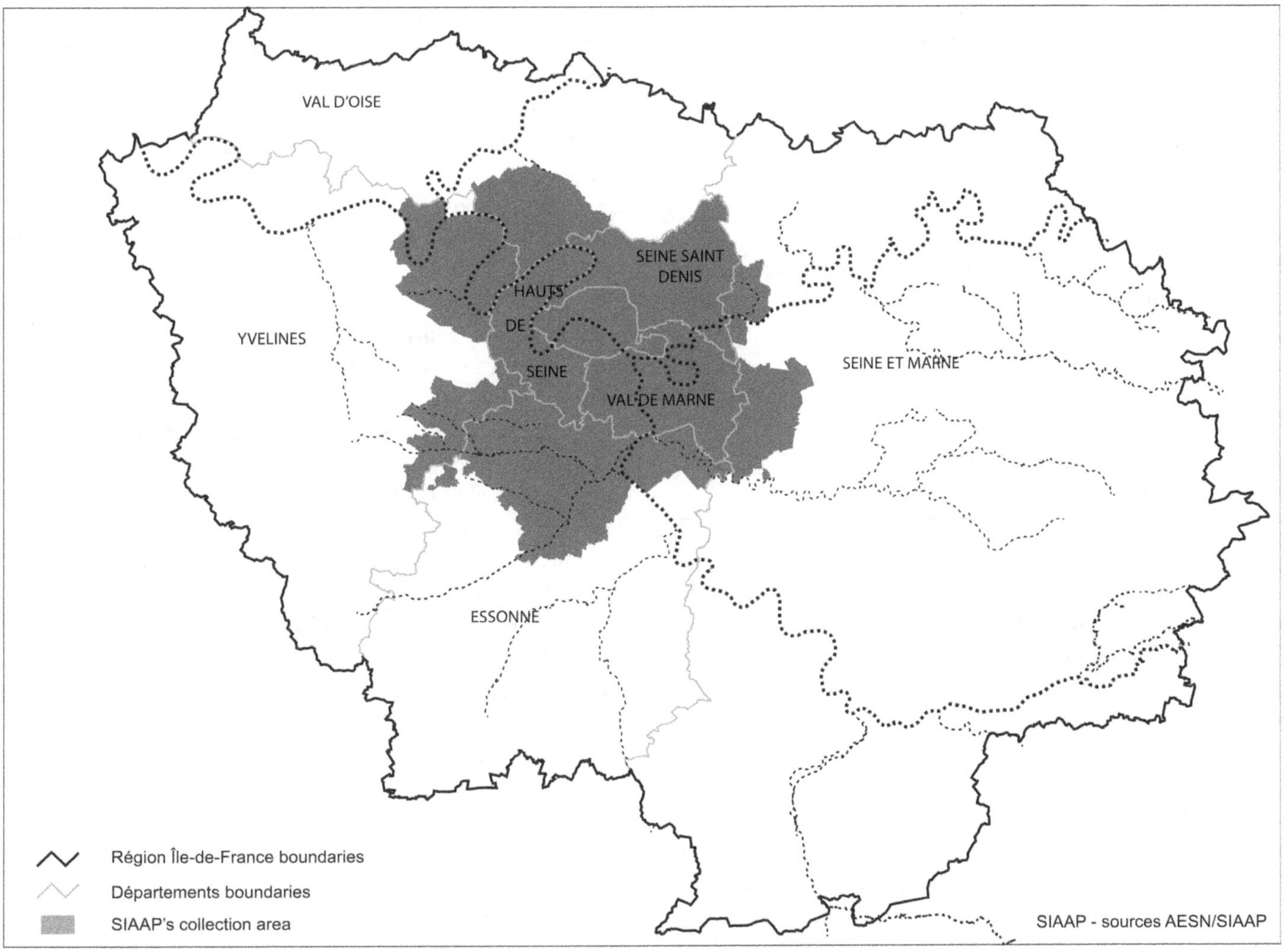

Figure 9.1 SIAAP's collection area and departments of the Île-de France Region

Table 9.1 *The surfaces of the main tributaries of the Seine*

Name	Oise	Marne	Yonne	Eure	Aube	Loing	Risle	Essonne	Epte	Yerres
Surface area (km^2)	16,667	12,920	10,836	6,017	4,660	4,150	2,300	1,870	1,490	1,030

This assessment was carried out; thanks to the supply of data from the sanitation departments of Paris (STEA), Hauts-de-Seine (DE 92), Seine-Saint-Denis (DEA 93) and Val-de-Marne (DSEA 94) as well as those of the SIAAP.

9.2 DESCRIPTION OF THE HYDROGRAPHIC NETWORK

The catchment of the Seine, covering an area of 78,650 km^2, concerns nearly 30% of the population of France. Its length is 776 km. It takes its source at an altitude of 446 m.

In Paris, its altitude is 26 m above the sea level, and the upstream catchment is 43,800 km^2. The catchment upstream from the Seine at Alfortville, which is a few kilometres upstream from its confluence with the Marne, is 30,800 km^2.

The surfaces of the main tributaries of the Seine are given in Table 9.1.

In the formation of the floods of the Seine, it is usually the Yonne that plays an important role.

However, in the case of the June 2016 flood, it was the Loing that played this role.

The hydrometric station used as a reference for the management of floods in the Paris area is that located at the pont d'Austerlitz in Paris. The reference level is 25.92 m.

9.3 GENERAL DESCRIPTION OF THE OPERATION OF THE SANITATION SYSTEM DURING A FLOOD PERIOD

The sanitation system of the central zone of Île de France is characterised by a main 400-kilometre framework comprised of the SIAAP transport networks and its sewage treatment plants (STP, Figure 9.2). This main network is supplied by the transport networks, owned by the *départements* around Paris and the City of Paris, but also the associations of communes (*Syndicats intercommunaux*) and the Territories connected to the SIAAP (Figure 9.2). The networks are meshed and are managed in real time with three objectives: protecting the natural environment, limiting flooding via network overflow and carrying of water correctly to the treatment plants. The central zone of this system is sanitised with a combined sewer system, while at the periphery, major portions are sanitised with a separate system.

In a normal situation, the stakeholders have a sanitation management policy intended to prevent discharges in dry weather and to limit as much as possible polluting discharges during rainy weather. In the present case, SIAAP has reconciled as much as possible the fight against pollution and against flooding while the *départements* have focussed their management on the fight against flooding via network overflows. Indeed, the *département*-level networks play a major role in the protection against flooding via network overflow in the municipalities and their territories.

In a flood situation, the flow of water in the networks is highly affected by the downstream constraint exerted by the elevation of the water level of the Seine and of the Marne rivers. Sanitation managers must in the first place limit massive intrusions of water from the Seine and Marne in the sewer networks in particular via combined sewer overflows (CSOs). For this, it is necessary to isolate the network from the river with gates. The excess water that cannot join its outlet via gravity is then pumped to the Seine or the Marne using flood-protection pumping stations. They play a capital role in protecting against flooding linked to the operation of the network, especially when it is raining on their catchment. Floods normally occur in the winter, the design of this type of structure focusses on low-intensity and long-lasting rainfalls. They are therefore not suitable for managing storms or high-intensity rainfalls that are often met in June.

The situation can be substantially aggravated by upstream inputs linked to water intrusions in the networks in flooded areas, to the rise in alluvial groundwater or to the increase in the pumping of mine water. As such, substantial infiltration into the sewer networks occurs in sectors where the overflows of the Seine and of the Marne take place, particularly in Val-de-Marne. The water then infiltrates via the gullies, manholes and other network accessories, leading to major hydraulic overloads in the sanitation system and to a strong dilution of the effluents.

Sewage treatment plants also play a major role in the operation of the system. Indeed, all of the wastewater arriving at these sites is pumped with lift stations that provide the indispensable function of emptying the sewage for the wastewater sanitation of upstream networks.

9.4 RAINFALL SITUATION

The studied hydrological episode that affected the operation of the sanitation system extended from 28 May 2016 to 24 June 2016; the total rainfall over this period was 135 mm. It was, however, preceded by a particularly rainy month in May since the total reached

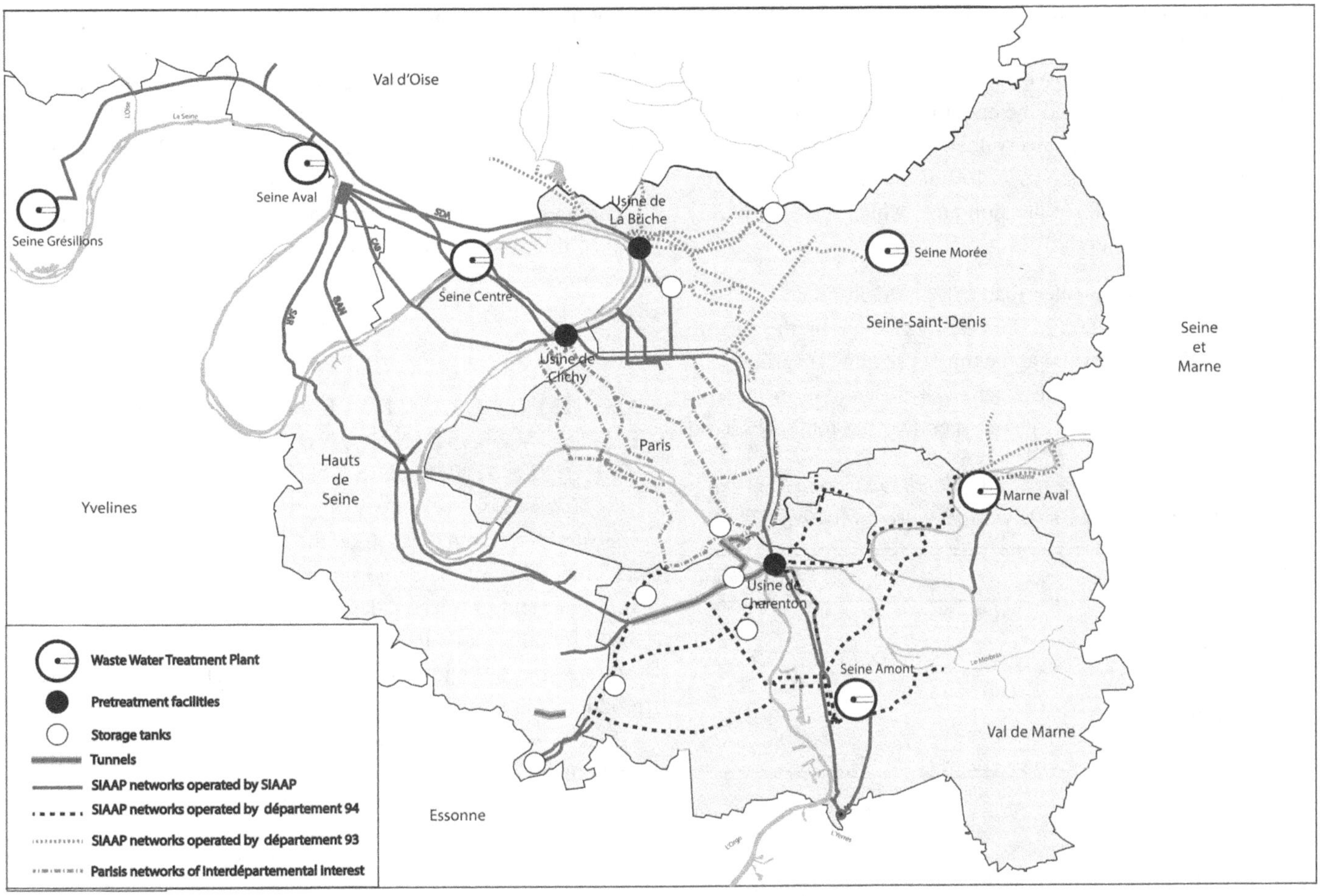

Figure 9.2 Location of the main infrastructures of the SIAAP

Figure 9.3 Flooding in Val de Marne (photos DSEA)

179 mm (measured at Paris Montsouris rain gauge) compared to 60 mm (average for 1951–1980) on average; this greatly influenced the hydrological behaviour of the basins of the Seine and of the Marne.

The following rainy episodes can be distinguished:

- From 9 to 12 May with a total of 31 mm.
- From 18 to 23 May with a total of 55 mm. In itself, this is already a substantial total since it is close to the monthly total.
- The rainy episode from 28 May to 2 June with a total of 97 mm with the particular episode on 30–31 May with a total only for those 2 days of 69 mm. This event, which occurred over a large portion of the Seine and of the Marne watersheds, is at the origin of the flood. However, before the flood actually occurred in the heart of the urban area (Figure 9.3), this rainy episode caused the most substantial discharges of the sanitation system as well as certain

subsequent malfunctions of which the flood has reinforced the effects and the duration.

- Then more local rainy episodes occurred between 12 and 20 June which affected the operation of a sanitation system that was hydraulically very loaded. Among them, we can mention the intense storm on 16 June that affected the south-east of the SIAAP's collection area, which was still highly affected by the flood.

With excess rainfall greater than 70%, Météo-France (French meteorological institute), in Île-de-France and in the Centre region, classified this period at the top of the rainiest springs.

Although this is not representative of the rainfall in the collection zone of the SIAAP, just to mention, the totals observed at the Paris-Montsouris weather station for different periods are summarised in Table 9.2 and Figure 9.4.

Table 9.2 *Summaries of the various rainfall totals from May and June – Météo-France data – Paris Montsouris – 00:00 h – 24:00 h*

Period	Totals mm
1/05 – 31/05	179.3
28/05 – 02/06	97.8
30/05 – 31/05	69.2
01/06 – 24/06	45.5
28/05 – 24/06	135.3

The estimation for the return period of the event varies according to the collection catchments. The SIAAP estimated the return periods of rain from 28 May to 1 June between 1 and 5 years according to the sectors. This rainfall situation combined with the flood situation of the Marne (Figure 9.3) and of the Seine confers an exceptional nature to the situation encountered.

9.5 HYDROLOGICAL SITUATIONS OF THE RIVER SEINE AND THE MARNE AND EFFECTS ON THE SANITATION SYSTEM

The substantial rain that fell on the Paris urban area affected the catchment area of the Seine and in particular one of its tributary the Loing and to a lesser degree the Marne. This rain was preceded by other rainfalls that were already substantial and which saturated the soil with water, making the catchment areas highly productive in terms of run-off.

Figure 9.5 shows the main characteristics of the behaviour of the Seine at Austerlitz bridge.

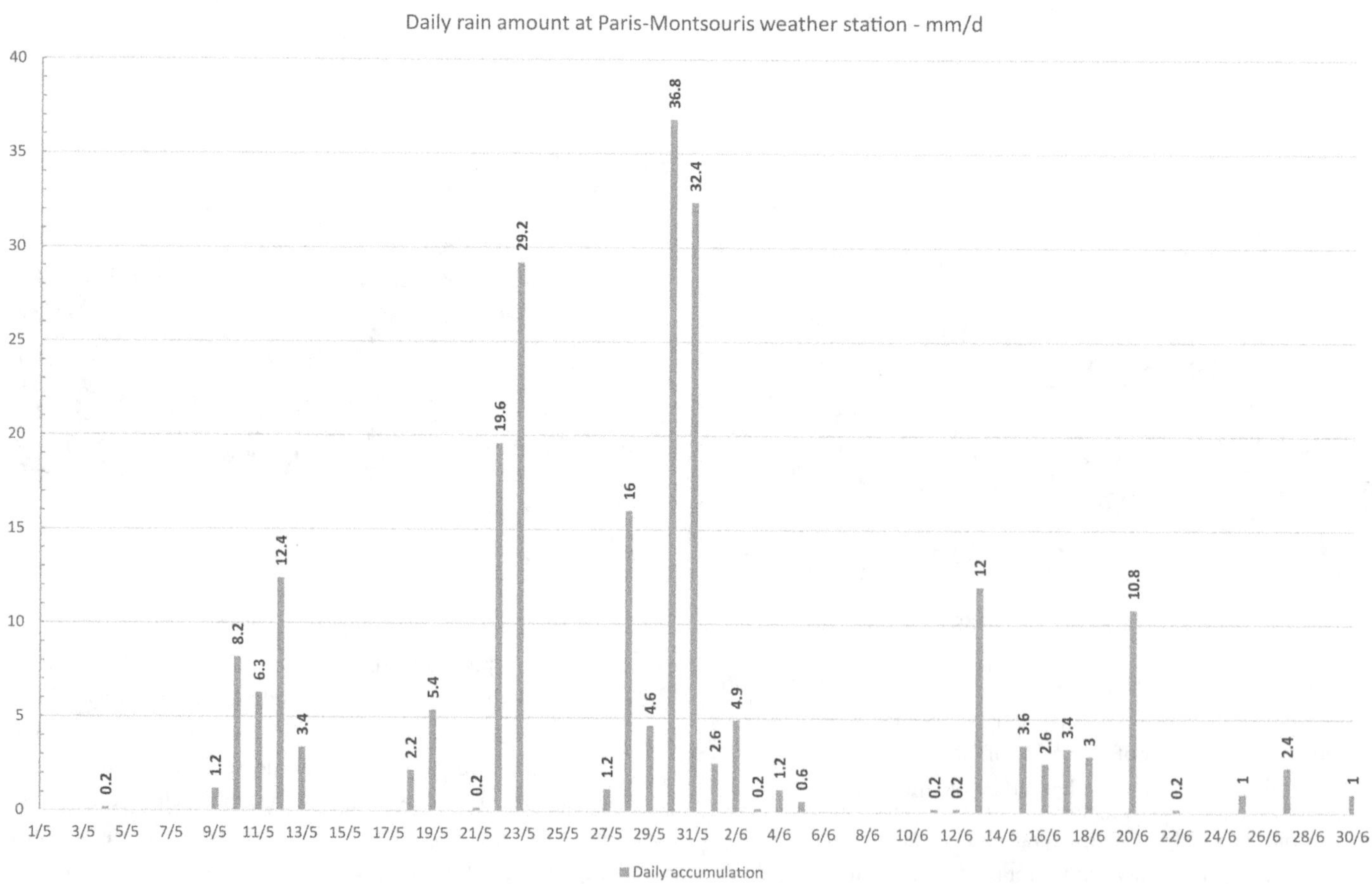

Figure 9.4 Rainfall during May and June – Paris Montsouris weather station

The rapid rise of water level in Paris, on 30 May 2016, can be explained by the proximity between the Loing – Seine confluence and the Paris area as well as by the quantity of the rain on the Essonne's and the Orge's catchment areas. The alert threshold defined by the State authorities was reached on 31 May at 06:00 h. The flood would develop to reach its maximum at pont d'Austerlitz on 4 June at 2:00 am. The water receding would be a long process that would take place in two phases. The first phase was rather fast and regular until 14 June. During the second phase, the flood worsened slightly due to rainfalls between 15 and 20 June. This episode would slow down the return to normal situation of the sanitation system's operation.

The comparison with other past great floods shows that the rise in the water was fast, similar to that of the 100-year flood in 1910 (Figure 9.7) and the level reached approaches that of the flood of 1982 (Figure 9.6) but with very different kinetics.

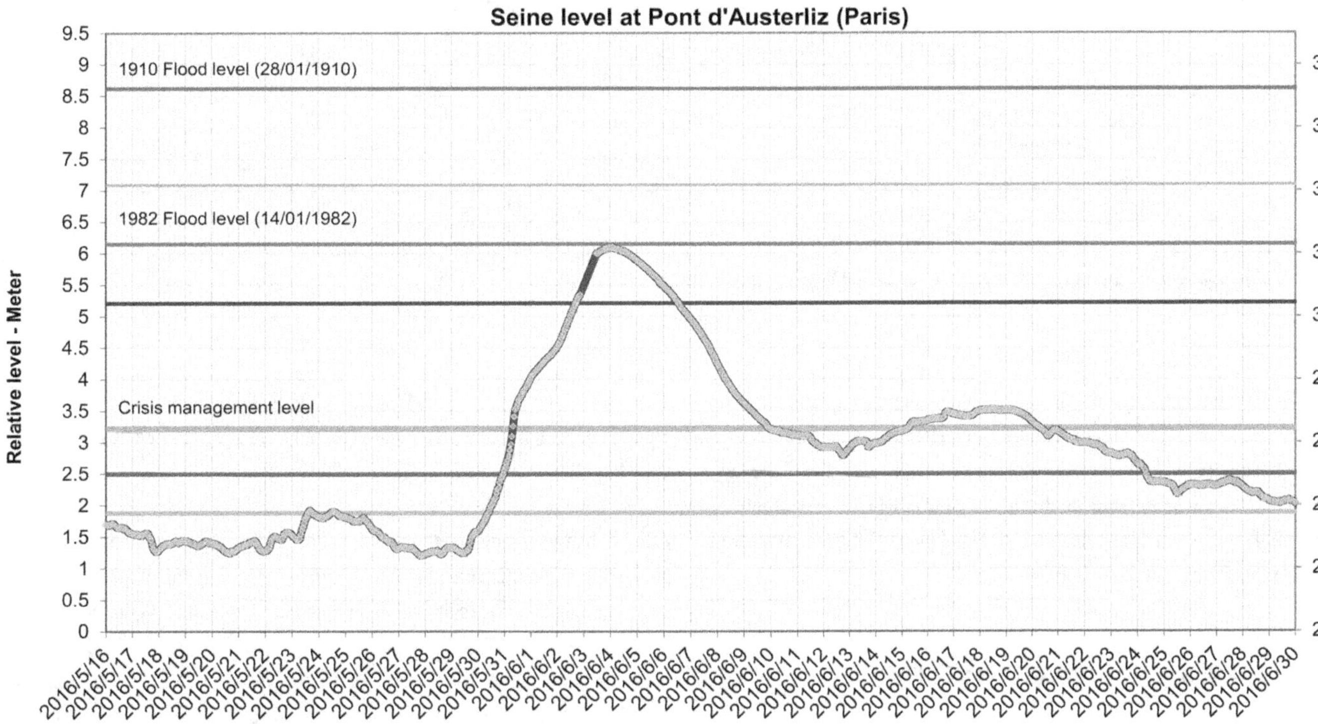

Figure 9.5 Hydrograph on the Seine at Pont d'Austerlitz (DRIEE data – SIAAP processing)

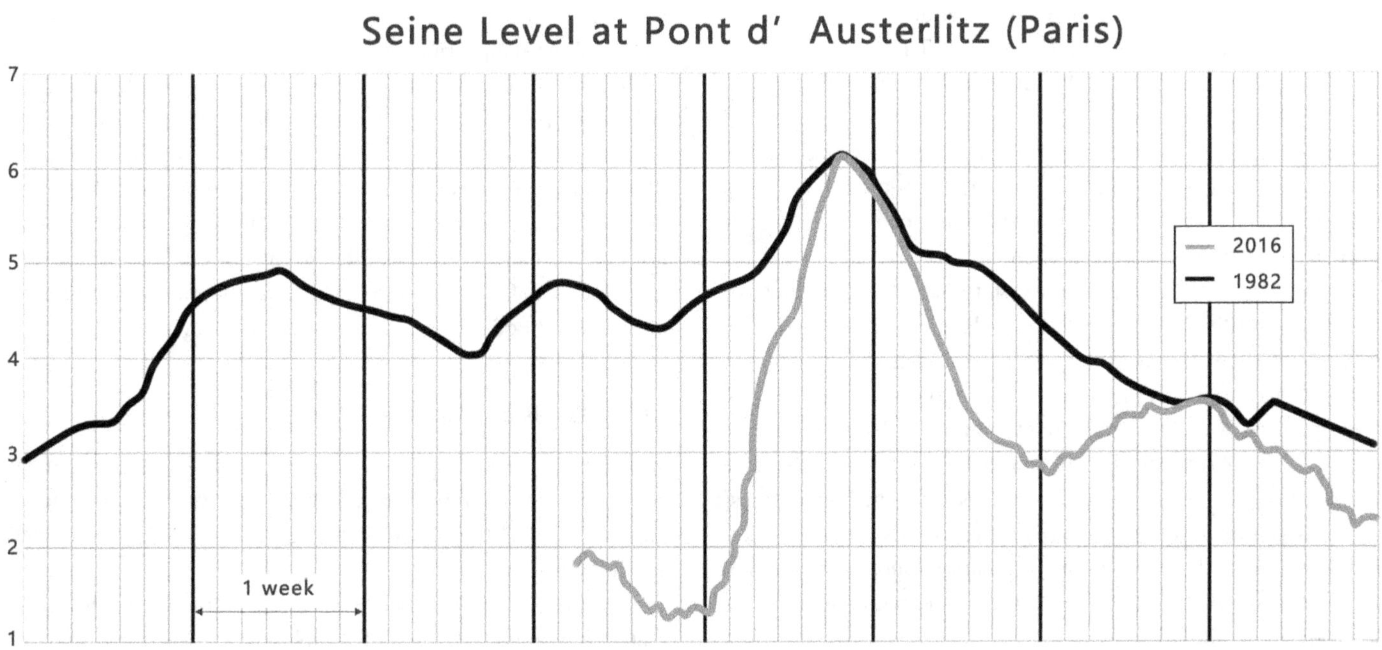

Figure 9.6 Hydrograph on the Seine at Pont d'Austerlitz (DRIEE data – SIAAP processing)

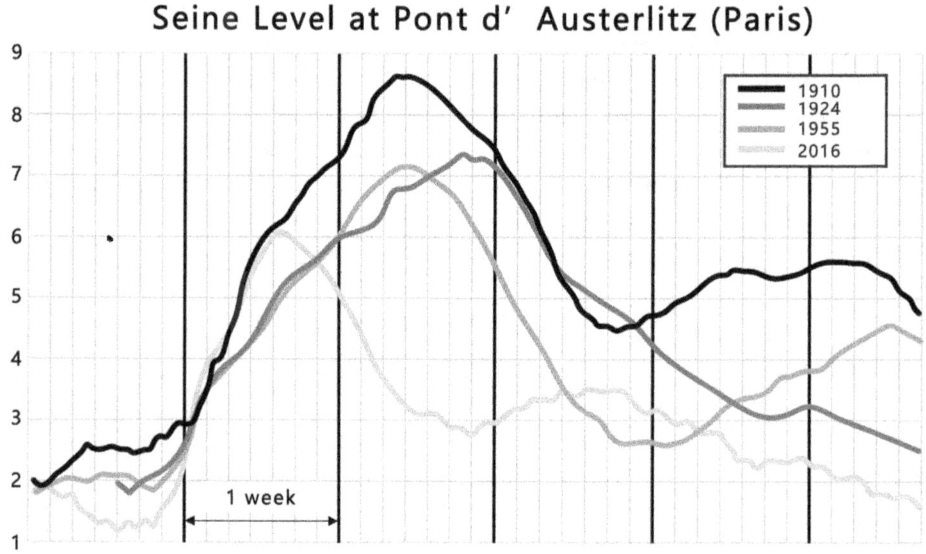

Figure 9.7 Hydrograph on the Seine at Pont d'Austerlitz (DRIEE data – SIAAP processing)

In light of the impact on the sanitation system's operation, we have considered the flood that occurred from 30 May to 24 June. With regards to the management of its facilities, the SIAAP has a first warning threshold at the level of 27.80 m at Austerlitz bridge; this level would be reached on 30 May at 09:30 h, and the situation would be over by 30 June.

9.6 GENERAL MANAGEMENT STRATEGY

The management of sanitation facilities by the *départements* or by SIAAP follows precise procedures that are put into application as the flood and the receding progress. They concern a very large number of actions and facilities for which details are not provided here.

Globally, in these particular conditions, the procedures aim to minimise the impact of the flood for the users of the sanitation system. They also aim to provide protection for equipment for the purpose of a quick re-start after the flood.

Management of the sanitation system has been the object of constant coordination between SIAAP and the *départements* as each of them instructions are closely linked. The management of this event was facilitated by several crisis exercises that were held in 2015 and at the beginning of 2016, in particular the SEQUANA exercise in March 2016. This major exercise involved all of the stakeholders in flood management: States' services and operators of public utilities but also private stakeholders such as those of goods distribution whose contribution is solicited during such events to provide certain supply functions for the populations.

9.7 GLOBAL ASSESSMENT OF THE TRANSPORTED AND DISCHARGED VOLUMES

The volumes generated by the sanitation system and managed during the initial phase of the flood, over the rainy period from 29 to 31 May, were exceptional since the total volume of the discharges exceeded 11 Mm^3/d compared with the 2.2 Mm^3/d of daily dry weather flow arriving in the SIAAP STP. The general profile of the volumes processed and discharged is presented in Figure 9.8 that particularly highlights the effects on the sanitation system of the 30–31 May event at the origin of the flood. At the end of the period, we also see here the effects of the rain from 14 to 21 June. The various volumes are summarised in Table 9.3.

The total volume discharged during the rain event from 29 to 31 May represents 21% of the total volume discharged over the entire flood period, from 29 May to 24 June, while the total rainfall from 29 to 31 May represents 55% of the total rainfall over the period from 29 May to 24 June. This difference between the share of the rainfall and the share of the volume discharged is linked to the intrusions of water from rivers in the networks during the high phase of the flood which caused overflows elsewhere.

9.7.1 The Rainy Episode (29 May to 31 May)

During this event, all of the SIAAP's transport and treatment facilities were hydraulically saturated. It was so exceptional that almost all of the SIAAP's discharge facilities were solicited simultaneously.

Over the three days that formed this rain phase, the STPs treated a volume of 13.5 Mm^3. Network discharges reached

Table 9.3 *Volume assessments for various periods of the flood*

	Total volume discharged m^3	STPs volume inlet m^3	Networks volumes discharged m^3	Combined networks m^3	Stormwater networks m^3
29–31/05/2016	26,801,200	13,510,000	13,291,300	7,170,500	6,120,400
29/05–24/06	131,254,200	93,252,900	38,001,200	15,461,200	22,539,700
Rain from 29–31 /Total	20%	14%	35%	46%	27%

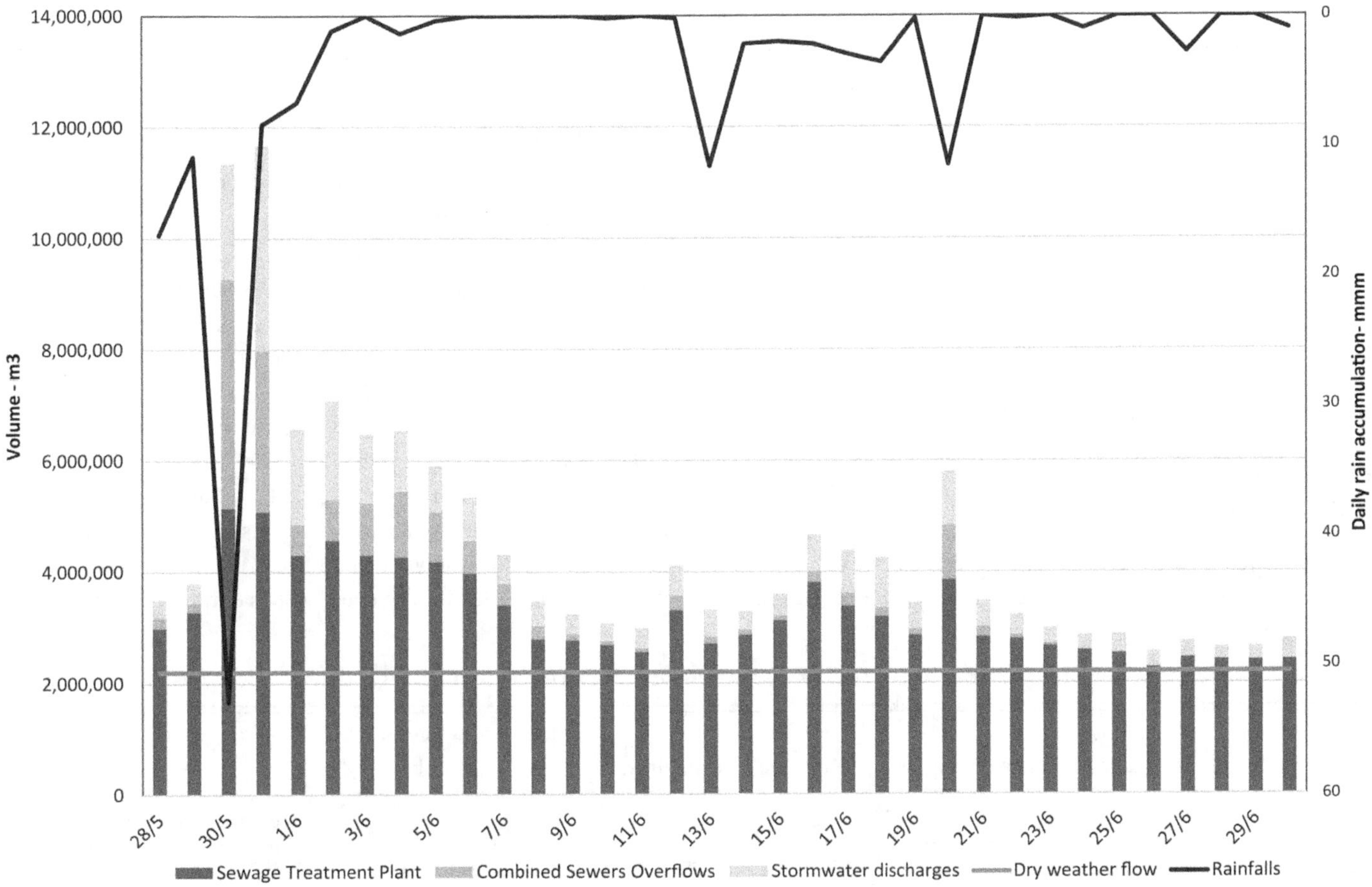

Figure 9.8 Change in the volumes discharged by the SIAAP plants and all networks taken as a whole (SIAAP, départements, rain (inc. rivers), CSOs)

13.3 Mm^3 of which 7.2 Mm^3 for the overflows of the combined sewer networks and 4.2 Mm^3 for separate stormwater discharges.

9.7.2 Discharges over the Flood Period (29 May to 24 June)

The volume assessments of the flood during this period are summarised in Table 9.3.

9.7.3 Assessment of the Pollutant Loads Discharged

The chosen method to carry out the assessment of the loads of the pollutants discharged all throughout the flood consisted in using reference concentration values defined during the update of the SIAAP's sanitation master plan. However, it was necessary to take into account the dilution caused by the strong and quick succession of rainfalls. For this, for each parameter, daily dilution coefficients were calculated on the observed concentrations basis at the inlet of the SIAAP's plants. These coefficients were then applied to the reference concentrations. However, the level of uncertainty surrounding these estimates remains high.

The discharged loads are summarised in Table 9.4. The data is differentiated between the rain period loads from 29 to 31 May and the total flood period loads from 29 May to 24 June.

Note that the loads discharged by the networks very clearly exceeded the STPs' loads discharges during the event from 29 to 31 May.

To allow for comparison with the usual data, the daily average loads were calculated. As reference data, for the purposes of comparison, the daily average of discharges for 2015 are also mentioned. These loads provide an idea of the extent 29 to 31 May events (Table 9.5).

Table 9.4 *Flows discharged by the various types of facilities over the various phases of the event (tonnes)*

Tonnes	Total flow discharged	Treatment plant discharges	Networks	Combined networks	Stormwater networks
		29–31/05/2016			
SS	1,550	480	1,070	840	220
BOD	560	190	370	350	20
COD	2,260	740	1,520	1,400	120
TKN	198	103	95	89	6
NH_4^+	102	62	40	38	2
Tot P	23	11	13	12	1
		29/05–24/2016			
SS	5,450	1,760	3,690	2,720	970
BOD	1,670	830	840	750	90
COD	7,470	4,030	3,450	2,930	520
TKN	728	520	208	183	25
NH_4^+	456	356	101	88	13
Tot P	106	79	28	24	4

9.8 ASSESSMENT OF THE SEWAGE TREATMENT PLANTS OPERATION

The STPs fulfilled their de-polluting function despite the altogether unique operating conditions.

These unique conditions concerned the volumes, the concentrations and the loads of pollutants which varied between the extremes, both in the maximums and in the minimums.

9.8.1 Treated Flows

The volumes treated in the SIAAP plants, whether over the entire flow or during the intense rain at the end of May, reached highly exceptional values. In particular at *Seine-Aval.* Table 9.6 shows the average daily volumes for various flood's periods. It also compares these values to the average daily values during all weather conditions taken over the year 2015 as well as the maximum flows observed during the flood.

The STPs treated very high volumes of water.

The volumes treated at *Seine-Aval*, *Seine-Amont* and *Marne-Aval* were very high: the average daily values over the flood's entire duration correspond to values usually observed during heavy rain. In 2015 only 11 to 13 times the average daily flows values exceeded the daily average values during the flood.

Table 9.5 *Daily average flows discharged by the various types of facilities over the various phases of the event (tonnes/day)*

t/d	Total flow discharged	Plant discharges	Networks	Combined networks	Stormwater networks	
			29–31/05/2016			Treatment plant average 2015
Volume – m^3/d	6,700,300	3,377,500	3,322,825	1,792,625	1,530,100	***2,352,000***
SS – t/d	388	120	268	210	55	***43.6***
BOD – t/d	140	48	93	88	5	***24.7***
COD – t/d	565	185	380	350	30	***119.8***
TKN – t/d	50	26	24	22	1	***21.9***
NH_4^+	26	16	10	9	0.6	***15.1***
Tot P – t/d	6	3	3	3	0.3	***2.4***
			29/05–24/06/2016			
Volume – m^3/d	4,687,650	3,330,460	1,357,190	552,190	804,990	
SS – t/d	195	63	132	97	35	
BOD – t/d	60	30	30	27	3	
COD – t/d	267	144	123	105	19	
TKN – t/d	26	19	7	6.5	0.9	
NH_4^+	16	13	4	3.1	0.5	
Tot P – t/d	4	3	1	0.8	0.1	

Table 9.6 *Average daily volumes treated at the SIAAP plants during the month of various periods*

	Seine-Aval	Seine-Centre	Seine-Grésillons (*)	Seine-Amont	Marne-Aval	Seine-Morée	Total
			29–31/05/2016 – m^3/d				
Volume	3,189,300	432,000	11,800	760,500	91,800	17,900	**4,503,300**
			29/05–24/06/2016 – m^3/d				
Volume	2,350,900	260,700	10,000	736,200	78,500	17,600	**3,453,800**
			Average *2015* – m^3/d				
Volume	*1,412,137*	*164,532*	*248,966*	*443,736*	*47,035*	*11,924*	*2,328,329*
			Max flow (m^3/d) observed during the flood event				
Max	3,664,000	495,569		870,517	110,714	18,070	
			Design capacities				
Dry weather flow	1,210,000	240,000	300,000	600,000	52,000	50,000	
Reference flow	2,300,000	400,000	300,000	800,000	75,000		
Peak flow in rainy weather	2,900,000	400,000	300,000	1,200,000	120,000		

* Plant operating in reduced mode due to work on its main supply facility.

Table 9.7 *Average concentrations at the inlet of the STPs*

	From 29 to 31 May 2016	29 May to 24 June 2016	Average 2015
SS mg/l	148	188	*261*
BOD mg/l	84	120	*191*
COD mg/l	223	302	*471*
TKN mg/l	21	31	*50*
NH_4^+ mg/l	12	21	*34*
Tot N mg/l	21	31	*51*
Tot P mg/l	3	3	*6*

9.8.2 Water Quality at the Inlets of Sewage Treatment Plants

The influents are characterised by their very strong dilution. This process was initiated at the first rains in the month of May. The very low concentrations encountered are entirely uncommon.

For the purposes of information, the average concentrations per period were calculated. These average concentrations are the weighted values with respect to the loads of different STPs.

We see here the dominant effect of the dilution on the concentration of the pollutants measured at the inlet of the STPs (Table 9.7). Finally, the concentrations at the peak of the flood are two to three times lower than the average concentration over the year.

Figure 9.9 shows the change in the water dilution with respect to the average concentration observed during the first week of May 2016. This dilution particularly affected the *Seine-Amont* plant which, in addition to the drop-in concentration, would see its load cut in half. In these conditions, the compliance with 70% removal of the total nitrogen required by the urban wastewater treatment directive was not able to be met as the concentration at the inlet was too low: so the observed output for June was 58%.

9.8.2.1 LOADS AT THE INLET OF THE STPS

A summary of the flows collected is given in Tables 9.8 and 9.9.

The loads are reported both as total tonnage per period (Table 9.8) and as average daily tonnage per period (Table 9.9).

Table 9.9 shows that as a whole the loads conveyed to the STPs are roughly retained. In line with the particular operating conditions of the sanitation system, losses of effluents occurred in various sectors of the supply on the south-east of SIAAP's catchment area. These losses of pollutants loads are the result of complex phenomenon linked to the substantial intrusions of water caused by the flooding of many sectors close to the banks of the Seine, Marne, Yerres and Orge rivers.

The details per STP, which are not presented here, give a more contrasted view. On some of the STPs, we can see a load deficit during the main flood period. This can be particularly seen on the *Seine-Amont* plant as shown in Figures 9.10 and 9.11, which show a load deficit at the inlet of the plant between 31 May and 9 June (Figure 9.10 and Figure 9.11). This is due to the fact that the catchment area of this plant was particularly impacted by the flood.

Table 9.8 *Summary of the total loads per period of observation – tonnes*

Tonnes	29 to 31 May 2016	29 May to 24 June 2016
Volume	13,509,960	93,252,950
SS	1,997	17,498
BOD	1,137	11,184
COD	3,009	28,159
TKN	278	2,884
NH_4^+	163	1,969
Tot N	288	2,931
Tot P	35	325

Table 9.9 *Summary of the daily average loads per period of observation – tonnes/day*

Tonnes/d	29 May 31 May 2016	29 May to 24 June 2016	Daily flows 2015
Volume	4,503,320	3,453,810	*2,328,329*
SS	666	648	*607*
BOD	379	414	*445*
COD	1,003	1,043	*1,097*
TKN	93	107	*117*
NH_4^+	54	73	*80*
Tot N	96	109	*118*
Tot P	12	12	*13*

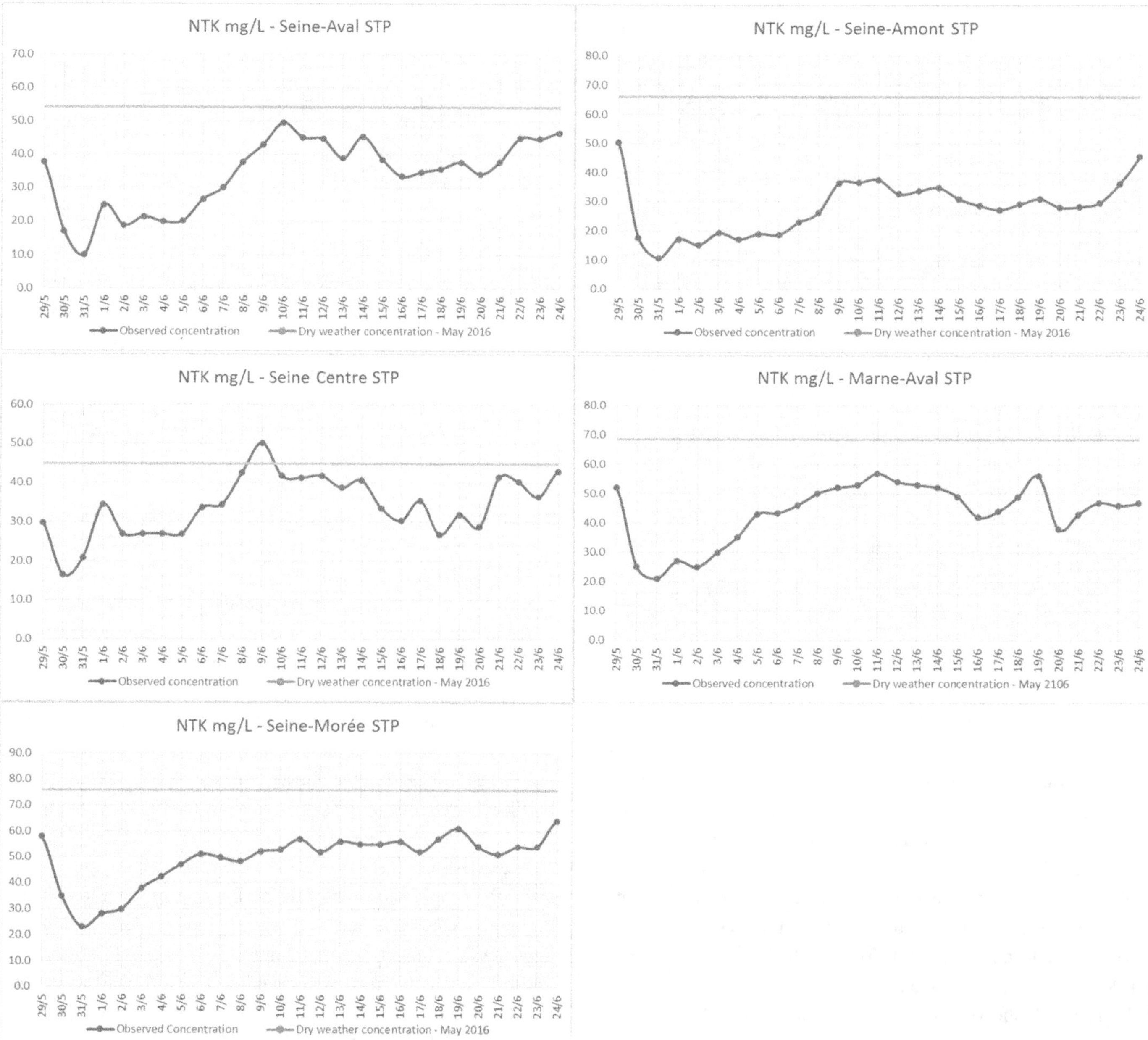

Figure 9.9 Change in TKN at the inlet of the SIAAP plants and comparison with the dry weather average

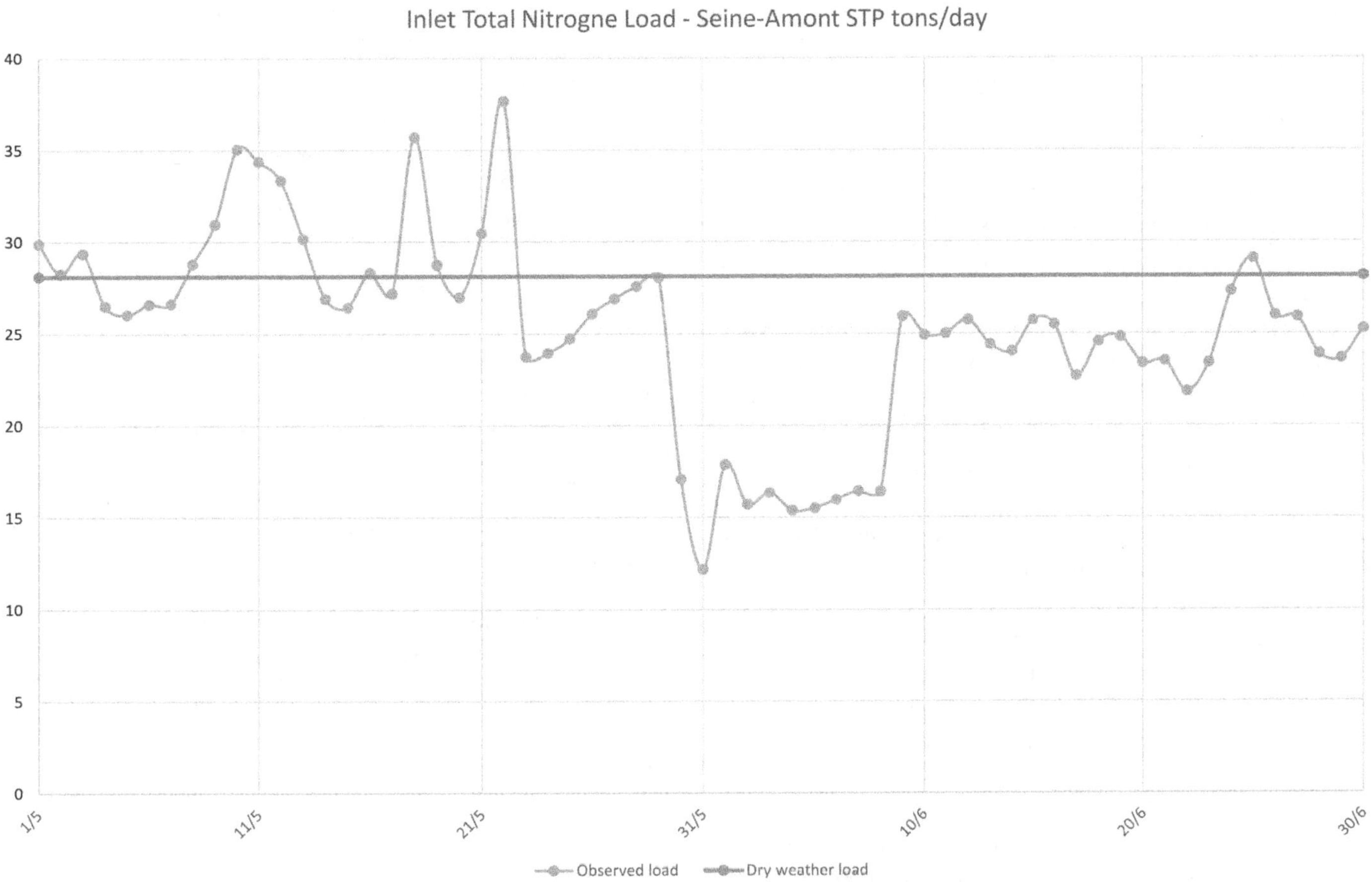

Figure 9.10 Change in the loads of total nitrogen at the inlet of Seine-Amont in tonnes/day from 1 May to the 30 June

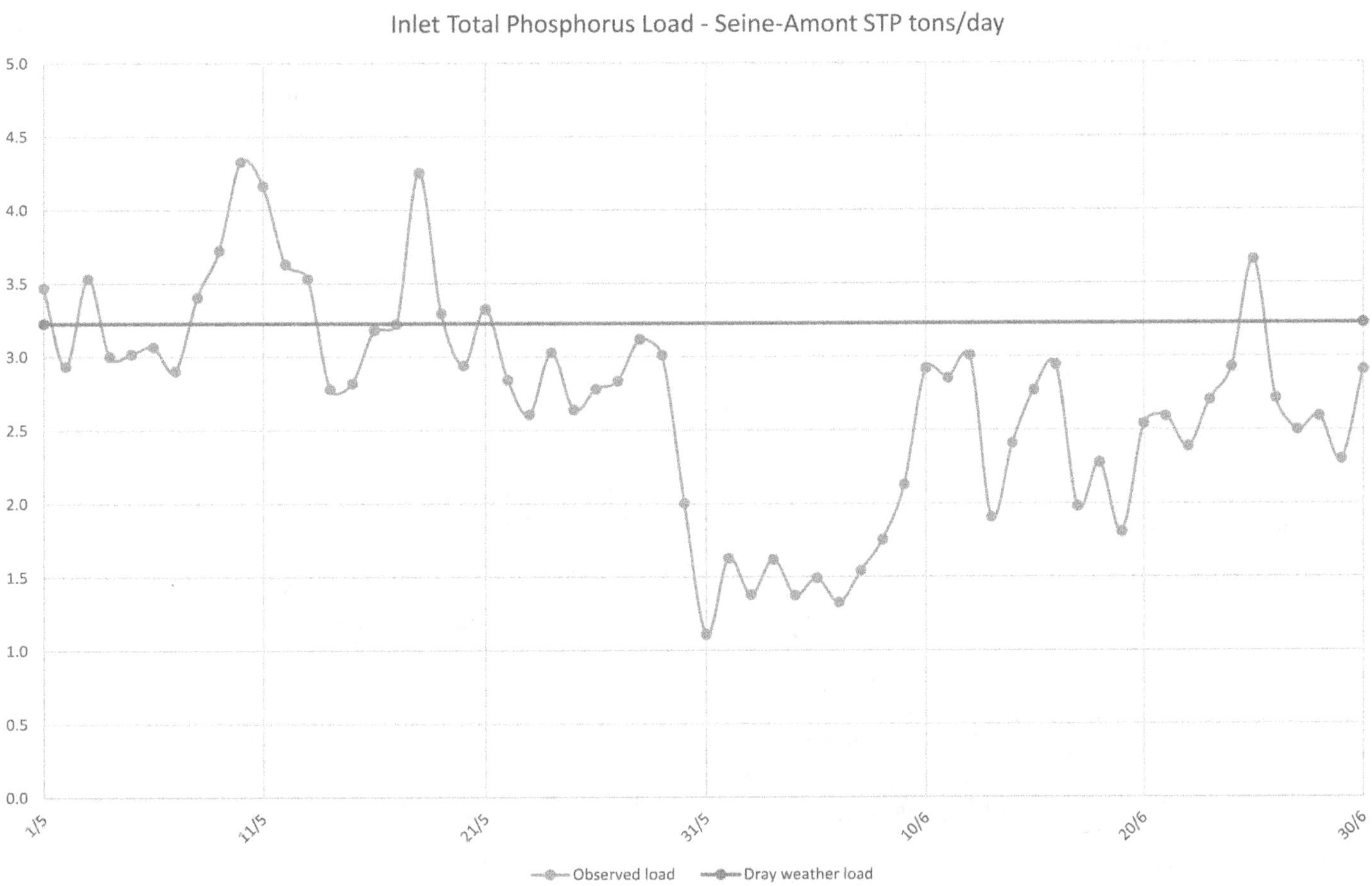

Figure 9.11 Change in the loads of total phosphorus at the entry of Seine-Amont in tonnes/day from 1 May to the 30 June

9.8.2.2 ASSESSMENT OF WATER INFILTRATION, WASTEWATER LOSSES AND WATER VOLUME BALANCE

As was indicated above; in times of flooding, substantial volumes of water can infiltrate into the networks. These volumes of infiltrated water were estimated at 29.6 Mm^3 or 1.2 Mm^3/d between 1 and 24 June.

On the contrary, the data shows a reduction of the loads of pollutants resulted by a loss en route of a portion of these loads. They were able to be assessed on the basis of the losses of total nitrogen loads observed in the supply catchment area of *Seine-Amont.*

Over the period of 30 May to 8 June corresponding to the load-deficit period at the inlet of the *Seine-Amont* plant, and in comparison to the average dry weather load, the total deficit was 126 tonnes of TKN. These calculations lead to a hydraulic assessment summarised in Figure 9.12.

The loads of pollutants and the calculations based on the effluent dilution lead to the infiltration water volume assessment over the period of 30 May to 8 June of approximately 10 Mm^3. The losses of wastewater volumes can be evaluated at about 6 Mm^3.

As a reminder the *Seine-Amont* sewerage network is a separate system. It therefore has a transport capacity in rainy weather that is more limited than that of the "west system" designed as a combined system.

9.8.3 Discharge Quality

In terms of discharge quality, Tables 9.10 and 9.11 provide a global vision of the concentrations at the outlet and the results obtained during the various periods under consideration. The percentages remained entirely correct except during the episode from 29 to 31 May when they were subjected to significant drops, drops partly due to the incoming loads.

Table 9.10 *Concentrations in mg/l at the outlet at different periods and comparison with the average 2015*

	29–31/05/ 2016	29/05–24/06/2016	Average 2015
SS	36	19	16
BOD	14	9	9
COD	54	43	47
TKN	8	6	9
NH_4^+	5	4	6
Tot. N	13	13	20
Tot. P	1	1	1

Table 9.11 *Removal percentage and comparison with the average 2015*

	29–31/05/ 2016 (%)	29/05–24/06/2016 (%)	Average 2015 (%)
SS	76	90	*93.8*
BOD	83	93	*95.1*
COD	76	86	*89.9*
TKN	63	82	*82.0*
NH_4^+	62	82	*81.8*
Tot. N	39	57	*60.2*
Tot. P	70	76	*81.9*

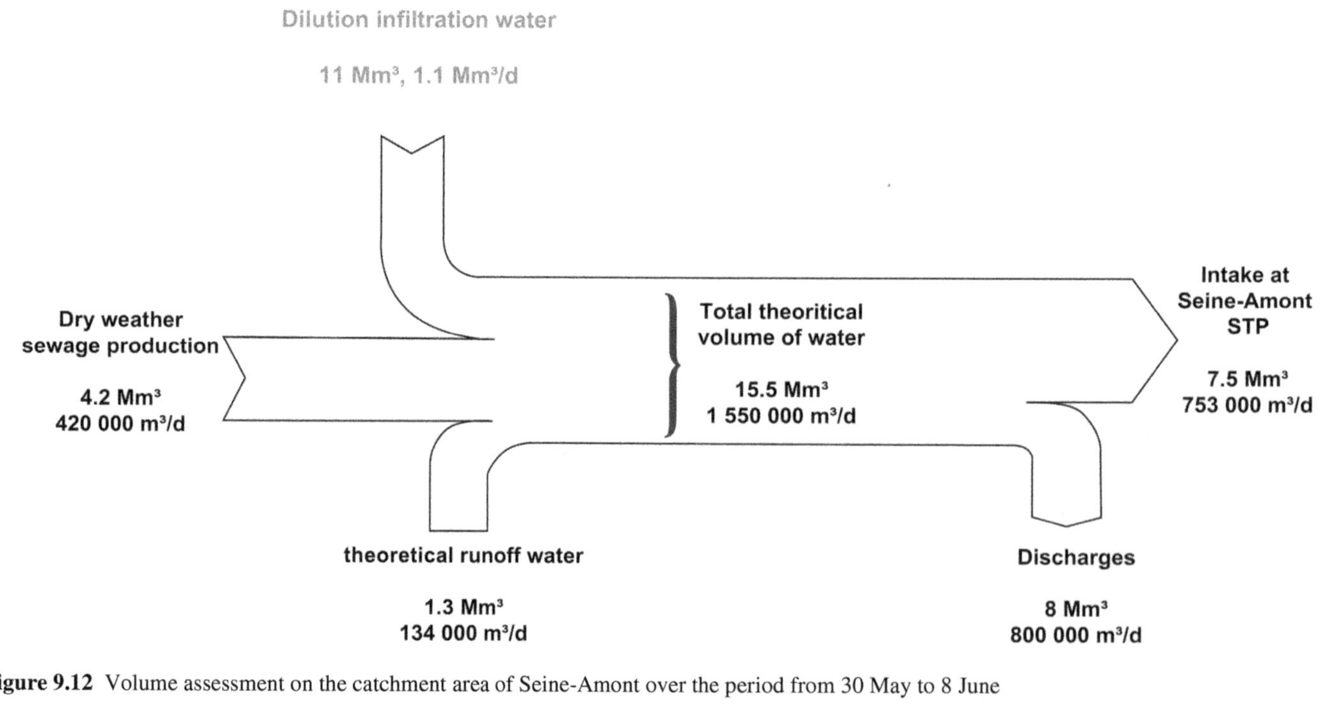

Figure 9.12 Volume assessment on the catchment area of Seine-Amont over the period from 30 May to 8 June

9.9 GLOBAL DE-POLLUTING ASSESSMENT

Table 9.12 summarises the global de-polluting rates for various phases of the flood. The term global de-polluting rate refers to the ratio between the discharged pollution and the total pollution produced. The total pollution produced is the sum of the loads at the inlet of the plants and of the network discharges.

During the rainy episode of 29 to 31 May, the de-polluting rate was about 50%, which would be considered as a mediocre rate. Naturally, the exceptional nature of this event explains this result and particularly the extent of the network loads discharges which are much higher than those of the STPs. The total flow produced – the sum of the flows at the inlet of the plants plus the loads discharged by the networks – is also very high.

In the flood phase, from 1 to 24 June that followed this event, the de-polluting rates were satisfactory in light of the treated flows and volumes.

Table 9.12 *Global de-polluting rate*

	De-polluting rate	
Tonnes	29 to 31 May 2016 (%)	29 May to 24 June 2016 (%)
Volume	51	73
SS	50	75
BOD	63	86
COD	51	77
TKN	70	90
NH_4^+	50	78
Tot P	52	70

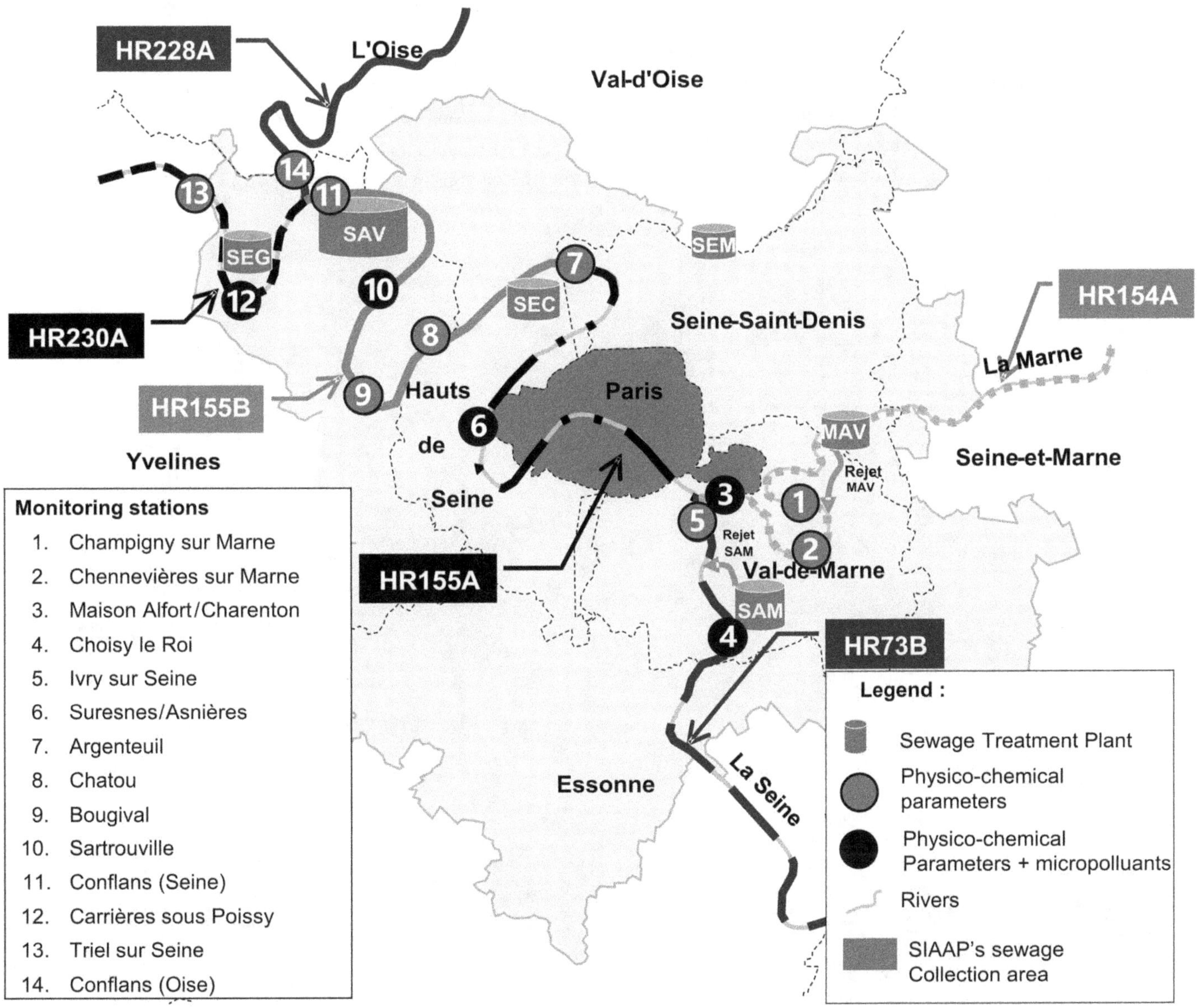

Figure 9.13 Seine stations used for the impact study

As such, globally, the effectiveness of the sanitation system has made it possible to eliminate about 80% of the pollutants load produced by the Paris urban area. The excellent behaviour of the TKN treatment can be underlined.

In conclusion, high volumes have been treated by the STPs with a degraded level of performance during the intense rainy episode. But over the flood period, as a whole, the obtained results were satisfactory. The volumes and loads discharged directly by the sewer networks were high and substantially affected the global sanitation de-polluting system yield.

9.10 ENVIRONMENTAL IMPACTS ASSESSMENT

9.10.1 Context

The self-monitoring network MeSeine (Measurements in Seine, Figure 9.13) has made it possible to assess the flood impact on the natural environment. This network was deployed in order to supervise the permanent discharge at the STPs as well as the stormwater discharges. It makes it possible to know the physico-chemical quality of the Marne and of the Seine.

The study of the flood impact on the natural environment is based on the results provided by the weekly sampling from the monitoring network which makes the physico-chemical quality assessment possible. As a complement to this weekly monitoring, a continuous monitoring allows for in-situ measurement of dissolved oxygen in the Seine.

9.10.2 Flood Impact on the Quality of the Seine and Marne Rivers

This section presents the results of the self-monitoring of the environment, using the analyses collected from samples taken from the Seine, for ammonia, nitrites, nitrates and continuous in-situ measurements, for dissolved oxygen. Note that the sampling upstream of the Paris urban area (e.g., Champigny, Alfortville, Choisy-le-Roi and Ivry-sur-Seine) was not able to be taken on 7 June following a social movement.

9.10.2.1 NITROGEN

9.10.2.1.1 Ammoniacal Nitrogen (NH_4^+)

Figure 9.14 makes it possible to view the change in NH_4^+ by plotting the change in the concentration of the weekly samples in

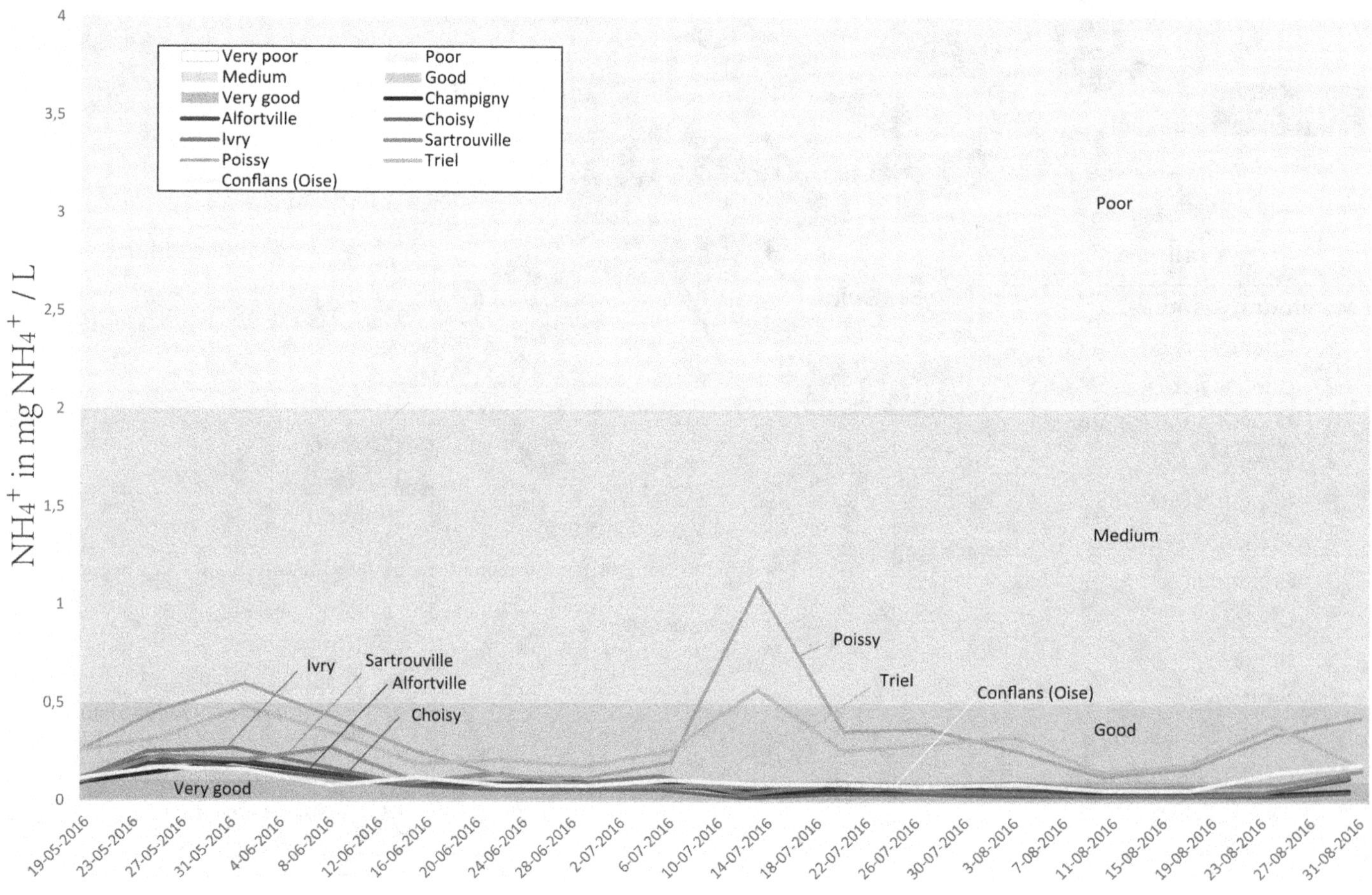

Figure 9.14 NH_4^+ in Seine, Oise and Marne

NH_4^+ measured in the upstream and downstream zones of Paris before the flood until the end of August. The colour code indicates the limit quality concentrations stemming from the order of 27 July 2015 on the evaluation of the quality of water bodies in liaison with the Water Framework Directive.

During the flood, the zone from Choisy-le-Roi to Alfortville is of the same magnitude of about 0.2 mg NH_4^+/L. We do, however, observe an increase between Sartrouville and Poissy. Only the value of NH_4^+ measured at Poissy would lead for a portion of the flood to a one-off classification, as average condition of the environment for this parameter directly concerned by the WFD. The situation returns to normal on all of the measuring points from mid-June.

An average condition is recorded downstream from *Seine-Aval* at Poissy and Triel-sur-Seine on 13 July following rainy weather.

9.10.2.1.2 Nitrous Nitrogen (Nitrites NO_2^-)

Figure 9.15 makes it possible to view the change in the concentrations of NO_2^- measured in the upstream and downstream zone of Paris from before the flood until the end of August. During the flood, the zone from Choisy-le-Roi to Alfortville is of the same magnitude of about 0.15 mg NO_2^+/L. We do, however, observe an increase between Sartrouville and Poissy. Among the various measurement stations, only the value of NO_2^- measured at Poissy would lead for a portion of the event studied to a one-off classification as average condition of the environment for this parameter. The situation returns to normal on all of the measuring points starting in mid-June.

An average quality is recorded on 13 July and 31 August downstream from *Seine-Aval* following rainy weather.

9.10.2.1.3 Nitric Nitrogen (Nitrates NO_3^-)

Figure 9.16 makes it possible to view the change in the concentrations of NO_3^- measured in the upstream and downstream zone of Paris from before the flood until the end of August. During the flood, the stations of Champigny and Alfortville on the Marne have the same behaviour with a concentration that on the whole is lower than on the Seine starting from the end of the flood, at the beginning of July. On the measuring stations on the Seine, we observe a dilution in nitrates during the flood phase. The Seine and the Marne are in good conditions with regards to the nitrate parameter.

9.10.2.2 ORTHOPHOSPHATES (PO_4^{3-})

Figure 9.17 makes it possible to view the change in orthophosphates, by plotting the concentrations of the weekly samples

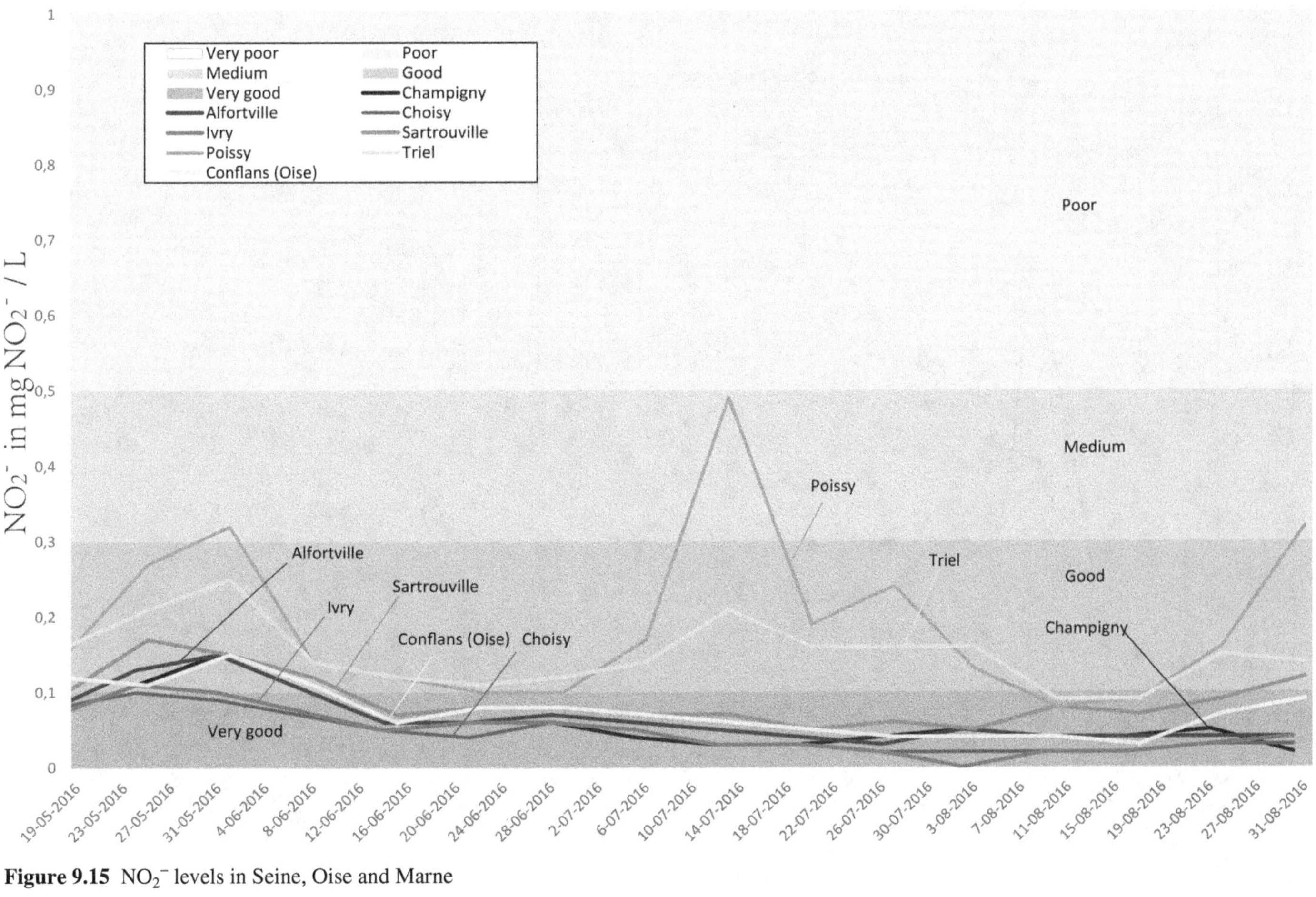

Figure 9.15 NO_2^- levels in Seine, Oise and Marne

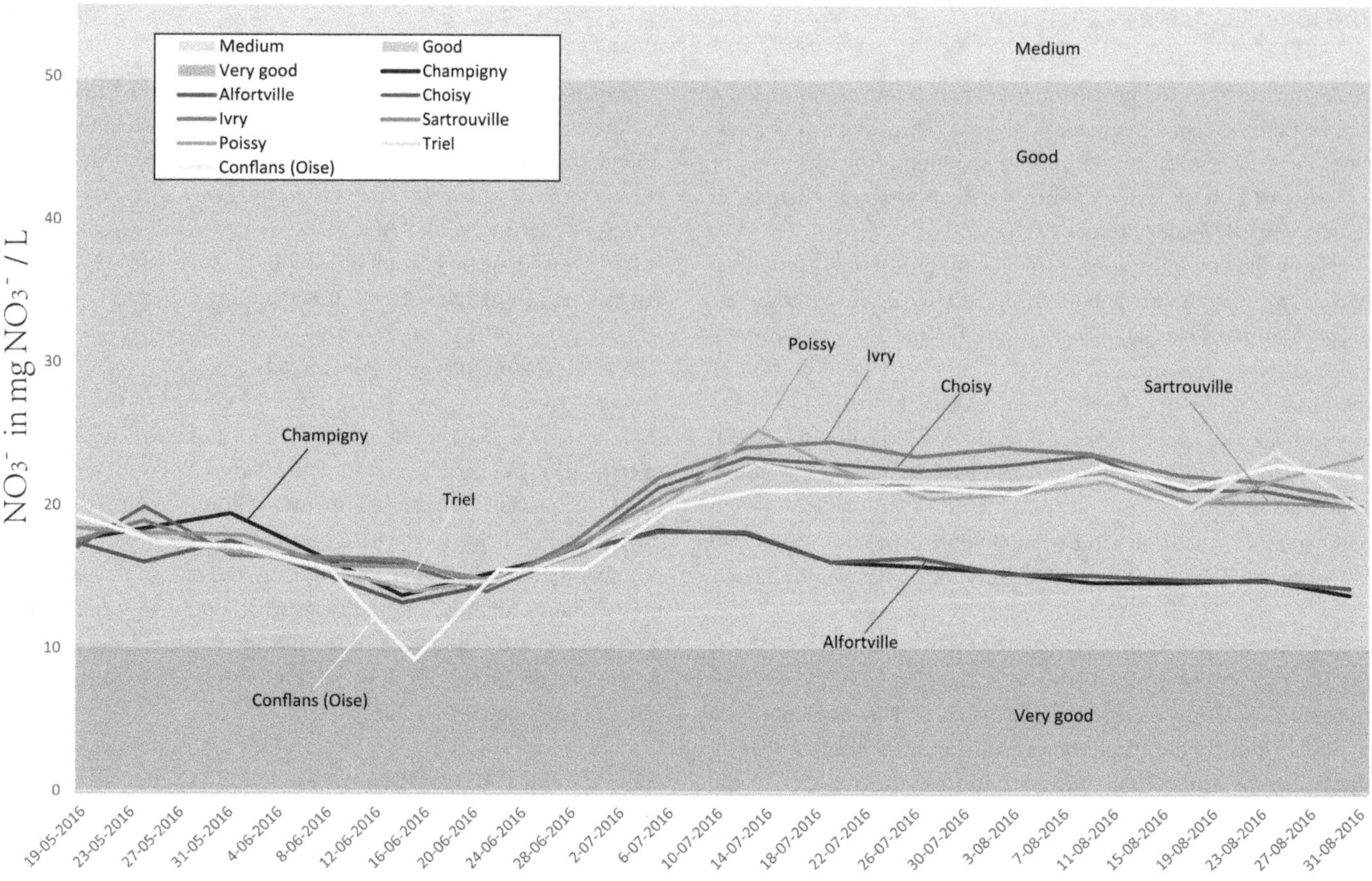

Figure 9.16 NO_3^- levels in Seine, Oise and Marne

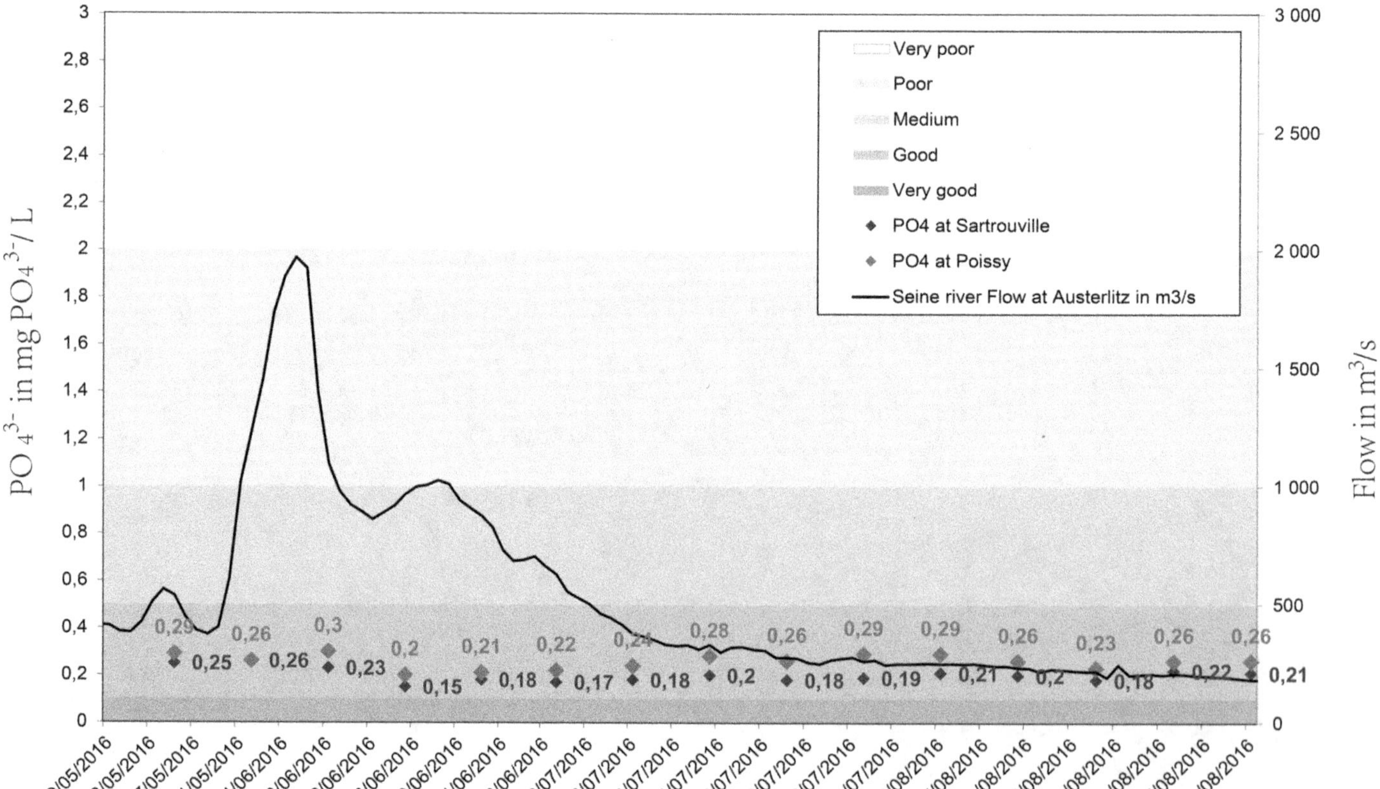

Figure 9.17 PO_4^{3-} levels upstream and downstream from Seine-downstream

taken at Sartrouville (upstream from *Seine-Aval* – orange), and at Poissy (15 km downstream from *Seine-Aval* and downstream from the confluence with the Oise – brown).

The quality on the orthophosphate parameter remains in good condition.

9.10.2.3 DISSOLVED OXYGEN

Figure 9.18 reports the content in dissolved oxygen measured continuously, from the Port à l'Anglais to Méricourt.

Due to the relatively low temperatures and the substantial flows, dissolved oxygen remains in a good condition during the flood and until the end of the period studied.

9.10.2.4 BACTERIOLOGY: *E. COLI* AND INTESTINAL ENTEROCOCCI

Figure 9.19 shows *Escherichia coli* in the Seine, from Choisy-le-Roi to Triel-sur-Seine. *E. coli* in MPN/100 ml are on a logarithmic scale. For information, the limit for sufficient bathing quality is indicated in a dotted line (900 MPN/100 ml).

Figure 9.20 shows the intestinal enterococci in the Seine, from Choisy-le-Roi to Triel-sur-Seine. The intestinal enterococci in MPN/100 mL are on a logarithmic scale. For information, the limit for sufficient bathing quality is indicated in a dotted line (330 MPN/100 ml).

The self-monitoring on bacteriology is carried out every 15 days, except at Poissy and Triel-sur-Seine where it is carried out every 7 days. We do not have any measurements at the highest point of the flood however.

The situation is globally equivalent before and during the flood period, due to the high variability that is naturally observed on these two parameters. The period following the flood appears to be lower than the rest of the year on certain measuring points, but these parameters are down regularly during the summer season.

9.10.3 Conclusions

The self-monitoring carried out by the SIAAP, using the analyses and samples and continuous measurements, has made it possible to reveal the impact of the flood on the quality of the river. However, it was not possible to conduct additional campaigns. Therefore, we do not have any data at the beginning and at the highest point of this exceptional situation.

With regards to *E. coli* and intestinal enterococci, we have a situation that is globally equivalent between pre-flood and post-flood.

For nutrients, what comes to light is that this quality was substantially affected only for ammonia and nitrites.

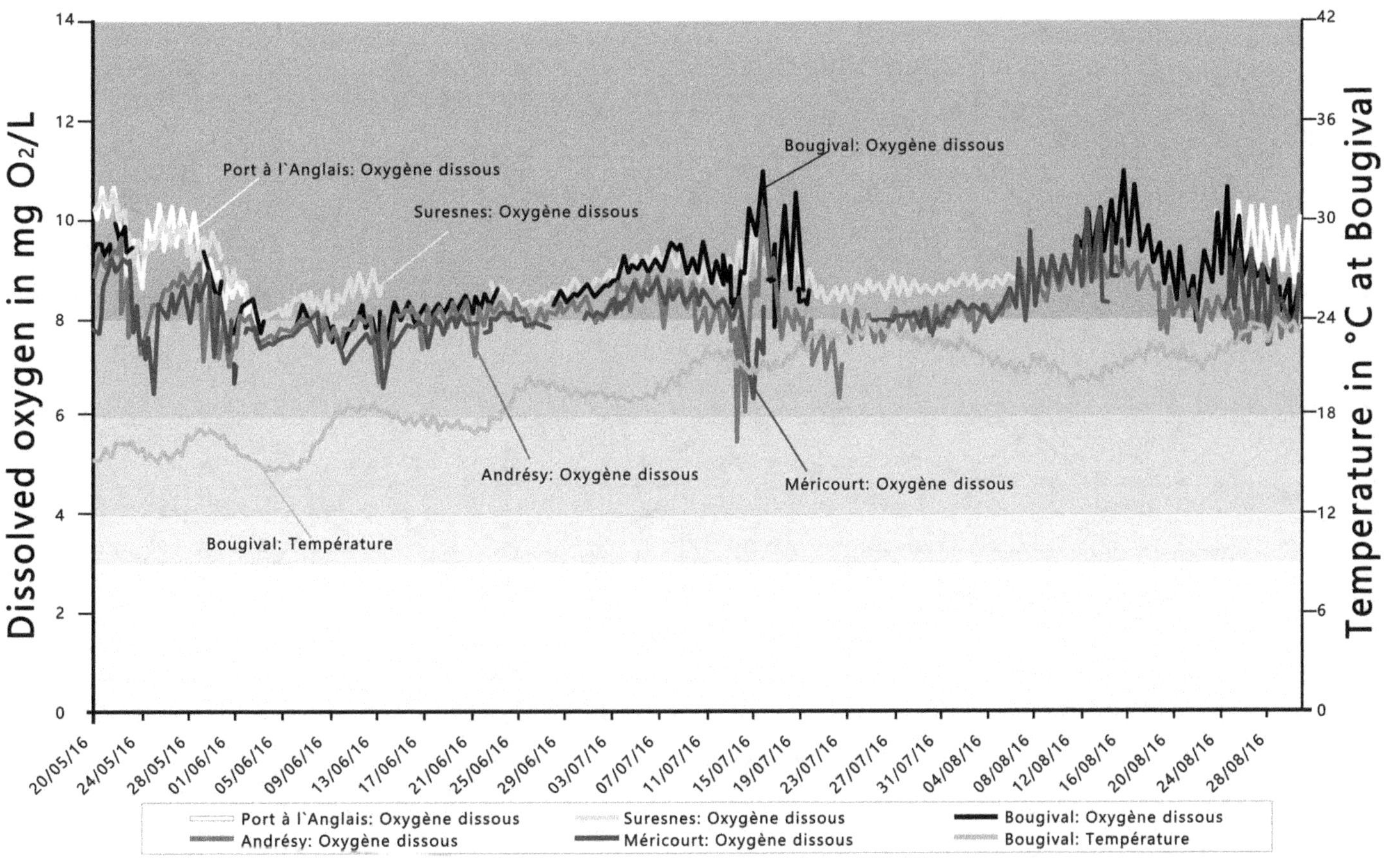

Figure 9.18 Dissolved oxygen and temperatures levels in the Seine

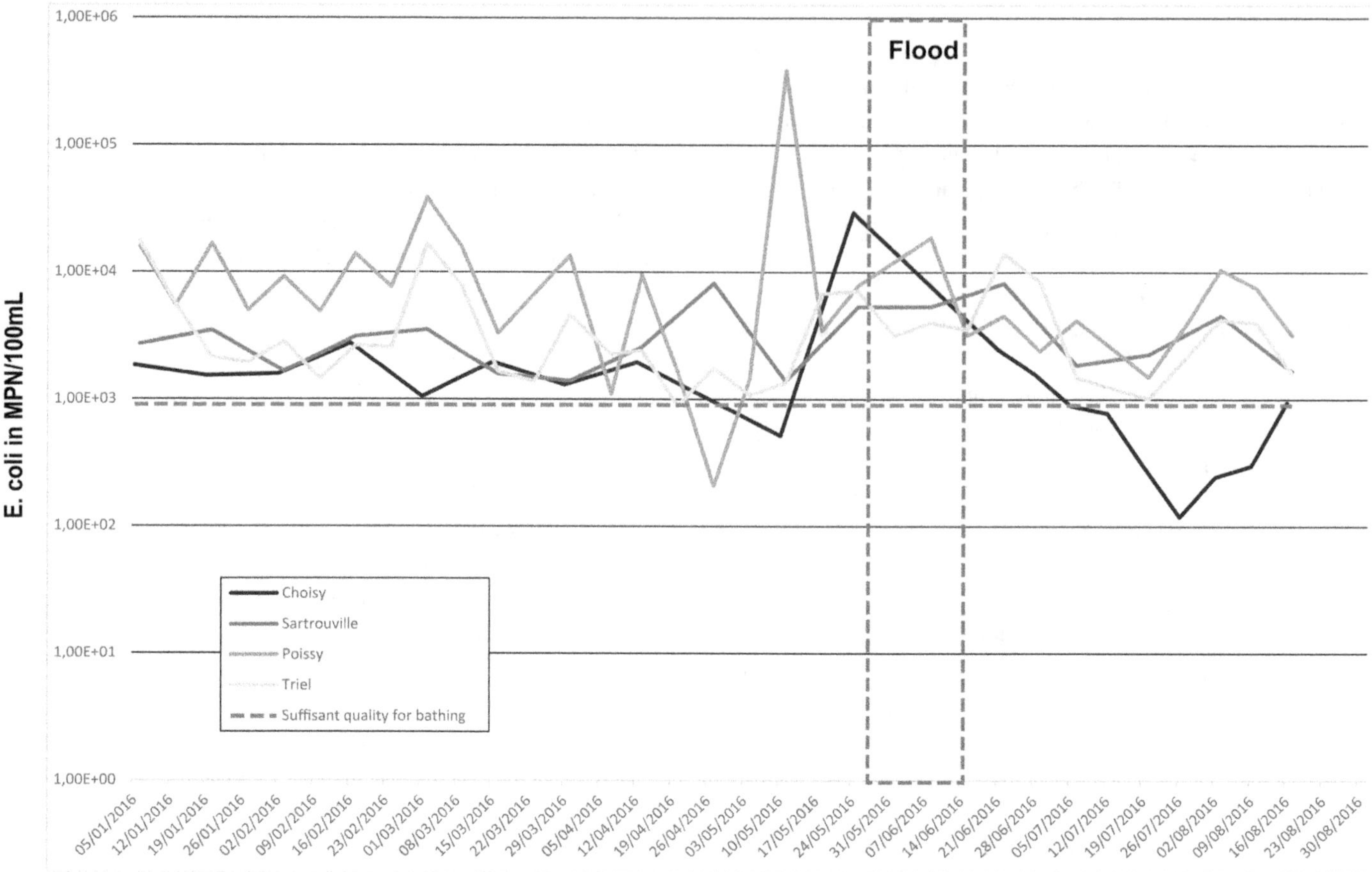

Figure 9.19 *E. coli* in Seine

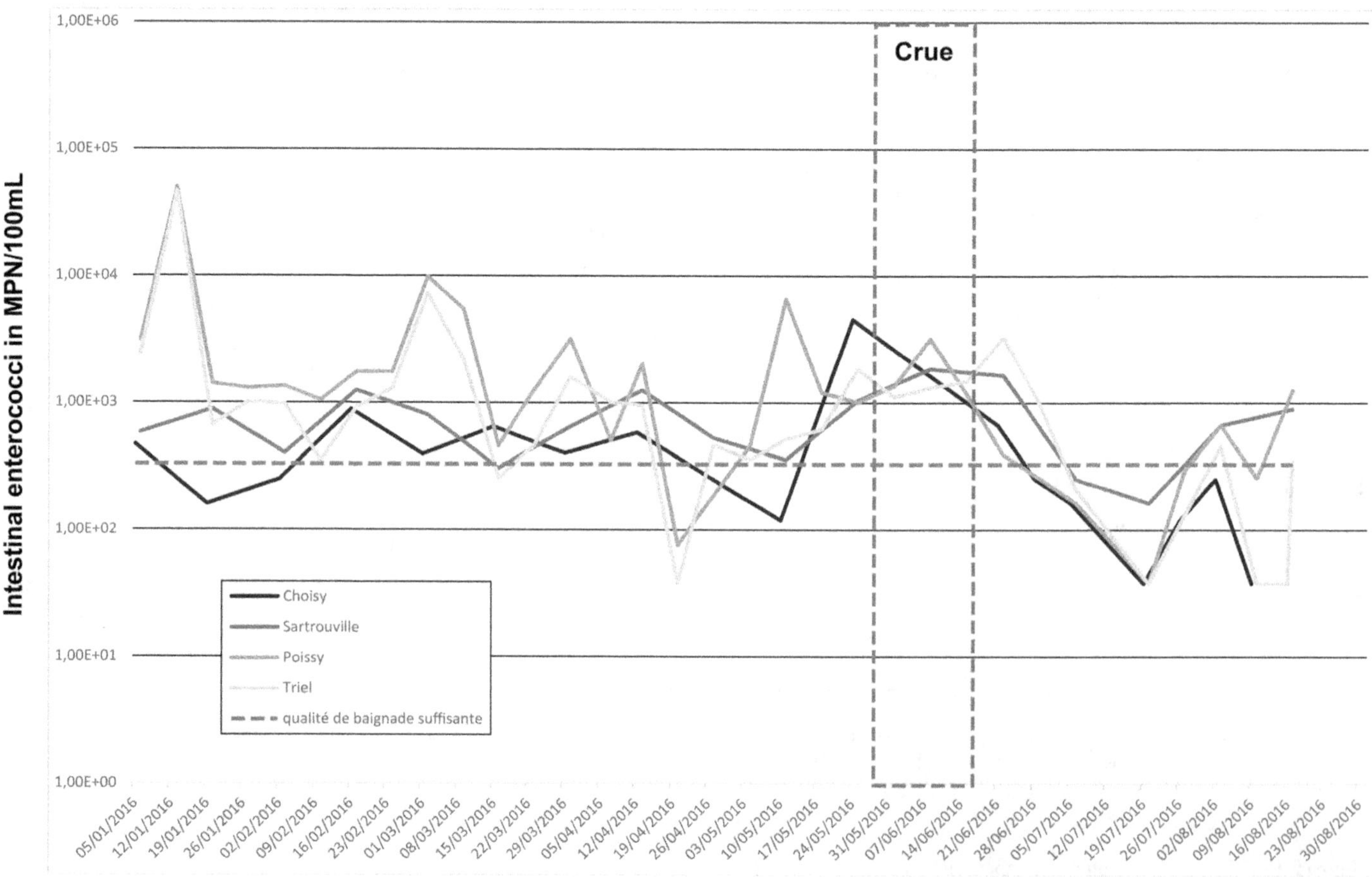

Figure 9.20 Intestinal enterococci levels in Seine

Nitrogen:

- Ammonia (NH_4^+) is downgraded to average condition.
- Nitrites (NO_2^-) are downgraded to average condition.
- Nitrates (NO_3^-) remains at good condition.

Phosphorus:

- Orthophosphates remain at good condition.
  The content in the Seine observed during the receding of the flood returned to good condition. The situation returns to normal in mid-June.

Dissolved oxygen:

- Oxygenation remains in good condition.

9.11 GENERAL CONCLUSION ON THE ASSESSMENT OF THE FLOOD

The exceptional situation was caused by substantial rain events from 29 to 31 May. It had a very strong impact on the operation of the sanitation system and forms, from a regulatory standpoint, an unusual situation. This rainy event has, moreover, given rise to major flooding of the Seine which also had consequences on the operation of the sanitation system.

Beyond the regulatory aspect, a coordinated and effective management between the sanitation operators must be retained from these events. Indeed, the exchanges were constant during these four weeks and the SEQUANA exercise largely contributed to the effectiveness of the management of this event. It also made it possible to streamline communication between all of the stakeholders.

The treatment systems made it possible to treat substantial volumes of water with purification performance at a good level which results in a limited impact on the environment. The extent of the flows in the Seine and Marne also contributed to limiting the impacts. The major overflows during the episode from 29 to 31 May 2016 occurred in an environment with flows that were already high. The effects – although visible in the conventional physico-chemical parameters – remained limited and have only a limited impact on the general quality of the environment. The high flows linked to the flood substantially increased the dilution capacity of the environment, as such contributing to limiting the impact on the environment, while, at the same time, the flows discharged by the sanitation system remained of the same magnitude as the daily averages. The measures taken did not concern micropollutants. No assessment on the discharges of the sanitation system on these parameters was conducted by SIAAP.

One must consider the robustness of the west sanitation system formed by the *Seine-Aval*, *Seine-Centre* and *Seine-Grésillons* plants with regards to the volumes transited.

Finally, the major contribution of the developments for more than 35 years of real-time control systems in each one of the *départements*, just like at SIAAP, must be underlined. These are major tools for coordinating actions and the effectiveness of interventions.

However, several points of attention must be underlined:

- Major difficulties in the south-east sector particularly linked to major pumping: the Crosnes pumping station, the only outlet of the urban wastewater for the valleys of the Orge, Yvette and Yerres to the *Seine-Upstream* plant;
- The high sensitivity of the territory of Val-de-Marne to the flood by the Seine and the Marne;
- The quantities of parasite water coming into the system and the volumes discharged are indicators of an operation that can be improved. The behaviour of the groundwater is still largely unknown, although research is currently being conducted.

Despite the damage in particular for the local population and the major disturbances caused by the flood, it should be underlined that several contextual elements were rather favourable. The following elements will be retained:

- As already indicated, the recent SEQUANA exercise prepared the teams well and insight had already been drawn from it, especially in terms of communication;
- The absence of rainy events at the same time as the peak of the flood. This point is particularly important because in June intense storms are frequent, yet the flood-protection pumping stations are not sized to handle this type of rain;
- The flood peak was of short duration. This limited the solicitation time of all of the equipment, some of which could have experienced difficulties over a longer duration. This in particular concerns the STPs. The event was also short enough to avoid causing any management issues concerning human resources;
- Finally, it was observed that the level at 6.10 m at Austerlitz seems to form a limit from which the difficulties could worsen substantially in particular concerning power supply, telecommunications and the mobilisation of staff in line with the difficulties concerning transport and the rest time required for intense activity.

10 "The Sponge City" Initiative in China

Haifeng Jia, Te Xu, and Zheng Wang

10.1 BACKGROUND

Ancient China has always respected Heaven and Man. Harmony with nature reflects the wisdoms of ancient Chinese. However, the past decade witnessed the rapid urbanisation in China. China's urban population has grown to 52.4% in 2015 from 42.5% in 2005, and the build-up areas have increased by 17,252 km^2. This roughly equates to an addition of 165 million people dwelling in urban areas in a decade. At the end of 2023 the urbanization rate in the country was 66.2% and is expected to reach 75–80% by 2035 (China: urbanization 2023. Statista. Retrieved 30 January 2024). The rapid urbanisation in China has led to a worsening "city syndrome" situation in recent years, including urban flooding and stormwater runoff pollution (Jia et al., 2017).

Frequent urban flooding is the first concern of the public. Pervious natural lands are gradually subsumed by the dramatically increasing impermeable concrete surface that prevents any water infiltration. Between 2008 and 2010, 62% of Chinese cities were flooded. And 39% of 351 cities in the country reported at least three serious floods in that time (Gaines, 2016). Beijing built-up area, for instance, has lost 16.4% of its green space since 1983 (Zuo et al., 2008). The result is a drastic increase in runoff generation. The situation is worsened by the continental, arid climate of Beijing with highly concentrated rainfall in summer. On 21 July 2012, these factors colluded with the insufficient stormwater infrastructure and consequently caused an extraordinary rainstorm disaster that cost 79 lives and a direct economic loss of 11.64 billion RMB (Su et al., 2015).

Stormwater runoff pollution including nonpoint source (NPS) pollution and combined sewage overflow (CSO) pollution is another extraordinary problem (Figure 10.1). For NPS pollution, many domestic researchers have carried out a lot of field monitoring and published their monitoring data since the new century. However, NPS pollution has not drawn Chinese government's attention until recent years and still remains to be solved. In 2008 and 2016, researchers from Tsinghua University have monitored the NPS pollution in the same urban area in Beijing. Results were disappointing that no improvement had been witnessed. Surface runoff often carries considerable pollutants that exceeds the national surface water standard (GB3838–2002) of Class IV (i.e., COD and TN are no more than 30 mg/L and 1.5 mg/l, respectively). That is, COD and TN in surface runoff are approximately 20–370 mg/l and 0.5–12.5 mg/l, respectively. Tianjin local government along with Tianjin University has also carried out a series of field monitoring before 2015. Results showed that COD and TN in surface runoff are approximately 17–508 mg/l and 2.5–5.5 mg/L, respectively. Particularly for the first flush runoff, the pollutant concentrations almost reach the pollution level of general sanitary sewage. Apart from NPS pollution, frequent CSO pollution during rainstorm also aggravates deterioration of aquatic ecosystems. With the promotion of reclaimed wastewater treatment plant during "the 13th Five-Year Plan," point source pollution would be well controlled. Consequently, NPS pollution and CSO pollution gradually become the main sources of pollution for urban water bodies.

Significant loss of life and property, as well as environmental pollution, modern urban stormwater management has attracted much attention of the Chinese government, professional experts, and the public. Inspired by the concept of low impact development (LID) and best management practices (BMPs) (Fletcher et al., 2015) prevalent in the developed countries, the construction of "Sponge City" was promoted in 2013 by Chinese government to alleviate the adverse impacts caused by rapid urbanisation. Compared with LID-BMP measures, the conventional grey infrastructures such as rainwater pipe network are relatively rigid, meaning that frequent updates and expansions to keep pace with the rapid urbanisation are almost impossible. Hence, a concept transition from the conventional "rapid draining" to the current mainstream of urban stormwater management begins to happen in China, an undertaking presented in an outlook paper published in "Nature" (Gaines, 2016). "The Sponge City" is an organic fusion of wisdoms of ancient Chinese and the advanced concepts of Western developed countries.

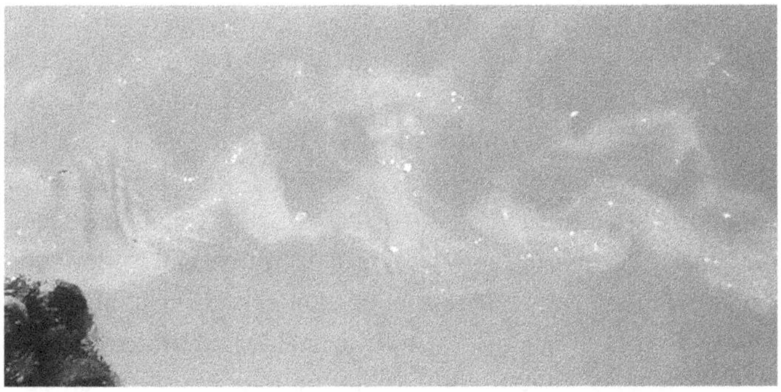

Figure 10.1 Grease and suspended solids levels on Beijing Moat after the rainstorm on 22 June 2017

10.2 THE SPONGE CITY INITIATIVE

10.2.1 Concepts of the Sponge City

Initially, the purpose of the Sponge City is defined as "natural storage, natural infiltration, and natural purification." In other words, the Sponge City is intended to mitigate urban flooding and purify stormwater runoff. In 2014, the issued "Technical Guide for the Construction of the Sponge City (Draft)" (MOHURD, 2014) (referred to as "the Guide") and the published "The meanings, approaches, and prospects of the Sponge City (LID)" (Qiu, 2015) gave the Sponge City a substantially clear definition. That is, the city can function like a sponge. During wet period, urban stormwater will be captured and detained by LID measures, while during dry period, the stored and purified stormwater could be recycled for scenic environment use and street sprinkling. Deviating from the traditional "rapid-draining" approach, the new paradigm calls for the use of natural processes such as soil and vegetation as part of the urban runoff control strategy.

Since 2015, a few proverbs, which further complemented the concepts of the Sponge City, have begun to appear in many official documents. The best-known two are the "four-phrase" objectives, that is, "no ponding during sprinkle, no flooding during rainstorm, no urban black-and-malodorous, and alleviated heat island effect," and the "six-word" principles including infiltrate, detain, store, purify, reuse, and drain.

The "four-phrase" objectives comprehensively summarise the multiple functions that the Sponge Cities are required to achieve. Apart from the essential functions such as stormwater runoff control, the Sponge City is also beneficial for ecological restoration and microclimate improvement, thereby providing local residents with life full of convenience and comfort as well as equipping the city with flexibility to deal with uncertainty of future developments (Xu et al., 2017a), such as the increasing flooding with the rapid urbanisation or continuous climate change (Liu et al., 2016).

The "six-word" principles form the guidelines for urban stormwater management. These principles are similar to those under the LID-BMPs paradigm. "Infiltration" can be interpreted as absorbing water just as a piece of ordinary sponge does, which, to some extent, makes the restricted hydrological cycle restored to the predevelopment status. Infiltration process effectively reduces the generation of surface runoff and increases groundwater recharge. "Infiltration" corresponds to the original concept of the Sponge City and thus has been given the first priority among the six words. The basic idea of "detention" is to "trade space for time." That means by extending flow path and time of flow routing, the peak time point can be significantly delayed and the peak flow intensity can decrease considerably. Different from infiltration that focusses on source control, "storage" lays emphasis on process control. Temporary storage of stormwater runoff is able to further reduce runoff volume and to put off peak time point. Because of the combined effects of infiltration, detention, and storage in the Sponge Cities, the burden of the conventional rainwater systems including the pipe network and the pumping station, which may have been overwhelmed by a sudden influx of stormwater, can be alleviated. Different from storage, "reuse" requires more complex rainwater collection systems that gather not only the stored water but also the infiltrated water. "The Sponge Cities" contain three kinds of rainwater collection systems, that is, systems in residential, industrial, and municipal public areas. Certainly, "purification" measures or facilities including natural treatments such as constructed wetland and intensified treatments such as filtrator are indispensable before qualified for reuse. The last but often ignored by a few city planners is "drainage." Considering the limited capacity of LID-BMP measures for infiltration and storage, it is the conventional rainwater systems that must be depended on to drain the excessive stormwater during extreme rainfall to avoid traffic paralysis and casualties. In this sense, the concept of the Sponge City is not purely "green" but about the integration of green and grey infrastructures (Jia et al., 2017).

10.2.2 The Key Movement of Sponge City Implementation: Pilot City Programme

After the emergence of Sponge City term in China 2013, there are a great advance. In October 2014, the China Ministry of Housing and Urban-Rural Construction (MHURC) issued a draft technical manual on Sponge City construction. In October 2015, the State Council of China announced a major expansion of the Sponge City Initiative and issued "Guiding Opinions on Advancing the Construction of Sponge Cities" (General Office of the State Council, 2015) (referred to as the Sponge City Plan (SCP)), which is being implemented nationwide. The SCP put forward the overall goals of "Sponge City" construction. That is, more than 20% of the built-up urban areas are built to meet the control targets by 2020, and more than 80% of the built-up urban areas are built to meet the control targets by 2030. The use of LID-BMPs will then be required for all new development and retrofit sites, science and commercial parks, green spaces, vehicle roads, pedestrian walkways, and so on.

In order to push forward the SCP smoothly, China Ministry of Finance (MOF), with support from MOHURC and the Ministry of Water Resources (MWR), have selected 30 cities among hundreds of applicants as national pilot sites under the SCP. The very strict pilot city screening was taken two batches. The first batch screening was conducted in April 2015. And 16 pilot cities were selected. The second batch screening was conducted in April 2016. And 14 pilot cities were selected. These 30 cities are located in 22 different provinces and the 4 municipalities, that is, Beijing, Shanghai, Chongqing, and Tianjin, as representatives of construction of "Sponge Cities" under different natural and social-economical circumstances of the five controlling zones. Accordingly, an assessment system for the pilot Sponge City has been established based on the "six-word" principles.

The pilot city programme helps obtain the experiences of Sponge City construction in different typical cities and thus support the national policy on Sponge City implementation. At the same time, the provincial governments responded to the SCP actively that many provincial pilot cities have appeared in all the 31 provinces in mainland China. Zhejiang Province, for example, selected Shaoxing, Quzhou, Lanxi, and Wenling as provincial pilot sites in 2016, in addition to the national pilot "Sponge Cities" Jiaxing and Ningbo. According to the SCP, many of the national and provincial pilot "Sponge Cities" have introduced their own policy and guides to specify the basic principles, targets and tasks, and key measures to support planning and design of the green and grey infrastructures. For example, "Technical Guidelines for the Sponge City Construction in Tianjin" have come into effect on 1 June 2016. This technical guideline specifies the four control targets suggested by the Guide (e.g., the annual runoff capture rate is divided into several small targets for different types of land use with distinct status of urban development), includes some useful monitoring data (e.g., rainfall patterns in different regions of Tianjin, event mean concentrations of COD, SS, TN, and TP on roads, and rooftops and green spaces), and provides the key design parameters and material lists for designing common LID-BMP measures.

"The Guide" (MOHURD, 2014) has given estimations on the cost of different "Sponge City" components in China (Figure 10.2). And 100 to 150 million RMB (around 19 to 26 million US dollars) is demanded per square kilometre (CPC Central Committee, 2014; General Office of the State Council, 2015). Estimations on the annual total investment in the Sponge Cities are also given in Table 10.1, which shows that nearly 1.8–2.5 trillion RMB or 264–360 billion US dollars is demanded for the 2020 goal. So where does such a huge amount of money come from? Public–private partnerships (PPP) are encouraged. In China, central finance will cover approximately 15–20% of the total investment. The rest has to be covered by local finance or private sectors. During operation phase, the profits of the private investors are mainly from government grants. The government imposes levies on stormwater

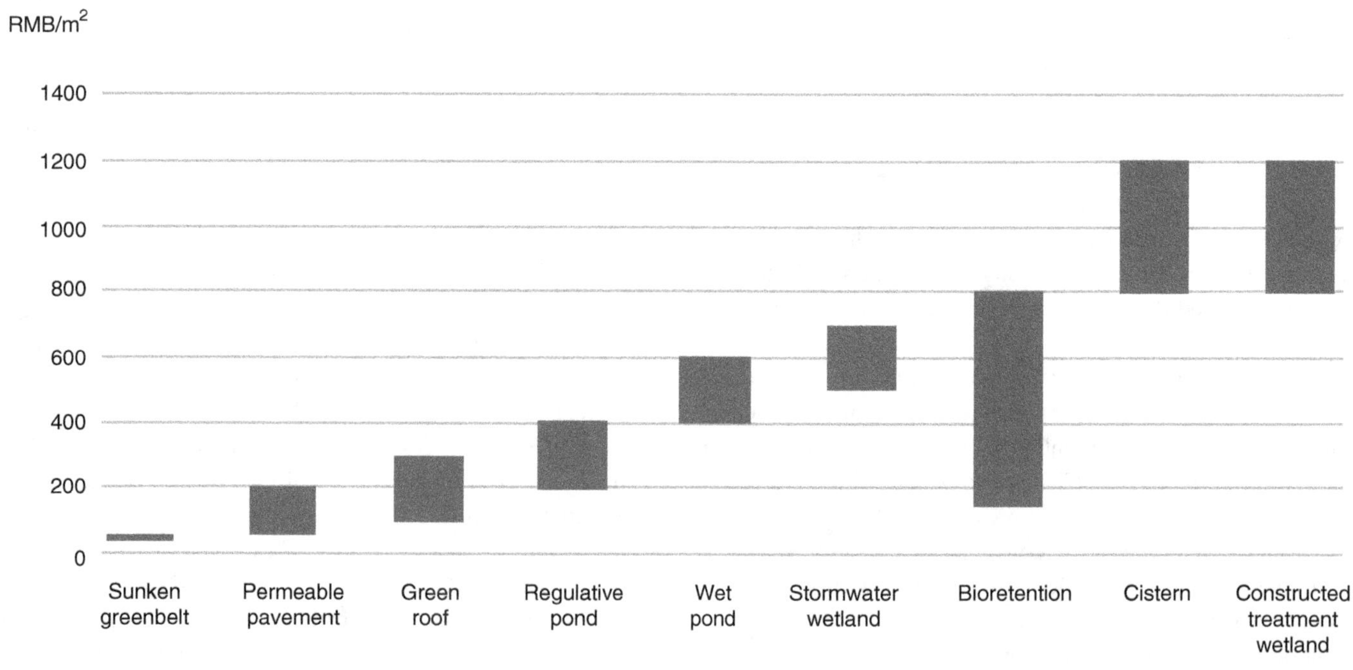

Figure 10.2 Cost of different "Sponge City" components

infrastructure from local residents. Then, the government pays the private investors according to the performance of stormwater runoff control and rainwater recycling. For local residents, improvement of living environment and increase in real estate value will be a good compensation for increase in tax incidence. In this sense, infrastructure PPP is not only an effective financial tool for government but also an incentive for private sectors. Take the national pilot cities as example. The total investment of the 30 national pilot "Sponge Cities" are estimated to be about 42.3 billion RMB or 6.35 billion US dollars. Each municipality, provincial capital, and other city will receive 600, 500, and 400 million RMB (around 88, 74, and 59 million US dollars) annually from the central government for three years, respectively. Cities will receive a 10% bonus from the central government if the PPP contribution exceeds a certain percentage of the overall budget. In most pilot projects, PPP is often combined with build-operate-transfer (BOT) mode. That is, the profits during the first 10–25-year operational phase (referred to as concession period) belong to the private sectors. After the concession period, the ownership of the project should be transferred to the local governments.

Table 10.1 *Estimation of total investment on Sponge City from 2015 to 2020*

	2015	2016	2017	2018	2019	2020	Total
Built-up urban areas (km^2)	51,764	53,834	55,987	58,227	6556	62,978	/
Sponge City ratio of built-up areas (%)	/	2	4	9	15	22	/
Sponge City area (km^2)	/	77	2239	5240	83	13,855	/
Highest total investment (100 million)	350	1588	2093	5402	6917	8589	24,939
Lowest total investment (100 million)	300	1100	1512	3901	4996	6203	18,012

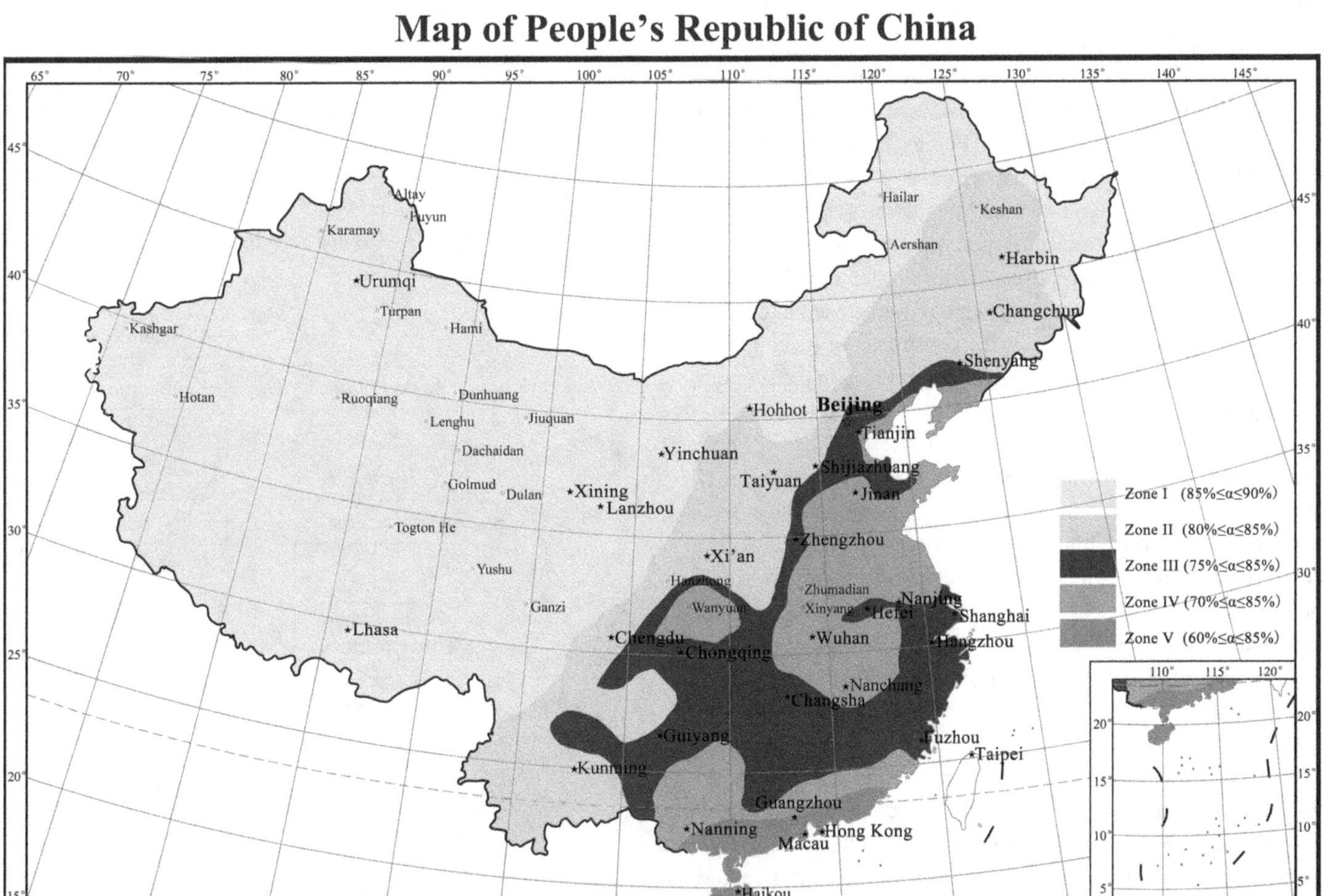

Figure 10.3 Zoning map of annual runoff capture rate on mainland China

10.3 THE KEY POINTS FOR SPONGE CITY IMPLEMENTATION

10.3.1 Technical Guide for Zoning Runoff Control Targets

"The Guide" (MOHURD, 2014) was the first and foremost technical guide for the subsequent sponge cities to determine runoff control targets, conduct LID-BMPs layout planning, and design urban stormwater management system. The Guide proposed four kinds of control targets. They are about runoff volume control, peak flow control, NPS pollution control, and rainwater reuse.

The target of runoff volume control is always mandatory, not only because of the initial functional positioning of the Sponge City, but also because of the significant correlations with the other three control targets. Annual runoff capture rate is one basic index to evaluate the performance of runoff volume control. The Guide provides city planners with the relationships between total annual runoff capture rate (%) and the corresponding designed rainfall (unit: mm) based on the statistical analysis of daily rainfall database of nearly 200 cities from 1983 to 2012 in China. China is vast. Considering the distinct climate conditions, hydrological characteristics and degree of urban development, the Guide classifies the mainland China into five zones with different annual runoff capture rate (Figure 10.3). For example, north-western China (Zone I) is located in the temperate continental climate zone. The arid or semiarid climate leads to the demand for rainwater reuse. Some areas such as the Loess Plateau have thin and loose soil, which is susceptible to stormwater runoff erosion. Moreover, the economic development and urbanisation of Zone I are relatively backward. Thus, the constraints of placing LID-BMP measures in Zone I are not so tight as those in the economically developed regions. As a result, the Guide set a relatively high annual runoff capture rate (i.e., at least 85%) for Zone I. On the contrary, southern China (Zone V and part of Zones III and IV) is located in the tropical or subtropical monsoon climate zone where extreme rainfall events are frequent. Under these circumstances, "drainage" is the most urgent among the "six-word" principles. On the other hand, most of the cities in southern China are economically developed with substantially high opportunity costs for placing LID-BMP measures especially on retrofit sites. That is, the Sponge City will be constructed at the expenses of high land acquisition costs or at the sacrifice of more profitable investment projects. In this sense, proposing LID-BMPs for use in the largest scale possible may succeed in terms of a high annual runoff capture rate, but would be costly (Xu et al., 2017a). Therefore, the Guide set a relatively low annual runoff capture rate (i.e., below 75% is acceptable) for southern China.

As for the other three control targets, the Guide does not specify the objectives that the Sponge Cities must satisfy but makes some instructive suggestions. For instance, suspended solid (SS) is suggested as the pollution index for the target of NPS pollution control because SS is often significantly correlated with other conventional pollutant indicators such as COD in stormwater runoff. Researches (Bedan and Calusen, 2009; Jia et al., 2015b; Mao et al., 2017) indicated that LID-BMPs are capable to remove 35–75% of SS loads annually from stormwater runoff. The local governments could set a proper objective within this range with sufficient consideration of the actual conditions, especially the safety of the entire water environment. Similar to the "six-word" principles, the Guide also recognises the important role that "drainage" plays in "the Sponge Cities." It points out that only LID-BMP measures make an incomplete "Sponge City" and mandates the integration of green and grey infrastructures to establish a complete urban stormwater management system from the source to the end. More specifically, the designed return period and runoff coefficients of the conventional rainwater pipe network and pumping station should still conform to the national standard "Code for Design of Outdoor Wastewater Engineering" (GB50014–2014).

10.3.2 Selection of Appropriate LID-BMP Measures

The first task of the Sponge City implementation is LID-BMPs selection and siting for each subcatchment. Site suitability should be given the first priority including site conditions, soil characteristics, groundwater characteristics, topography, catchment properties, and space requirement (Figure 10.4). In

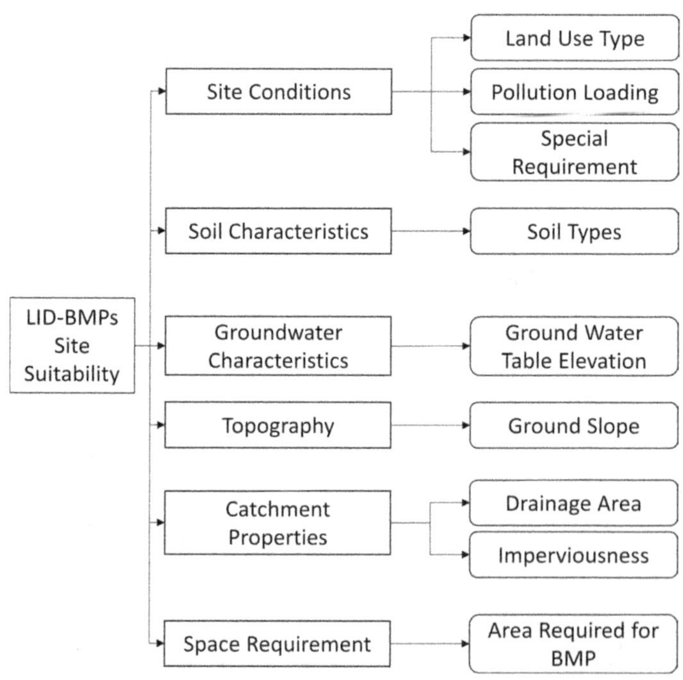

Figure 10.4 Indicators of LID-BMPs site suitability (Jia et al., 2013)

Table 10.2 *Comparison among the most frequently applied LID-BMP measures in China (MOHURD, 2014)*

	functions					Control targets			Treatments		Costs		
LID-BMP measure	Storage	Groundwater recharge	Peak flow reduction	Purification	Conveyance	Runoff volume	Peak flow	Runoff pollution	Decentralised	Centralised	Construction	Maintenance	Aesthetic effect
Permeable pavement	○	●	◎	◎	○	●	◎	◎	√		○	○	○
Permeable concrete	○	○	◎	◎	○	◎	◎	◎	√		●	◎	○
Permeable asphalt	○	○	◎	◎	○	◎	◎	◎	√		●	◎	○
Green roof	○	○	◎	◎	○	●	◎	◎	√		●	◎	●
Sunken greenbelt	○	●	◎	◎	○	●	◎	◎	√		○	○	◎
Bioretention	○	●	◎	●	○	●	◎	●	√		○	○	●
Wet pond	●	○	●	◎	○	●	●	◎		√	●	◎	●
Rain garden	●	○	●	●	○	●	●	●	√	√	●	◎	●
Cistern	●	○	◎	◎	○	●	◎	◎	√		○	○	○
Regulative pond	○	○	●	◎	○	○	●	◎		√	●	◎	◎
Conveyance-based swale	◎	○	○	◎	●	◎	○	◎	√		○	○	◎
Vegetative swale	○	●	○	◎	●	●	○	●	√		○	○	●
Grass strip	○	○	○	●	/	○	○	●	√		○	○	◎
Initial split-flow equipment	◎	○	○	●	/	○	○	●	√		○	◎	○

●-*strong*, ◎-*moderate*, ○-*weak*

addition, relevant national standards and policies should also be considered. For example, porous pavements are suitable for patios, parking lots, sidewalks, and even driveways, but the fire lane must not be occupied during construction; rain gardens are suitable for residential lands, but enough spacing must be maintained from the buildings thereby avoiding foundation erosions; and so on. After identifying potential LID-BMP measures, other factors such as runoff control efficiency, costs of construction and maintenance, aesthetic effects, system reliability, and the priority control target of the city should be considered for city planners to make judicious decisions. The Guide has made a comprehensive comparison among a total of 22 types of LID-BMP measures in several aspects (MOHURD, 2014). Table 10.2 lists a part of them, that is, the considerably detailed information about the most common LID-BMP measures in China. Then, a mechanism for ranking the LID-BMP measures is required to determine the most satisfactory ones for placement in each subcatchment. Analytical Hierarchical Process is adequate to fulfil the requirement (Young et al., 2010). Jia et al. (2013) have devised BMPSELEC that integrates the aforementioned processes of LID-BMPs selection and siting into one software program. In addition, the System for Urban Stormwater Treatment and Analysis Integration (SUSTAIN) has a built-in BMP siting tool (Shoemaker et al., 2009), which is also accessible for "Sponge Cities." Foshan was the first case in China to successfully use these tools (Jia et al., 2013, 2015a; Mao et al., 2017).

10.3.3 Decomposition of Control Targets

The second task of the Sponge City implementation is to determine the area or volume of each LID-BMP measure in each subcatchment. In essence, this task is about decomposition of the priority control target. "Bottom-up" method is recommended by the Guide. Take the target of runoff volume control as an example to demonstrate the basic idea of the bottom-up method. The runoff volume that is required to be eliminated on site can be calculated as:

$$V = 10H\varphi F \tag{1}$$

where H is the designed rainfall, mm, which is determined by the relationships between total annual runoff capture rate and designed rainfall; F is the area of the entire planning region, m^2; φ is the average runoff coefficient of the entire planning region, which is the weighted average of φ_i and F_i of each subcatchment. Hence, the runoff volume required to be eliminated by each subcatchment can be calculated as:

$$V_i = 10H\varphi_i F_i \tag{2}$$

Then, for each subcatchment, the assigned runoff volume is decomposed into several parts that will be treated by each LID-BMP measure in planning. It is worth mentioning that subjective judgements will be inevitably involved in this step because of unavailable formulas for decomposition. Finally, the area or volume of each selected LID-BMP measure can be determined according to its burden. Although the bottom-up method seems easy to master, it would be rather unfeasible when too many LID-BMP measures are to be placed in the same subcatchment. The reason lies in the unreliable subjective judgements. The results maybe deviate from a cost-effective planning scheme. In this situation, "up-down" method is a good alternative.

The basic idea of the up-down method is multi-objective optimisation (Xu et al., 2017a). The area or volume of each selected LID-BMP measure is the decision variable of optimisation, while the annual runoff capture rate and the total cost are the objective functions. The prediction of the annual runoff capture rate must rely on stormwater runoff modelling, such as Storm Water Management Model (SWMM) and SUSTAIN. The total cost may include construction cost, operational cost, maintenance cost, and even opportunity cost. A life-cycle cost is also recommended (Xu et al., 2017b). Results of optimisation are a set of trade-off solutions, namely Pareto solutions. Non-dominated Sorting Genetic Algorithm (NSGA-II) is the most popular algorithm for LID-BMPs layout optimisation as NSGA-II presents the fastest convergence speed and good quality of Pareto solution set (Deb et al., 2002). The model SUSTAIN has a built-in optimisation tool that is exactly established on NSGA-II (Shoemaker, 2009). Figure 10.5 shows an example of LID-BMPs layout optimisation in Foshan, China using SUSTAIN. In the figure, the X axis represents the total cost of LID-BMPs implementation, the unit is million CNY (RMB). The Y axis represents the annual runoff capture rate of LID-BMPs planning schemes. We can find the best solution of LID-BMPs layout planning, which is the "knee-of-curve" point in the figure.

10.3.4 Integration of Green and Grey Infrastructure

Integration of green and grey infrastructures is the core aspect of design of urban stormwater management system. Take Tianjin, one of the pilot "Sponge Cities," as an example. In 2016, Tianjin's local government has proposed a sketch of integration of green and grey infrastructures as shown in Figure 10.6 that is worthy of reference for other "Sponge Cities." Urban stormwater runoff comprises of off-site runoff and on-site runoff. The former means runoff generated from the urban area that is not suitable for implementing LID-BMPs, whereas the latter means runoff generated from the urban area that is suitable for LID-BMPs placement. For the runoff off site, the first flush should be split into sanitary sewage pipeline by initial split-flow equipment to avoid the first flush pollution. For the runoff on site, the runoff from traditional rooftops and impervious areas should be firstly treated by LID-BMP measures in series, referred to as LID-BMP chain. Common LID-BMP

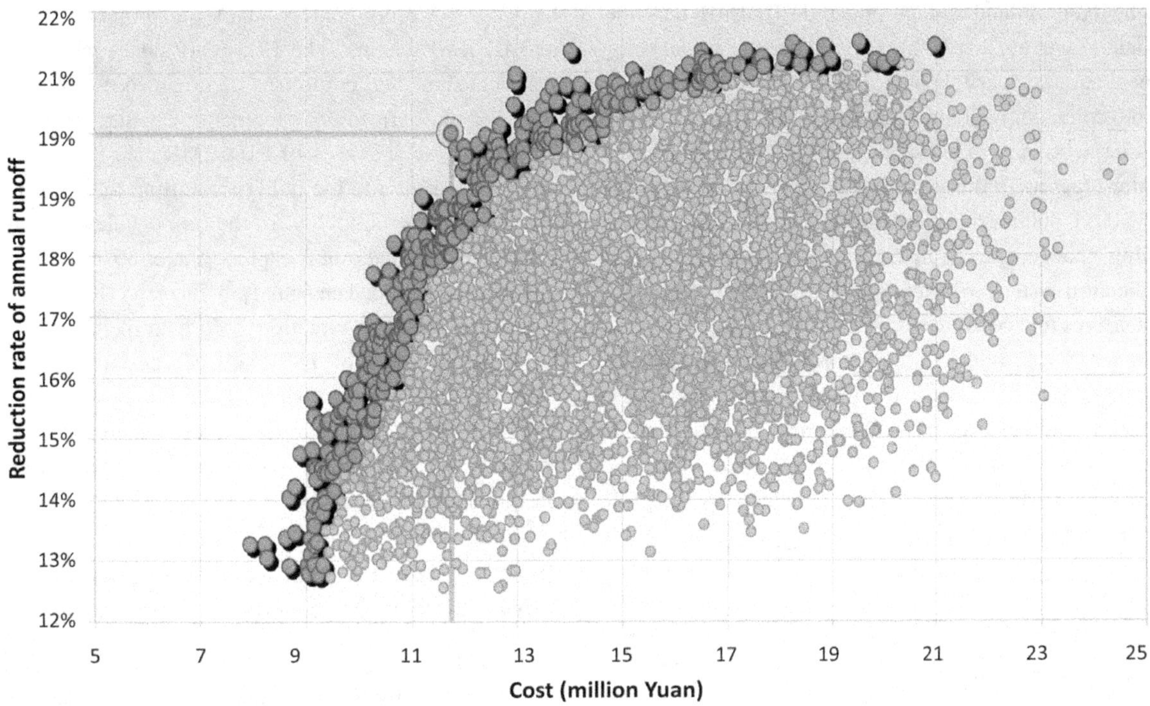

Figure 10.5 Example of Pareto solution set after LID-BMPs layout optimisation in Suzhou, China (Jia et al., 2015a)

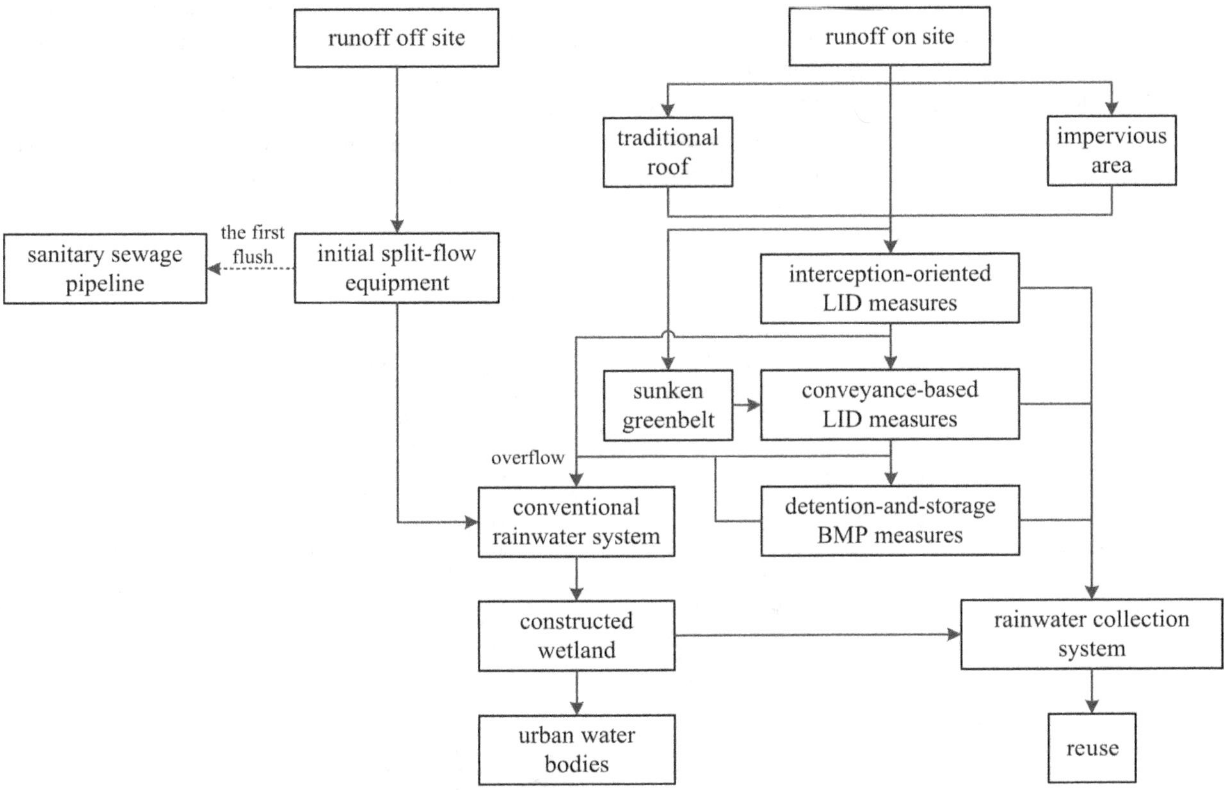

Figure 10.6 Diagram of the integration of green and grey infrastructures

measures can be categorised under interception-oriented LID (e.g., green roof, cistern, and permeable pavement), conveyance-based LID (e.g., vegetative swale), and detention-and-storage BMPs (e.g., rain garden and wet pond) according to their main functions. It was recommended that the interception-oriented LID-BMPs be deployed upstream, the conveyance-based LID-BMP be in the middle, and the retention-and-storage-oriented LID-BMP be at the end of the chain (Xu et al., 2017a, b). The runoff after initial split-flow equipment and the overflow part of LID-BMP chains will be discharged into conventional rainwater system. This part of runoff still carries a large amount of pollutant loads. Therefore, the outlet of rainwater pipeline in this system is constructed wetland for natural purification, but not rivers or lakes just as the conventional "rapid draining" concept does. In addition, the aforementioned rainwater collection system will gather the purified rainwater from LID-BMP measures and constructed wetland for reuse.

So we can see that a modern urban stormwater management system combines green infrastructures (LID-BMP measures, sunken greenbelt, and constructed wetland) and grey infrastructures (initial split-flow equipment and conventional rainwater system) effectively to bring the "six-word" principles into full play, but not a simple connection.

10.4 CASE STUDIES OF SPONGE CITY CONSTRUCTION

10.4.1 Shenzhen: A Pilot City to Implement LID-BMPs in Series

Shenzhen was selected as the national pilot Sponge City in the second screening batch. However, it was one of the earliest cities in China to implement LID practice. In the performance assessment of pilot Sponge City organised by MOHURD in 2016, Shenzhen won the first place in the second batch of pilot "Sponge Cities." The prominent point of the project is the application of LID-BMP chains. The LID-BMP chain in Guangming New District, Shenzhen comprises of green roof, rain barrels, raised flower bed, permeable pavement, vegetative swale, and rain garden. These six types of LID-BMP measures are deployed in series. According to the field monitoring, the LID-BMP chain is able to reduce around 70% of the runoff volume annually, which satisfies the annual runoff capture rate of 60–85% (Zone V) stipulated by the Guide. The peak flow intensity decreases by 55–88% compared with the situation before implementing LID-BMPs. Besides, over 80% of TSS loads can be removed. In summary, the LID-BMP chain maximises the utility of each independent unit and makes the entire rainwater system substantially robust to resist a heavy rainfall, whose intensity may even exceed the designed return period.

10.4.2 Tianjin: A Pilot City to Integrate Green Infrastructure with the Grey Infrastructure

Tanggu Bay, a coastal urban area in Tianjin, has established a stormwater management system near Haihe River (Figure 10.7(a)). The core aspect of designing such a system is exactly the integration of green and grey infrastructures (Figures 10.7(a) and (b)). The on-site runoff will firstly be treated by the LID-BMP measures and then overflow into the rainwater pipe. The off-site runoff will firstly be treated by initial split-flow equipment and then discharged into the rainwater pipe. The pumping station helps to raise up the collected rainwater into the constructed wetland for further storage and

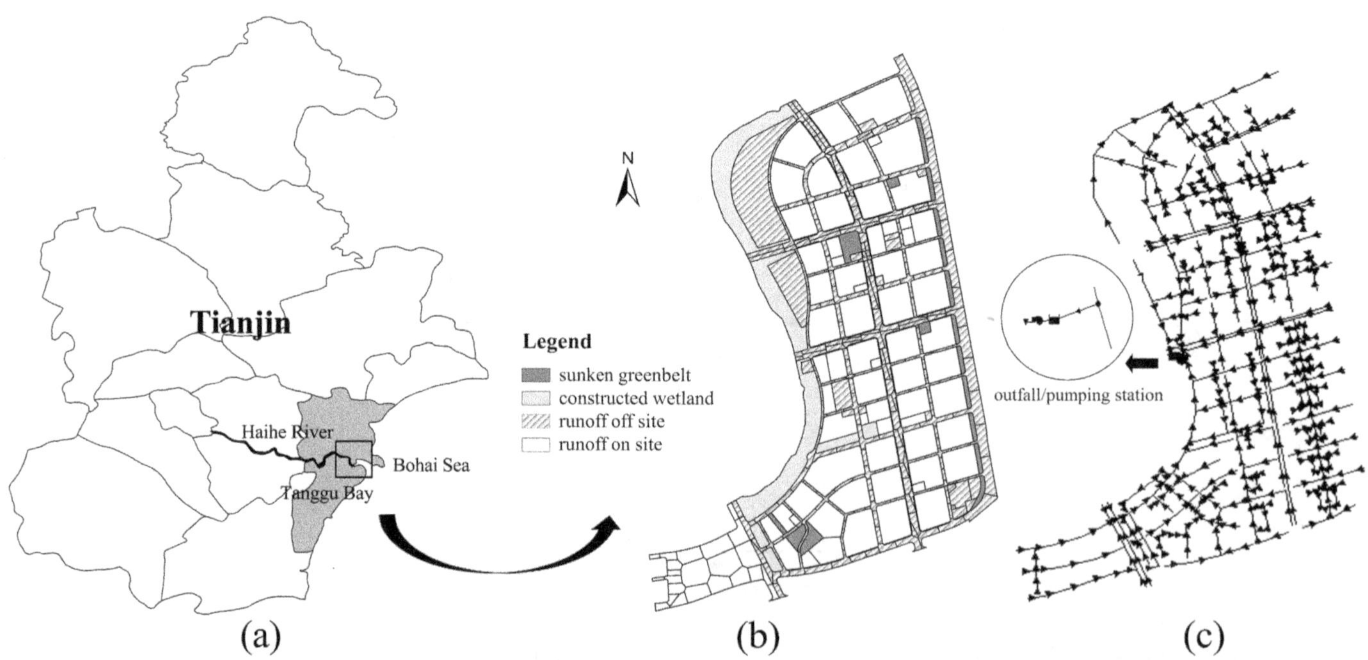

Figure 10.7 An example of integration of green and grey infrastructures in Tanggu Bay, Tianjin. (a) Location of the study site; (b) distribution of green infrastructures (SWMM); (c) distribution of grey infrastructures (SWMM) (Xu et al., 2017a)

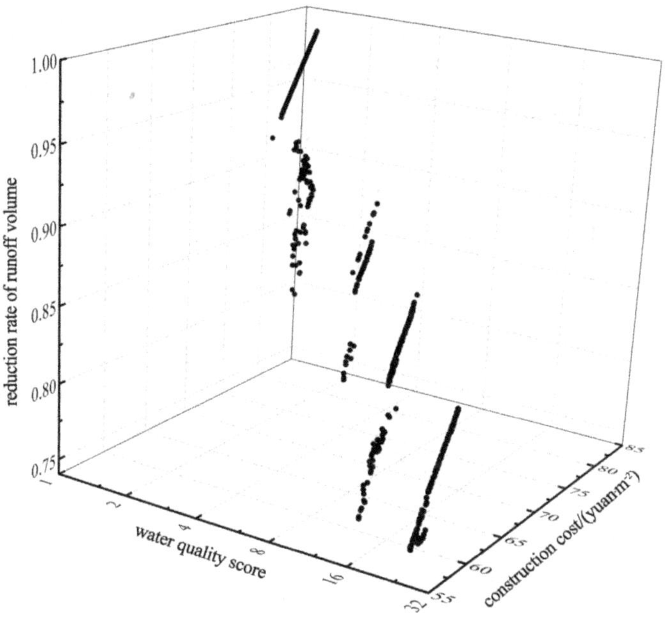

Figure 10.8 Results of rainwater system optimisation in Tianjin, China (Xu et al., 2017a)

purification. The system was optimised by coupling NSGA-II to SWMM. Different from Figure 10.5, this case optimised three objective functions, that is, runoff volume control, runoff water quality control, and construction cost (Figure 10.8). After optimisation, the system is able to meet the annual runoff capture rate of 70–85% (Zone IV), to reduce 22–46% peak flow intensity, and to remove COD, TSS, TN, and TP loads by 32–49%, 34–50%, 56–64%, and 45–56%, respectively.

10.4.3 Xi'an: A Pilot City to Implement Rain Gardens over Underground Garages

In the course of "Sponge City" practice, the Chinese engineers have devised a few new structures and construction methods of common LID-BMP measures thereby expanding their applicable areas. For example, areas with high water table, shallow bedrock depth, poor soil conditions, or underground garage is not suitable for implementing rain gardens because of restricted infiltration ability. Xi'an is such a city with restricted infiltration ability. Fortunately, by adding a drainage layer, which consists of gravel and blind pipes, and a drainage plate at the bottom of an ordinary rain garden (Figure 10.9), the structural improved rain gardens can drain the infiltrated rainwater into rainwater collection system in time. Consequently, the limited infiltration ability could be maintained for a longer period of time. Many infiltration-restricted areas in Xi'an, especially residential districts with underground garage, have adopted this type of rain garden as illustrated in Figure 10.9. Certainly, this type of rain garden has been extensively applied in many other areas with restricted infiltration ability in China.

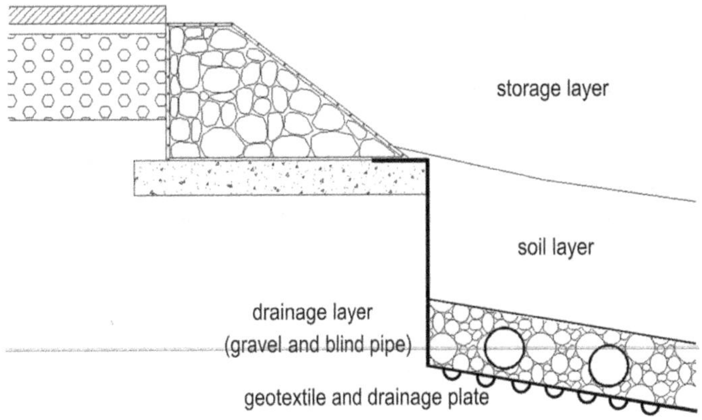

Figure 10.9 Structural improvement of rain garden in infiltration-restricted areas

References

Bedan, E., & Clausen, J. (2009). Stormwater runoff quality and quantity from traditional and low impact development watersheds1. *JAWRA Journal of the American Water Resources Association*, 45. 998–1008. DOI: 10.1111/j.1752-1688.2009.00342.x.

China: urbanization. (2023). Statista. Retrieved 2024-01-30

CPC Central Committee, the State Council. (2014). The National New Urbanization Plan (2014–2020).

Deb, K., Agrawal, S., Pratap, A., & Meyarivan, T. (2002). A fast and elitist multiobjective genetic algorithm: NSGA-II. *IEEE Transactions on Evolutionary Computation*, 6, 182–197.

Fletcher, T. D., Shuster, W., Hunt, W. F., Ashley, R., Butler, D., Arthur, S., Trowsdale, S., Barraud, S., Semadeni-Davies, A., Bertrand-Krajewski, J.-L., Mikkelsen, P. S., Rivard, G., Uhl, M., Dagenais, D., & Viklander, M. (2015). SUDS, LID, BMPs, WSUD and more – The evolution and application of terminology surrounding urban drainage. *Urban Water Journal*, 12(7), 525–542, DOI: 10.1080/1573062X.2014.916314

Gaines, J. (2016). Flooding: Water potential. *Nature*, 531, S54–S55. https://doi.org/10.1038/531S54a

General Office of the State Council. (2015). Guiding Opinions on Advancing the Construction of Sponge Cities.

IWA (International Water Association). (2017). The hard road to "soft" cities. The Source, 7.

Jia, H., Yao, H., Tang, Y., Yu, S. L., Zhen, J. X., & Lu, Y. (2013). Development of a multi-criteria index ranking system for urban runoff best management practices (BMPs) selection. *Environment Monitoring and Assessment*, 185, 7915–7933.

Jia, H., Yao, H., Tang, Y., Yu, S. L., Field, R., & Tafuri, A. N. (2015a). LID-BMPs planning for urban runoff control and the case study in China. *Journal of Environmental Management*, 149, 65–76.

Jia, H., Wang, X., Ti, C., Zhai, Y., Field, R., Tafuri, A. N., Cai, H., & Yu, S. L. (2015b). Field monitoring of a LID-BMP treatment train system in China. *Environmental Monitoring & Assessment*, 187(6), 1–18.

Jia, H., Zhen, X., Clar, M., & Yu, S. L. (2017). China's Sponge City construction: A discussion on technical approaches. *Frontiers of Environmental Science & Engineering*, 11(4), 18. DOI: 10.1007/s11783-017-0984-9

Liu, Y., Theller, L. O., Pijanowski, B. C., & Engel, B. (2016). Optimal selection and placement of green infrastructure to reduce impacts of land use change and climate change on hydrology and water quality: An application to the Trail Creek Watershed, Indiana. *Science of the Total Environment*, 553, 149–163.

Mao, X., Jia, H., & Yu, S. L. (2017). Assessing the ecological benefits of aggregate LID-BMPs through modeling. *Ecological Modelling*, 353, 139–149.

Ministry of Housing and Urban-Rural Development. (2014). Technical Guide for the Construction of the Sponge City (Draft). Beijing. Retrieved from www.mohurd.gov.cn/wjfb/201411/W020141102041225.pdf

Qiu, B. (2015). The meanings, approaches, and prospects of the Sponge City (LID). *Water & Wastewater Engineering*, 41(3), 1–7. (in Chinese)

Shoemaker, L., Riverson, J., Khalid, A., Zhen, J., Sabu, P., & Rafi, T. (2009). SUSTAIN, a framework for placement of best management practices in urban watersheds to protect water quality. Report EPA/600/R-09/095, USEPA. Washington, DC, USA.

Su, Y., Zhao, F., & Tan, L. (2015). Whether a large disaster could change public concern and risk perception: A case study of the 7/21 extraordinary rainstorm disaster in Beijing in 2012. *Natural Hazards*, 78(1), 555–567.

Xu, T., Jia, H., Wang, Z., Mao, C., & Xu, C. (2017a). SWMM-based methodology for block-scale LID-BMPs planning based on site-scale multi-objective optimization: A case study in Tianjin. *Frontiers of Environmental Science & Engineering*, 11(4), 1.

Xu, C., Hong, J., Jia, H., Liang, S., & Xu, T. (2017b). Life cycle environmental and economic assessment of a LID-BMP treatment train system: A case study in China. *Journal of Cleaner Production*, 149, 227–237.

Young, K. D., Younos, T., Dymond, R. L., Kibler, D., & Lee, D. H. (2010). Application of the analytical hierarchy process for selecting and modeling stormwater best management practices. *Journal of Contemporary Water Resources and Education*, 146, 50–63.

Zuo, J., Liu, C., Zheng, H., & Chen, Y. (2008). Countermeasures and characteristics of Beijing urban rainwater utilization. *Resources Sciences*, 5(30), 990–998. (in Chinese)

Conclusions

Blanca Jiménez Cisneros and Alexandros K. Makarigakis

URBANISATION

We live in a world that has begun to experience an unprecedented amount of uncertainty due to the effects that global changes have.

The United Nations estimate that from approximately 8.2 billion people living worldwide in 2024, the global population could grow to around 9.7 billion in 2050, and 10.3 billion in the mid-2080's (UNDESA, 2019 and UNDESA 2024).

People living in urban areas (3.42 billion) outnumbered for the first time the global rural population (3.41 billion) by the middle of 2009 and since then the world has become more urban than rural (UNDESA, 2010). The urban population of the world has grown rapidly from 751 million in 1950 to 4.2 billion in 2018.

It is projected that by 2050, the percentage of people living in urban agglomerates will increase to 68%, a substantial increase to the 55% of the urban population in 2018 (UNDESA, 2019). The majority of this increase in the size of the world's urban population is expected to take place in just a few countries such as India, China, and Nigeria, which are projected to account for 35% of the projected growth between 2018 and 2050.

Today, the most urbanised regions include Northern America (with 82% of its population living in urban areas in 2018), Latin America and the Caribbean (81%), Europe (74%) and Oceania (68%). The level of urbanisation in Asia is now approximating 50%. In contrast, Africa remains mostly rural, with 43% of its population living in urban areas, whereas in certain regions, such as Latin America, the urbanisation percentage has reached 80% since the past century.

URBAN WATER CYCLE

The urban water cycle is a relatively new concept used to describe water problems in cities (Jimenez, 2009). It is useful in identifying the disruption of the water cycle at a city level and, in particular, the conventional and nonconventional sources of pollution. In the context of urban floods, it is important to understand these differences, in order to be able to replace the conventional global approach used to solve water problems, with a local approach based on the size and within the context of cities. Examples to address water quality under this framework are provided by Jimenez (2011).

The COVID-19 pandemic demonstrated in the context of cities, the importance of having both enough water to wash hands and to sanitise homes, to help reduce the spread of the virus. Cleanup practices using water reduces the transmission of respiratory diseases by 25% (Jimenez, 2020), which in turn translates into a reduction of 25% in hospitalisation and deaths.

During the pandemic, an overall increase on water demand and a different demand distribution across the various users in cities have been observed and reported. The demand for municipal water increased substantially, whereas the water demand for commercial and industrial purposes dropped, especially during the first months of countries imposing related lockdowns.

INCREASING URBAN FLOODS

Countries, are well aware and concerned with the effects of the global increase in temperature resulting in climatic variability and change, affecting particularly water in cities in quantity and quality (Jimenez et al., 2014; Caretta et al., 2022). This temperature increase along with the high density of buildings and the often limited green spaces result in cities experiencing a heat island effect (HIE), which in turn contributes to an increasing amount of extreme rain events and flash floods. Indeed, data and information of more than four decades have shown that HIE induces localised, extremely strong rainfall events in cities, which are the origin to many urban floods (Jauregui, 1997). Indeed, the HIE is increasing temperature in cities that reaches up to 10°C, and the induced localised showers often surpass the cities' sewers capacity, which are designed following conventional criteria and therefore do not account for this weather extreme (Marsalek et al., 2008; Jimenez, 2009) resulting in urban flash floods. In addition, the presence of more intense rain events caused by climate change (Jimenez, 2014) reinforces this situation.

Floods have accounted for 44% of all disaster events from 2000 to 2019, averaging 163 events per year and have affected more than 1.6 billion people globally comprising the highest figure of occurrence for any type of disaster (UNDRR, 2020).

However, the figures aggregate both plain and urban floods. There is need to characterise each separately in the future to have additional information for policymakers to take decision on where and how it is more relevant to act. In particular, flooding events need to support the basin–city link, so that proposed solutions can be designed and implemented optimally.

Flood events are more common in Asia, which constitutes 41% of floods worldwide, followed by Africa and the Americas. India with 17 floods per year and China experiencing 20 floods annually on average are the two most affected countries in the past 20 years, with over 1.2 billion people affected. The death toll, although as not high as other disasters (like earthquakes or tsunamis) can be significant, surpassing 6,000 souls in June 2013, when the deadliest flood event was experienced in India (UNDRR, 2020). EM-DAT data show that floods comprise the second most costliest disaster type causing financial losses of more than 650 billion USD in the past twenty years (storms being the costliest). In an urban setting, flooding is the most frequent disaster, causing widespread economic costs, environmental degradation, material and physical damages, and loss of human lives. Therefore, integrated urban water and flood risk management have become important fields of study for sustainable development (e.g., Jha et al., 2011; Tucci, 2007).

Thus far, the responsibility for planning and acting before, during and after a flood has fallen to the appropriate authorities without involving citizens. Social hydrology and citizen science are trying to push for a paradigm shift in search of sustainable solutions and further research on this theme should be encouraged and financed.

Furthermore, new approaches, such as the Climate Risk Informed Decision Analysis or CRIDA, are needed to provide a framework in identifying potential water security risks and in offering a stepwise planning process to identify and address the challenges for sustainable water resources management (Gilroy, 2018).

Until enough planning and investment have been realised to minimise the risk from flooding, countries and cities are requested to manage extreme events preventing disasters from happening and minimising the harm to their citizens. Planning though at city level often does not suffice as the origin of the challenges to manage is often outside the political jurisdiction of the elected officials and the technical personnel. In such cases, a basin-wide approach should be encouraged with strong links to city officials and structures.

URBAN FLOODS PREVENTION

Prevention often relies on the development and implementation of early warning systems, which mainly address fluvial flood events (e.g., Buchholz et. al., 2013), which are still missing for pluvial (rainfall) flood events in urban areas. Early warning systems for urban areas have to be based on a prediction of a pluvial flood and include rain forecast, prediction of flooded areas, and prediction of transport paths of contaminants. Prevention further relies on the planning of cities considering elements of the urban water cycle behaviour and effects (Jimenez, 2007, 2009; Jimenez et al., 2014; Marsalek, 2007).

The effects of flooding, especially in an urban setting, have been mainly studied and researched with a view to prevent and minimise large amounts of water from causing damage to infrastructure and related to the volume of water, leaving aside those for water quality.

To date, significant research examining the effects of such events in an urban setting to the quality of water bodies receiving the overflow has been conducted. However, there are still insufficient data to provide guidance at global level to decision-makers, despite the demonstration of their significant effects which reach beyond the area of flooding and for a time period exceeding that of the event itself (Brabec et al., 2009; Cardonha et al., 2004; EPA, 2003; Jimenez and Rose, 2009; Lee and Bang, 2000; Marsalek et al., 2007).

The impacts of flooding on the quality of groundwater used for urban water supply require, according to the geology, time to manifest as often the sites where pollution is introduced are far from the sites where water is abstracted. This allows policymakers to adapt the operating conditions of drinking water systems (Jiménez Cisneros et al., 2017), which unfortunately is not the case for surface water sources, where impacts are manifested almost immediately.

Thus, controlling water quantity in urban stormwater management came from the needs of urban life as well as the wishes to avoid urban flooding, whereas the emergence and development of stormwater quality management are mainly driven by strict norms, standards, and the gradually improved legal and regulatory systems (Shaver et al., 1995).

Urbanisation patterns and elevation of the overall standard of living (quality of life) are related to an increase in consumption of goods, which quite often associated to the increase of pollutants in a city. Whether it is an increased level of atmospheric pollution, dust, spilled motor oil, tire residue on the streets, or other types of pollutants, higher concentrations and amounts can be found. These elements affect the stormwater quality during extreme rain events and flooding.

Furthermore, in the cases where the sewage system is combined (both wastewater and storm water collected), the health threat increases as inadequate sewage and drainage systems cannot handle the sudden large flows generated, releasing pathogens and in some cases toxic chemicals into the water system. The spatial and temporal scale of these exchanges of pollutants remains largely unknown due to the wide range of pollution sources, the complexity of the interactions themselves, and lack of integrated sampling data of all sources (e.g., Jimenez and Rose, 2009).

The majority of the focus to deal with urban flooding and address the issue on water quality degradation is structural solutions/systems such as the SuDS. These include permeable surfaces such as green roofs and pervious pavements, infiltration trenches and basins, swales, filter drains and strips, detention basins, retention ponds, and wetlands. However, the design of SuDS is such that its focus, like that of conventional, piped drainage, is on attenuation of the storm peak. The research has therefore focused on the water quantity reduction capabilities of SuDS systems, and less on their abilities to improve water quality particularly their performance over the long term (Kirby, 2005). The new concepts, such as the one of 'the Sponge City' combining both the planning of cities with innovative ways to manage drainage seems to be the future, but are still to find ways to be implementable in already build cities.

Moreover, factors causing an impact on the urban water quality are both structural and non-structural and are highly context specific. A number of case studies found in [Chapter 9] of this book present environmental, social and economic implications of water pollution by urban floods ranging from harming the aquatic life and ecosystems, to soil degradation, biodiversity loss, the spreading of diseases, beach closings, microbial resistance, infrastructural damage and loss of crops and fish, as examples of indirect consequences. The case studies also show that the application of measurements depend on the context.

There is a need for low cost and efficient adaptation measures, not only to prevent flooding, but also to protect the water quality in times of extreme rain events. Research towards developing customised solutions context appropriate, ensuring the capacity to maintain them so that they can offer their services in a sustainable manner should be financed.

These systems need to be part of a holistic approach of Water Sensitive Urban Design (WSUD), or Low Impact Development (LID), that consider water as a "liquid asset" (Semadini-Davies et al., 2008) rather than a "wasted waste" and viewing it as "opportunity water" (Chocat et al., 2004) instead. UNESCO's second volume of the Global Water Security Issues series, focused on "Water reuse within the context of circular economy" (UNESCO and UNESCO i-WSSM, 2020) to promote this way of thinking among decision-makers.

NEED FOR ADDITIONAL INFORMATION AND RESEARCH

We need to understand and assess factors and mechanisms of urban floods as well as their impacts on water availability in both quantity and quality. More information is needed on the quality aspect than on the quantity one. Contributing to the understanding of the diffusion of pollutants during urban floods causing water quality degradation such as inadequate sewage and drainage systems, rainfall characteristics such as intensity and duration, but also the previous days rain patterns, seasonal characteristics (both in the rainfall and in the catchment) (Liu, 2013), state of cleanliness (both in the sewer and in the catchment), the type of flooding, the degree of impervious surfaces, topography, the amount of rainfall, human displacement, population density and climate is relevant.

There are shortcomings concerning water quality management due to a lack of a thorough understanding of the degree and the processes of water pollution during urban floods. The non-point character of the pollution makes it difficult to trace back all the pollutants to their source and calls for an interdisciplinary approach including sample data analyses and modelling.

One of the "low hanging fruit" vis-à-vis future research is the collection and reporting of data on the impacts of urban floods on the quantity and quality of water for cities. Furthermore, research is required to develop a common methodology in assessing water quality during and after urban flooding and extreme rain events in different urban regions across the world.

The importance of institutional, policy and educational considerations in protecting the water quality during extreme rain events and flooding in cities should not be underestimated. Soft measures of this sort that strengthen the governance-related capacity are often more crucial in providing solutions than infrastructure-based solutions. Enhancing the comprehension and encouraging citizens to keep streets and storm drains clean, to avoid dumping various pollutants and dispose of them properly cannot and should not be underestimated.

There are a number of challenges to reflect upon when designing realistic research plans as large flood events are infrequent (Walling, He and Nicholas, 1996) and sampling and recording of parameters' values during and immediately after flood events is very difficult (Zak, Rohovec and Navratil, 2009; Zonta et al., 2005). Similarly, long-term monitoring projects are usually not attractive to the donor community that seeks instant gratification with results that can be presented to their own constituencies and which demonstrate to the layman the impact of their contribution.

UNESCO'S IHP ROLE

Sustainable development requires cities, which are resilient to global changes. Flooding events have increased both in frequency and intensity due to the effects of such changes. The capacity to estimate risks and impact associated with urban flooding on water quality is crucial for the development of efficient mitigation and adaptation strategies and plans. It is with the implementation of such plans that the road to sustainability will be paved.

In its Ninth Phase (2022–2029), UNESCO's Intergovernmental Hydrological Programme (IHP) has identified floods as a research area to tackle dedicating an output on "Scientific knowledge, methodologies and tools in addressing water-related disasters, such as flood and drought elaborated and/or enhanced towards timely forecasting." Furthermore, it will pursue research to achieve sustainable urban water management researching adaptive water management tools. A multidisciplinary science approach (hydrological, social socio-economics sciences etc.) will be used to provide a new holistic framework at all levels. Sharing scientific knowledge and understanding of water issues to all stakeholders, in urban areas and rural areas in a watershed will allow the development of a common vision and implementation of science-based solutions.

Integrated Urban Water Management is an approach to managing the entire urban water cycle in an integrated way ("Integrated Urban Water Management" in the UNESCO Urban Water Series, 2007–2012) and it could provide the foundation upon which the effects of urban floods could be incorporated. The Global Water Security Issues (GWSI) Series Issue No. 4 focuses on Water Security and Cities: Integrated Urban Water Management and outlines various IUWM options and applications to highlight the importance of integrated approaches to urban water management to make cities sustainable and resilient. An open call will be made to the scientific and research communities to support further work on the issue of water quality and urban flooding with a view of better understanding processes and to develop tools for decision-makers that will support science-based policymaking and infrastructure development.

References

Brabec, E., Shulte, S., and Richards, P. (2009). Impervious surfaces and water quality: A review of current literature and its implications for watershed planning, *Journal of Planning Literature* 16(4): 499–514.

Cardona, O. D., van Aalst, M. K., Birkmann, J., Fordham, M., McGregor, G., Perez, R., Pulwarty, R. S., Schipper, E. L. F., and Sinh, B. T. (2012). Determinants of risk: Exposure and vulnerability. In: *Managing the Risks of Extreme Events and Disasters to Advance Climate Change Adaptation* [Field, C. B., V. Barros, T. F. Stocker, D. Qin, D. J. Dokken, K. L. Ebi, M. D. Mastrandrea, K. J. Mach, G.-K. Plattner, S. K. Allen, M. Tignor, and P. M. Midgley (eds.)]. A Special Report of Working Groups I and II of the Intergovernmental Panel on Climate Change (IPCC). Cambridge University Press, Cambridge, UK, and New York, NY, pp. 65–108.

Caretta, M. A., Mukherji, A., Arfanuzzaman, M., Betts, R. A., Gelfan, A., Hirabayashi, Y., Lissner, T. K., Liu, J., Lopez Gunn, E., Morgan, R., Mwanga, S., and Supratid, S. (2022). Water. In: *Climate Change 2022: Impacts, Adaptation and Vulnerability. Contribution of Working Group II to the Sixth Assessment Report of the Intergovernmental Panel on Climate Change* [Pörtner, H.-O., D.C. Roberts, M. Tignor, E.S. Poloczanska, K. Mintenbeck, A. Alegría, M. Craig, S. Langsdorf, S. Löschke, V. Möller, A. Okem, and B. Rama (eds.)]. Cambridge University Press, Cambridge, UK and New York, NY, pp. 551–712, doi:10.1017/9781009325844.006.

Chocat, B., Ashley, R., Marsalek, J., Matos, M. R., Rauch, W., Schilling, W., and Urbonas, B.: Urban drainage: Out-of-Sight-Out-of-Mind. NOVATECH (2004). 5th Intl Conf on sustainable techniques and strategies in urban water management. Brelot, E., Chocat, B., Desbordes, M., Villeurbanne: 2004;2:1659–1690.

Earl, S., Maxted, J., Curtis, G., and Carter, D. (1995). Watershed protection using an integrated approach. In: *Stormwater NPDES Related Monitoring Needs: Proceedings of an Engineering Foundation Conference* [Torno, H. C. (ed.)]. American Society of Civil Engineers, New York.

Gilroy, K. (2018). Water resources planning and design for future uncertainties – Collaborative Risk Informed Decision Analysis (CRIDA), [Mendoza, G., J. Matthews, and A. Jeuken (eds.)]. Washington, USA, ICIWaRM.

Jauregui, F. (1997). Heat Island development in Mexico City, *Atmospheric Environment*, 31(22), November 1997: 3821–3831.

Jha, A., Proverbs, D., and Bhattacharya, N. (2012). *Cities and Flooding: A guide to integrated urban flood risk management for the 21st Century*, The World Bank, GFDRR.

Jimenez, B. (2008). Wastewater risks in the urban water cycle. In: *Urban Water Security: Managing Risks*. [Jimenez, B., and J. Rose (eds.)], pp. 151–200. Taylor and Francis Group. ISBN: 978-0-415-48567-8.

Jimenez, B. (2009). Risks and integrated management of the urban water cycle in megacities of the developing world: Mexico city. In: *Chapter in Water and Urban Development Paradigms*. [Feyen, S. J., and K. M. Neville (eds.)], pp. 387–397. CRC Press Taylor and Francis Group UNESCO, London, UK. ISBN: 978-0-415-483346.

Jimenez, B. (2011). Safe sanitation in low economic development areas, Treatise M.Sc. 82. In: *Chapter in Treatise on Water Science*, vol. 4. [Wilderer, P. (ed.)], pp. 147–201. Academic Press, Oxford.

Jiménez Cisneros, B. E., Oki, T., Arnell, N. W., Benito, G., Cogley, J. G., Döll, P., Jiang, T., and Mwakalila, S. S. (2014). Freshwater resources. In: *Climate Change (2014) Impacts, Adaptation, and Vulnerability. Part A: Global and Sectoral Aspects. Contribution of Working Group II to the Fifth Assessment Report of the Intergovernmental Panel on Climate Change* [Field, C. B., V. R. Barros, D. J. Dokken, K. J. Mach, M. D. Mastrandrea, T. E. Bilir, M. Chatterjee, K. L. Ebi, Y. O. Estrada, R. C. Genova, B. Girma, E. S. Kissel, A. N. Levy, S. MacCracken, P. R. Mastrandrea, and L. L. White (eds.)]. Cambridge University Press, Cambridge, UK and New York, NY, pp. 229–269. http://ipcc-wg2.gov/AR5/images/uploads/WGIIAR5-Chap3_FGDall.pdf

Jiménez Cisneros, B., Navarro González, I., and Montes Rojas, R. (2017). Water quality risks associated with climate change: Three Mexico City case studies, II Adaptation and Integral Manegement in Climate Change-Sensitive Cities: Challenges and Opportunities in Urban Governance, Chapter 3 in Climate Change-Sensitive Cities: Building Capacities for Urban Resilence, Sustainability, and Equity. Gian Carlo Delgado Ramos (Ed.). Climate Change Sensitive Cities PNUD. E-ISBN: 978-607-02-9973-5. 81–132 pp.

Jimenez, B. (2020). Nuevas direcciones para la investigación urbana y las políticas públicas en el ámbito del agua en Cities and Covid-19: New Directions for Urban Research and Public Policies. Delgado Ramos G and Lopes Garcia D. Editors Plataforma de Conocimiento para la Transformación Urbana, International Network for Government Science Advice, www.transformacionurbana.mx, ISBN: 978-1-64945-190-3, DOI: 10.5281/3894075E, pp. 92–97.

Kirby, A. (2005). Proceedings of the Institution of Civil Engineers – Municipal Engineering, Paper 14024, Issue ME01, (1–8).

Lee, J. H., and Bang, K. W. (2000). Characterization of Urban Stormwater Runoff. *Water Research*, 34, 1773–1780.

Liu, L., Xu, H., and Wang, Y. (2017). Impacts of 1.5 and 2 °C global warming on water availability and extreme hydrological events in Yiluo and Beijiang River catchments in China. *Climatic Change*, 145, 145–158. https://doi.org/10.1007/s10584-017-2072-3

Marsalek, J., Jimenez, B., Malmquist, P., Karamouz, M., Goldenfum, J., and Chocat, B. (2007). *Urban Water Cycle Processes and Interactions, Urban Water Series*, Taylor & Francis Group, 152 pp. The Netherlands. [152 pp en inglés ISBN: 978-92-3-104010-3 y 160 pp. y en Chino. ISBN: 978-7-5170-2583-2].

Semadeni Davies, A., Hernebring, C., Svensson, G., and Gustafsson, L.-G. (2008). The impacts of climate change and urbanisation on drainage in Helsingborg, Sweden: Suburban stormwater. *Journal of Hydrology*, 350(1–2), pp. 114–125.

Tucci, C. (2007). Urban Flooding Management. World Meteorological Organization. Cap-Net International Network for Capacity Building in Integrated Water Resources Management. WMO/TD No. 1372.

UNDRR, CRED: The human cost of disasters: an overview of the last 20 years (2000–2019), 2020.

UNESCO, UN-Water. (2020). *United Nations World Water Development Report 2020: Water and Climate Change*. Paris: UNESCO.

UNESCO and UNESCO i-WSSM. (2020). Water Reuse within a Circular Economy Context (Series II). Global Water Security Issues (GWSI) Series – No.2. UNESCO Publishing, Paris.

United Nations, Department of Economic and Social Affairs, Population Division (2019). *World Population Prospects 2019: Highlights* (ST/ESA/SER.A/423).

United Nations. (2024). World Population Prospects 2024: Summary of Results. UN DESA/POP/2024/TR/NO. 9. New York: United Nations.

United Nations, Department of Economic and Social Affairs, Population Division (2010). *World Urbanization Prospects: The 2009 Revision* (ST/ESA/SER.A/296). New York: United Nations.

United Nations, Department of Economic and Social Affairs, Population Division (2019). *World Urbanization Prospects: The 2018 Revision* (ST/ESA/SER.A/420). New York: United Nations.

U.S. Environmental Protection Agency. (2008). *EPA's Report on the Environment* (2003 Draft). Washington, DC: U.S. Environmental Protection Agency.

Walling, D. E., He, Q., & Nicholas, A. P. (1996). Interpreting particle size effects in the adsorption of 137Cs and unsupported 210Pb by mineral soils and sediments. *Journal of Environmental Radioactivity*, 30(2), 117–137.

Walsh, C. (2000). Molecular mechanisms that confer antibacterial drug resistance. *Nature*, 406, 775–781. https://doi.org/10.1038/35021219

Žák, J., Rohovec, T., & Navrátil, P. (2009). Fluxes of heavy metals from a highly polluted watershed during flood events: A case study of the Litavka River, Czech Republic. *Water, Air, and Soil Pollution*, 203(1), 343–358.

Zonta, R., Collavini, F., Zaggia, L., and Zuliani, A. (2005). The effect of floods on the transport of suspended sediments and contaminants: A case study from the estuary of the Dese River (Venice Lagoon, Italy). *Environment International*, 31(7), 948–958. https://doi.org/10.1016/j.envint.2005.05.005

Index